Woven Arch Bridge

This book focuses on the woven arch bridge, an arch-shaped structure that is one of the most extraordinary timber building traditions of the world. The woven arch bridge exists widely in different cultures and its specific nature is conceptualized by the author as a kind of "universal uniqueness," challenging widespread viewpoints on its origin and genealogy.

Taking this argument as its main thread, the book traces the histories of different woven-arch-bridge-cultures and investigates in particular the woven arch bridge in the mountains of Southeast of China from three angles, using both archaeological and anthropological methods. Resting upon these case studies, a definition of typology and a new theory of structural evolution are established, while the book also draws comparisons between western and eastern timber building cultures and offers new insights on the differences between East Asia and Europe.

The book also provides a large number of examples and illustrations of the bridge, and will be of great value and inspiration for architects and scholars studying the history of architecture, bridges, and construction, while also appealing to general readers interested in historical bridges and traditional construction technology.

Liu Yan, an architectural historian, earned the degree of Doktor-Ingenieur in *Bauforschung* (Building Archaeology) from the Technical University of Munich, Germany. He is a Member of the Koldewey-Gesellschaft and a Junior Fellow at the Society of Fellows in Liberal Arts, Southern University of Science and Technology and now works as a lecturer at Kunming University of Science and Technology, China.

China Perspectives

The *China Perspectives* series focuses on translating and publishing works by leading Chinese scholars, writing about both global topics and China-related themes. It covers Humanities & Social Sciences, Education, Media and Psychology, as well as many interdisciplinary themes.

This is the first time any of these books have been published in English for international readers. The series aims to put forward a Chinese perspective, give insights into cutting-edge academic thinking in China, and inspire researchers globally.

Titles in architecture currently include:

Vertical Urbanism
Designing Compact Cities in China
Zhongjie Lin, José L.S. Gámez

Woven Arch Bridge
Histories of Constructional Thoughts
Liu Yan

For more information, please visit
www.routledge.com/China-Perspectives/book-series/CPH

Woven Arch Bridge
Histories of Constructional Thoughts

Liu Yan

Routledge
Taylor & Francis Group

LONDON AND NEW YORK

First published in English 2021
by Routledge
2 Park Square, Milton Park, Abingdon, Oxon OX14 4RN

and by Routledge
52 Vanderbilt Avenue, New York, NY 10017

Routledge is an imprint of the Taylor & Francis Group, an informa business

© 2021 Liu Yan

British Library Cataloguing-in-Publication Data
A catalogue record for this book is available from the British Library

Library of Congress Cataloging-in-Publication Data
A catalog record has been requested for this book

ISBN: 978-0-367-61823-0 (hbk)
ISBN: 978-1-003-10934-1 (ebk)

Typeset in Times New Roman
by Newgen Publishing UK

Brücken bauen, statt Mauern.
(Build bridges, not walls.)
        – German saying

Contents

Figures

Introduction

Let's imagine some scenes from a movie trailer...

0.1 55 BC, Gaul, in today's Rhineland-Palatinate, Germany

The hilly wilderness of Western Europe.

In an army tent, five or six tribal leaders stand around, anxious and impatient. They look at each other and then one of their number, chosen to represent the group, speaks: "We've had enough of the Germans harassing us! All we ask is that you transport your army over the Rhine! We can provide enough ships."

Before them stands an officer with shining armour, sharp eyes, and a solemn expression. He opens his mouth to speak, waits for a moment and says: "Yes, the Roman Legion will go over the Rhine – but not by ship."

The camera follows the officer's eyes, taking in rapid changes of scene: dense forest on the hills, the edges of the woods being felled by soldiers in Roman armour... From the edge of a broad, fast-moving river, a bridge construction extends out into the middle of the river and huge tamping machines are hitting wooden stakes stuck in the water rhythmically.

0.2 1520s, Vicenza, Italy

Inside a mason's workshop.

The sound of the tamping machines turns into the same rhythm, but made this time by a wooden hammer, pounding a chisel held by a young but rough hand. Fragments of white marble chip off.

The speed slows and comes to a stop. We see a close-up of a boy's face, a far-away look in his eyes...

"Andrea, stop daydreaming! What are you thinking?" says a voice from the other side of the room.

"Caesar."

"Caesar who?"

"*Julius* Caesar, of course! The great Caesar!"

The boy leans down, picks up a pencil and a piece of paper, and begins to draw something.

0.3 1481, Milan, Italy

The camera pans from the boy's lines on a piece of crumpled paper, to neat lines on a flat paper surface, and slowly sweeps over some sketches of various installations, weapons, bridges…

There is the sound of a door opening, and the well-dressed young man who made the drawings quickly closes the sketchbook and stands up. He follows a servant into a grandly decorated hall, salutes and speaks:

"My most illustrious Lord, my name is Leonardo, from Vinci, applying to be your military engineer. I have many skills." He holds out his sketchbook.

0.4 1032, Qinzhou, China

From the face of an Italian lord to that of an Asian official, in a simply-furnished government office.

The candlelight flickers; the elder has a suspicious expression on his face. His long beard trembles as he utters his question, speaking slowly and solemnly: "Do you really mean to tell me that you can build a bridge over the Yang River?"

A man in a dirty uniform, that of a low-ranking prison warder, bows to him, his eyes cunning: "That is correct, Your Honour. I can build a bridge simply with these." He holds out his hand, which contains a bunch of chopsticks.

0.5 1857, Taishun, Zhejiang, China

Aerial view of endless mountains and forest.

The view quickly goes from the air down to a small town, flies across a market, hovers along a river, and slowly comes to settle on a construction site set up over the river. A huge wooden bridge appears to be half-finished, villagers crowd on the river bank.

Screams and shouts can suddenly be heard from the middle of the river, followed by the sound of splitting timber. The entire bridge structure and the half-detached scaffolding topple over, and with sudden loud rumbles, the wooden dragon breaks into hundreds of pieces and is taken away by the water.

0.6 1870s, Enshu, Japan

A temple garden in the morning light: neatly-arranged plants and buildings with an Asian aesthetic in the background.

A young novice is sweeping the courtyard. He slows his movements down and looks at the quiet and colourful autumn garden. The camera stops on his serene and joyful and face covered in sunlight.

0.7 1937, Shouning, Fujian, China

Deep green water in a mountain valley, enclosed by dense trees connected by a stone footpath.

Four giant wooden pillars stand in the water. They are connected to each other by tree trunks, and to the rocky abutments at the banks of the river.

An old man with silver hair, a square ruler resting at his neck, lights some incense and prays to the silent river.

A young worker steps forward, accepts a red envelope from a villager, grabs a bowl of liquor from a table, and quaffs it down in one. He puts down the bowl, shoves an axe into his waistband, and steps out onto the tree trunks towards the middle of the river.

0.8 1898, San Francisco, USA

Outside the window, we see busy, muddy streets.

The scene pans to the interior of the room, and we look around with store boys' eyes: it's full of Buddha statues, porcelain, East Asian-style wooden tables and chairs, and screens. The camera focuses on a figure who pushes open the door and steps inside.

He has an Asian face and is dressed in a well-worn but neat sailor's uniform. "My name is Toichiro Kawai, and I come to see Mr Victor Marsh. He told me I'd find him here," he says with a strong foreign accent, and then adds, "some years ago." The shot stops on the middle-aged man's face.

0.9 1913, San Marino, California, USA

We see a drop of sweat dripping down that same face. The camera pans out to show us a broader view, and we see the man kneeling on a curved bridge above a small pond. The arch-shaped wooden bridge in the centre of the scene is set in a garden with neatly-cut shrubs, Asian plants, and a small hill crowned with an Asian-style house.

0.10 1955, Wuhan, Hubei, China

The bridge in the centre of the previous scene changes into an image of a similar bridge, in a black-and-white drawing, on the page of a magazine. The camera pans out, to show a young man in a dark grey coat staring at the drawing in a concentrated way.

"Xiao Tang!"

Hearing the offscreen shout, the young man puts down the magazine. A colleague of his appears at the door, speaking rather excitedly: "Xiao Tang! Have you heard? The design for the Yangtze Great Bridge has been selected! And it's your design! Premier Zhou himself chose it!"

0.11 2002, Beijing, China

Late summer. In the centre of the campus, the leaves of the plane trees sway lazily. The camera pans over the busy flow of bicycles, through a window into a classroom on the ground floor.

A venerable professor on the stage is speaking in a rather vigorous voice: "But engineers from ancient times also had marvellous creations." He clicks, and a new image appears on the screen: it's the very drawing we saw in the last scene. "This Song Dynasty bridge, in *Qingming Shanghe Tu*, is a real historical and technical wonder. Even today, it's not easy for modern engineers to make the statics calculations for that bridge."

The camera goes from the platform down to the audience, sweeping over a hundred young faces.

0.12 Fade-out to darkness – and then a title appears on the screen

Stories of Woven Arch Bridges.

Welcome to my world, the world of woven arch bridges. And yet, this is not a movie: what you hold in your hands is a product of my decade-long research.

In 2015, around the time I started composing the first version of this book (my doctoral thesis), I watched a science-fiction movie, *Cloud Atlas*, directed by Tom Tykwer and the Wachowskis in 2012. It tells six independent stories, happens in six entirely different spaces and times in human history, and is cut in a montage sequence which, amazingly, shares the exact same leitmotiv.

After watching it I felt: "This is exactly the structure of my dissertation!" It's just a pity that the guidelines for academic writing don't include the montage technique!

The core of this book focuses on the woven arch bridge, a rather extraordinary structural form, rare in human history. Everywhere they appear, they are considered as an unusual or even one-of-a-kind creation, and yet, they appeared in multiple different spaces and times in adapted forms, in parallel and independently. Unique yet universal, the existence of woven arch bridges in all these different cultures is an unparalleled phenomenon in building history.

This book explores the history of each of these specific historical structures.

The first part consists of three stories from three different cultures. Each story is about a (group of) woven arch bridge(s) and covers a large span of time and/or space. The first spans from Ancient Rome to Renaissance Italy; the second recounts an overseas journey from Japan to the US; and the third covers the entire academic life of a Chinese scholar.

These three chapters are independent in narrative terms, like three links in a chain, and the reader is free to start from any chapter they like.

The second part also contains three chapters; however, the whole of Part Two is designed to be the fourth link, parallel with the first three chapters.

Its comparative length indicates the incomparable scale of this group of objects: the largest number of woven arch bridges is found in the mountains of Southeast China. Three different angles are provided by the three chapters, for which the keywords are: build, building, builders.

The four individual links appear to be very distinct in form, temperament, and methodology. They are arranged to be as independent as possible. However, just like the stories in *Cloud Atlas*, no matter how diverse they appear, they share the same core. The leitmotif of our stories focuses on the same questions: despite the peculiarity of the bridge type, how was each woven arch bridge created (i.e. actually designed and built) and how did they emerge in their respective cultures? In order to dig out the hidden history, I almost turned myself into a detective, incorporating all possible skills from other disciplines, and gathering evidence from the different corners of the world.

The strenuous trips were worth it. On our journey looking for this unique yet universal object, we experience in-depth forays into these different cultures, and reveal the unusual landscape that we can't hope to discover by following only a standard guidebook. And when we put together the pieces of treasures we find at all these destinations, they magically turn into a much grander view: a peculiar map of human history on constructional thinking.

Part I

Woven arch bridges

Three stories

1 From Caesar to Da Vinci

The woven structure's Italian root

1.1 Glory to Caesar!

1.1.1 Caesar and his Rhine Bridge

From 58 to 50 BC, during his conquest of Gaul, Julius Caesar (100–44 BC) led his Roman legion into a series of wars to quell the threat posed by the hostile Gallic tribes and other neighbouring forces.

In 55 BC, as a reaction to the constant harassment by some German tribes at the boundary of Roman Gaul, Caesar decided to lead his legion across the Rhine.

A natural boundary between Gaul and the German tribes, the Rhine was known for its great depth, width, and the rapidity of its flow. Although crossing by boat would have been less of a problem, to Caesar this was unsafe and undignified. He determined to make the effort to build a bridge – only then could he frighten the reckless and combative barbarians.

The exact location of the bridge is not clear today; most suggestions point to the area between Andernach and Koblenz, in a comparatively open part of the Rhine Valley. Accordingly, the length of the bridge is speculated to have been between 200 and 400 metres.

The bridge was built in ten days including the collection of timbers. Caesar led his legion over the river, took his revenge on the Germanic villages and struck fear into the Germans. He then led the army back to Gaul and destroyed the bridge. The action was a great military success.

In his *Commentarii de bello gallico* (Gallic Wars), the report written by his own hand in third-person narration, Caesar described the construction of the bridge with an engineer's precision and in great detail:

> Rationem pontis hanc instituit: tigna bina sesquipedalia paulum ab imo praeacuta dimensa ad altitudinem fluminis intervallo pedum duorum inter se iungebat. Haec cum machinationibus immissa in flumen defixerat festuculisque adegerat, non sublicae modo derecte ad perpendiculum, sed prone ac fastigate, ut secundum naturam fluminis procumberent, his item contraria duo ad eundem modum iuncta

intervallo pedum quadragenum ab inferiore parte contra vim atque impetum fluminis conversa statuebat. Haec utraque insuper bipedalibus trabibus immissis, quantum eorum tignorum iunctura distabat, binis utrimque *fibulis* ab extrema parte distinebantur. Quibus disclusis atque in contrariam partem revinctis tanta erat operis firmitudo atque ea rerum natura, ut, quo maior vis aquae se incitavisset, hoc artius inligata tenerentur. Haec derecta materia iniecta contexebantur ac longuriis cratibusque consternebantur. Ac nihilo setius sublicae et ad inferiorem partem fluminis oblique agebantur, quae pro ariete subiectae et cum omni opere coniunctae vim fluminis exciperent, et aliae item supra pontem mediocri spatio ut si arborum trunci sive naves deciendi operis causa essent a barbaris missae, his defensoribus earum rerum vis minueretur neu ponti nocerent.

(Caesar 1980, 195–8)

[He proceeded to construct a bridge on the following plan: He caused pairs of balks eighteen inches thick, sharpened a little way from the base and measured to suit the depths of the river, to be coupled together at an interval of two feet. These he lowered into the river by means of rafts, and set fast, and drove home by rammers; not, like piles, straight up and down, but leaning forward at a uniform slope, so that they inclined in the direction of the stream. Opposite to these, again, were planted two balks coupled in the same fashion, at a distance of forty feet from base to base of each pair, slanted against the force and onrush of the stream. These pairs of balks had two-foot transoms let into them atop, filling the interval at which they were coupled, and were kept apart by a pair of *fibulis* on the outer side at each end. So, as they were held apart and contrariwise clamped together, the stability of the structure was so great and its character such that, the greater the force and thrust of the water, the tighter were the balks held in lock. These trestles were interconnected by timber laid over at right angles, and floored with long poles and wattlework, And further, piles were driven in aslant on the side facing downstream, thrust out below like a buttress; and others likewise at a little distance above the bridge, so that if trunks of trees, or vessels, were launched by the natives to break down the structure, these fenders might lessen the force of such shocks, and prevent them from damaging the bridge.]

(Caesar 2006, 62)

In the text, Caesar described a form of trestle bridge with inclined posts in pairs on either side, holding a beam in between. Though written in a detailed and accurate manner, in places, the description is confusing to later readers. The most disturbing term is *"fibulis"* (plural of *"fibulae"* or *"fibula"*), which describes the elements used in pairs at the joints of the middle beam and the posts. Thanks to this particular device, the structure would be even more stable against the thrust of the river water.

There is no satisfactory contemporary definition of the word "*fibulis.*" A similar term, "*fibula,*" and its plural form "*fibulae,*" meaning brooch or pin for fastening garments, might be one of the closest definitions. In the legion, they may have symbolized specific ranks or positions in the Roman army.

Two millennia later, when modern scholars shone the spotlight on Caesar's Rhine Bridge again, they raised a series of questions including the location of the bridge, the geological environment of the site, the bridge form, the construction process, etc. Among these questions, the structure of the bridge was the topic most focused on; thus the explanation on the form and function of the *fibulis* served as the key.[1]

Although in the narration, Caesar is the bridge planner ("*nationem pontis hanc instituit*"/ "*He proceeded to construct a bridge on the following plan*") this bridge is commonly attributed to Mamurra, his *praefectus fabrum* (officer in charge of engineering) at that time, a man who held this position from 58 to 54 BC (McDermott 1983, 292–307).

Our knowledge about Mamurra today mainly comes from the description by the poet Catullus: his great fortune from the spoils of wars, his extravagant and lascivious life and – probably just a rumour – his homosexual relationship with Caesar. Despite these bad reputations, Mamurra was considered "by all means the best military engineer of his day" (Frank 1928, 157–9). Among his genius inventions was also a new kind of ship which enabled Caesar's second invasion of Britain.

There was some speculation as to whether Mamurra and Vitruvius were the same person (Thielscher 1961), although this opinion did not gain general acceptance (Ruffel and Soubiran 1962). The two do have some things in common. Vitruvius, the most well-known architect of antiquity, is the author of the first preserved architectural treatise – *De Architectura* (Ten Books on Architecture) –which is dedicated to his patron, the emperor Caesar Augustus. He also served in the Roman army as *praefectus fabrum* under Julius Caesar and was good at designing military constructions and machines. But among other factors that led to the refutation of identifying these two names as one man was the fact that Vitruvius apparently held a much less prominent position than Mamurra. Besides, the Rhine Bridge gives Vitruvius another "alibi." In his *De Architectura*, in the tenth book he wrote on military architecture and engines, mainly derived from his experience with Caesar in the civil wars, but made no mention about bridge-building in his work, neither military nor civil.

---

1 For a discussion on bridge construction, see: Cohausen (1867), Rheinhard (1883), Schleussinger (1884), Menge (1885), Zimmerhaeckel (1899), Schramm (1922), Saatmann and Thielscher (1939), Drachmann (1965), Bundgard (1965).

1.1.2 From a legend to an issue of architecture study

1.1.2.1 The Rhine Bridge in Caesar's legend

Even in the "Dark Ages," Caesar never lost popularity in the land of his rise to power and in the territories he had conquered. He enjoyed such a high reputation that he somehow became a legendary or mythical figure. He was adored as the first emperor of Rome, considered to be a man of the highest morality, a military genius, the conqueror of the French, and the founder of a group of German cities along the Rhine, many of which he had never actually set foot in.[2]

Several medieval manuscripts on Caesar's life and the wars he fought have survived. A number of them are illuminated, and there is no graph of the bridge. The reason for this lies partly with the public's interest, which focused on the legendary emperor and his dramatic life story, partly with the separation between the professions of scribe and illustrator. Book illustrations as part of the illumination served mostly for readers' convenience, serving to locate their reading progress; the drawings were mostly inserted according to a routine layout taught in the workshops. During the manufacture of a hand-copied book, spaces for illustrations were left blank while the scribes wrote, and only afterwards, when the copying was finished, did the illustrators start their work. Sometimes the illustrators might receive a hint from the scribes as to the content of the drawings through a shallow sketch in the margin or the like, but sometimes, the drawings had no direct connection with the text.[3]

This situation changed in the Renaissance. After Caesar and his texts were taken up by the humanists, he was finally recognized as the author of the Commentaries[4] and it was realized that he too had been an actual person and not a mythical figure. In his role as pure Latin author and model historian, he was at the top of the list of those whose works were read and translated in Europe, especially in Italy. With the newly-invented and ever more-widespread printing technique, Caesar's works were extensively published and read, and widely translated into vernacular languages. The first Italian versions came out in the first half of the fifteenth century; the first German version in 1507, Spanish in 1529, and English about the same time.

Thus it is no surprise that the earliest drawings of Caesar's Rhine Bridge only appeared in the Renaissance, as part of the book illustrations. The focus on the Rhine Bridge arose parallel to the concentration on his text. Only when the readers had sufficient linguistic and historical interests to dig into the text

---

2 The role of Caesar in the Middle Ages, see Griffin (2009) and Gundolf (1904).

3 For a general understanding of the method of production of medieval manuscripts, see Watson (2003), Alexander (1992), De Hamel (1992 and 2001)

4 Coluccio Salutati (1331–1406) was the first to realize that Caesar himself, instead of Julius Celsus, was the author of the commentaries. See Griffin (2009, 340).

rather than focus only on the stories, did the engineering of the Rhine Bridge catch their eye.

This gradual change in attitude was witnessed by some early Renaissance examples from outside Italy. The beautiful hand drawings of the various versions of *Faits des Romains* at the beginning of the fourteenth century and the woodcut illustrations of the *Les Commentaires de Jules César* in the latter half of the fifteenth century included no image of the bridge. When bridges finally appeared in the versions mentioned below, their description kept to a conventional flavour of biographical or moral interest. Thus, the illustrations showed the construction process of the bridge through the depiction of crane and builders. The bridges were set in the drawings as the stage for the story. Their structural forms clearly contradicted Caesar's description and were probably drawn according to contemporary common bridge forms, showing the illustrators' unconcern or even unfamiliarity with the text to which it related.

In the first German translation by Matthias Ringmann, *Julius der erst römisch Keyser von seinem Kriege* (1507), a trestle bridge with diagonal struts is depicted in the centre of the woodcut with a wheeled crane. The bridge (Figure 1.1) is almost identical to other images of bridges in the same book. In the text, the term "*fibulis*" is translated as "*Nageln*" (nails).

Another example comes from private documents from the French court, where the *Commentaires de la guerre gallique* (1519–20) was translated by François du Moulin, the tutor of the young king François I. This reworking of Caesar's text was used in the education of the young monarchs, and Caesar's original text was cut and rearranged into a question-and-answer form. The book was exquisitely illustrated. The bridge (Figure 1.2) is on an enormous scale, takes centre stage in the drawing. However, the construction is nothing like Caesar's description of it: the bridge is not even a trestle bridge, posts are singular and stand vertically, and beams run in a longitudinal direction.

1.1.2.2 The first light from the architecture discipline

In Renaissance Italy, Caesar's Rhine Bridge received its first scientific examination, resulting in numerous illustrated *Commentarii*. The first light to shine on the Rhine Bridge, however – thanks to the emergence of professional architects – was in Leon Battista Alberti's (1404–72) *De re aedificatoria* (On Architecture).

In the Holy Year of 1450, the main bridge linking the Vatican with the city of Rome, the *Ponte Sant'Angelo,* was damaged by overuse by pilgrims, and its parapets collapsed. Alberti was commissioned to repair the bridge.

The chapter on bridges in *De re aedificatoria* deals generally with the problems faced in this project.[5] Before Alberti engaged in the discussion of

---

5 For a brief biography and introduction of Alberti, see Rykwert (1998).

(a)

(b)

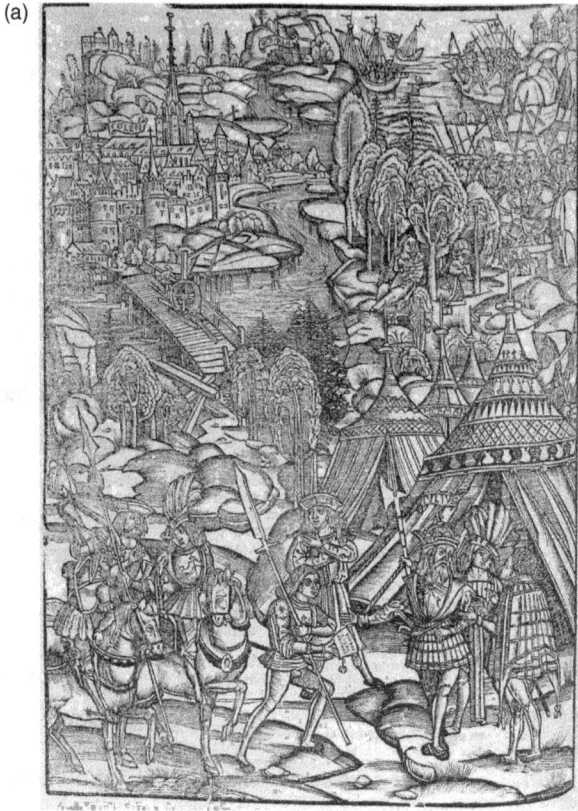

Figure 1.1 Caesar's Rhine Bridge in Matthias Ringmann's German version, 1507.
Source: Ringmann 1507.

Figure 1.2 Caesar's Rhine Bridge by François du Moulin, 1519–1520.

Source: the British Library.*

*Bridge from BL Harley 6205, f.23. Digital manuscript of the British Library, accessed: 20/7/2018www.bl.uk/catalogues/illuminatedmanuscripts/ILLUMIN.ASP? Size=mid&IllID=23179

the stone bridge, he briefly mentioned wooden bridges, using Caesar's Rhine Bridge as an example.

Alberti's writing – unlike the later tradition of architectural theory texts which were written for fellow artists, architects and craftsmen – was aimed at nobles and great merchants. He wrote in elegant Latin for the educated to read out loud. The original version of *De re aedificatoria* is text only, with no illustrations.[6] For the section on the Rhine Bridge, he copied Caesar's text almost word for word, with only tiny modifications to polish the expression. For the crucial expression *"fibulis,"* he kept Caesar's original term.

Although Alberti was the first author to take Caesar's Rhine Bridge as an object of architectural study, he did not add any new information or bring any personal understanding to the discussion.

Alberti's *De re aedificatoria* was first published in 1485, posthumously, in pure Latin. Both early Latin and Italian versions remained unillustrated. It was only in 1550, in the famous Italian translation entitled *"L'architettura di Leon Batista Alberti"* by Cosimo Bartoli – an enduringly influential version that was accompanied by an abundance of architectural woodcut drawings – that the illustration of the Rhine Bridge first appeared. But since Bartoli himself had no engineering background, the image of the bridge (Figure 1.3) merely reflected the decades-long discussion in the *Commentarii* illustration tradition. The *"fibulis"* was translated as *"legature"* following Popoleschi's example (see below) and was depicted simply by schematic rope knots at the connection of the beam and posts. This illustration is inherited in the modern English translation by Joseph Rykwert (1988), who used the term "bracket" for *"fibulae"* (108).

1.1.2.3   The first illustrator of the Rhine Bridge

The first study on the Rhine Bridge for the *Commentarii* is by Fra Giovanni Giocondo (1433–1515), a Roman Catholic monk. He was the illustrator of the 1513 Aldine edition of the *Commentarii* (in Latin). The drawing of the Rhine Bridge (Figure 1.4) appears among drawings of maps and military facilities.

To make the illustration of the bridge even clearer, he labelled the individual building elements using letters of the alphabet, with corresponding explanations. Although this drawing was not the first illustration of the Rhine Bridge—contrary to what his contemporary biographer Giorgio Vasari stated (2009, 5), the list of terms he added does make him the first to carry out a real study of the image of the bridge.

Here, the *"fibulis,"* denoted by the letter "D," are expressed as a pair of struts, forming a triangular connection between the beam and the inclined posts.

Giocondo also mentioned that he had consulted the architectural treatises by Vitruvius and Alberti to interpret such technical passages. In fact, he had not only consulted Vitruvius's writing; he was the producer of the first correct edition of Vitruvius's *De Architectura* (1511), a whole two years earlier even than his *Commentarii.* (Griffin 2009, 350–1)

His version of *De Architectura* was the first edition to be accompanied by a glossary and abundant illustrations, including an illustration of the so-called

---

6 For the reason and background for the absence of illustrations, see Carpo et al. (2007).

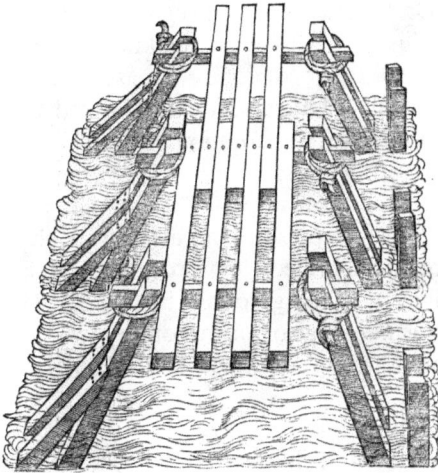

Figure 1.3 Illustration for Caesar's Rhine Bridge in the Italian translation of *De re aedificatoria* by Cosimo Bartoli (1550).

Source: Alberti 1550.

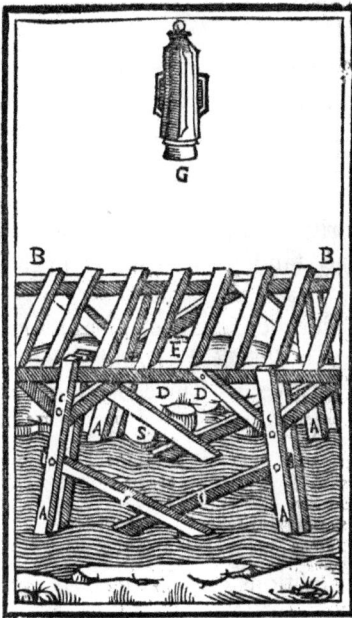

Figure 1.4 Rhine Bridge by Fra Giovanni Giocondo (1519).

Source: Caesar 1519.

"Vitruvius Man" (the proportions of the human body according to Vitruvius), made some 80 years earlier than Da Vinci's celebrated version.

However, as mentioned, Vitruvius never discussed any kind of bridges in his work, so his texts were of no help to Giocondo in this case. Alberti, too, had not really discussed Caesar's Bridge; but rather mainly copied Caesar's text. Thus, Giocondo was left to puzzle of the restoration of the Rhine Bridge in Caesar's text without references.

Unlike all the previous editors of Caesar's text, Giocondo was not only a fine scholar, with a mastery of Greek and Latin, but also a capable architect experienced in carrying out practical projects, many dealing with bridges and canals. One striking example, according to his biographer Giorgio Vasari (1511–1574), was that he was in charge of the restoration of the *Ponte della Pietra* bridge in Verona when the city was under the rule of Emperor Maximilian (1490–1516):

> It was seen to be necessary to refound the central pier, which had been destroyed many times in the past, and Fra Giocondo gave the design for refounding it, and also for safeguarding it in such a manner that it might never be destroyed again. His method of safeguarding it was as follows: he gave orders that the pier should be kept always bound together with long double piles fixed below the water on every side, to the end that these might so protect it that the river should not be able to undermine it.
>
> (Vasari 2009, 4)

We can see from this description that Giocondo's reinforcement of the piers with piles to keep them from being damaged by the water could be an effective imitation of the treatment by the Rhine Bridge. Therefore, the study of Caesar's text could have been of direct benefit to Giocondo in his constructional practices. And at the same time, Giocondo's qualifications in bridge construction would give his voice authority in the explanation of Caesar's bridge. His drawings were adopted as the standard illustrations in the centuries that followed, used in various versions and translations.

Giocondo was a friend of Aldus Manutius the Elder, the humanist, printer, and publisher, founder of the Aldine Press at Venice. From this family came later another version of the illustration of Caesar's Rhine Bridge, of which more later on.

1.1.2.4 Various Italian illustrations

Almost immediately following Giocondo, more illustrations on the Rhine bridge came out in Latin and vernacular editions of *Commentarii*, showing distinct differences from Giocondo's understanding of the "*fibulis.*"

In the 1517 Italian version *Comentarii di C. Iulio Cesare Tradotti* by Augustine Vrtica of Porta Genova, the "*fibulis*" is simply translated as "*fibule*," and illustrated by a pair of dowels (Figure 1.5).

Figure 1.5 Restoration of Rhine Bridge by Augustine Vrtica of Porta Genova (1517). *Source*: Caesar 1517.

While Dante Popoleschi's Italian version *Commentarii di Iulio Cesare* of 1518 rendered the "*fibulis*" as "*legature*" (plural form of "*legatura*," meaning "binding" or "ligature"), depicted (Figure 1.6) by a combination of rope knots (as the term "*legatura*" would indicate), and – perhaps by way of a compromise – a pair of dowels, probably because of the awareness of the illustrator, that the rope knots are far from strong enough for such a real structure.

Popoleschi's version of the commentary was praised as written in perfect Italian, the appropriate equivalent of Caesar's 'pure' Latin. Together with the awareness of their own language, it was also a period when the Italians found it necessary to acquire knowledge of French and German, the countries that had formerly been part of Caesar's empire, but which since 1494 had been waging war against Italy. "Caesar's Commentaries thus afforded Italians, depressed by their lack of military success against foreign powers in the Italian Wars, a consolatory backward glance at a happier military past" (Griffin 2009, 355).

In 1571, Aldus Manutius the Younger published a Latin version of *Commentarii, Caii Iulii Caesaris Commentariorum*, this edition including maps and military bridges, fortifications, siege towers, weapons, and some strange animals Caesar mentioned from the German forests. Thus by this time, Caesar's text was not only teaching pure Latin, virtues, military tactics, and engineering, but also European geography and natural history as well (Griffin 2009, 355).

The illustration of the Rhine Bridge provides a vivid image of the construction with a crane ramming on the posts. The structure is also marked with letters of the alphabet and an explanation of the terminology, following Giocondo's example. However, it seems that the illustrator of the woodcut (Figure 1.7) either wasn't paying attention to how a bridge is arranged

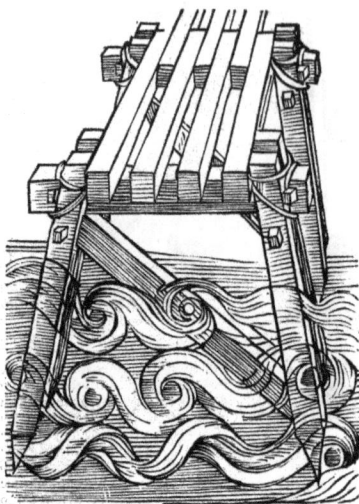

Figure 1.6 Restoration of Rhine Bridge by Dante Popoleschi (1518).
Source: Caesar 1518.

Figure 1.7 Restoration of Rhine Bridge in the Aldus Manutius the Younger' s edition
(1571/1574).
Source: Manutius 1574.

and constructed – or didn't really know, as the trestle is set by itself in the
water and in a wrong direction, and the *"fibulis"* (B) are simply expressed
as short horizontal wooden bars linking the posts. To go with the drawing,
the caption under it was 'adapted' from Caesar's original text to make it fit
better:

B. Trabes transversariae bipedales, quibus ea tigna iungebantur interuallo pedu duorum, ab vtroque latere fibulis infixis.

[B. Transverse beams of two feet, wherein the posts were joined in an interval of two feet, fastened from both sides by *fibulis*.]

Despite its many inaccuracies, the Aldus Manutius version was also reprinted several times in the sixteenth century.

Renaissance Italy has two traditions relating to the image of Caesar's Rhine Bridge: the translation and illustration of the *Commentarii,* and the new fashion for architectural treatise writing that had arisen. The most influential drawing of the bridge comes from a man who combined these two traditions – the architect Andrea Palladio (1508–1580).

1.1.3 Palladio's Rhine Bridge

1.1.3.1 I quattro libri dell'architettura

In 1570, continuing the tradition began by Alberti, Palladio published his great architectural treatise entitled *I quattro libri dell'architettura* (The Four Books of Architecture). Unlike Alberti, who had a preference for Latin, Palladio wrote in Italian and filled it with rich illustrations by his own hand.

In the section on bridges, in which wooden bridges were discussed first, he focused on Caesar's Rhine Bridge, and he mentioned proudly, that he imagined (*imaginai*) it in his youth when he first read the commentaries (Figure 1.8).

This fact is also reflected in his statements and attitude towards his and former studies: he felt that "(Caesar's Bridge) has been variously set down in designs (*disegno*) according to diverse inventions (*inuentioni*)," and that he is introducing here his "way" (*modo*) (Palladio 1570, 12; Palladio 1965, 63).

Palladio quoted Caesar's Latin text for the complete description of the bridge, and then translated it into Italian. In the translation he kept the word "*fibule*," which are expressed in his drawing as a pair of bolts, clamping the crossing of the beam and the posts and notched into them. Thus, they are able to interlock the main structural members and keep them from moving apart under load.

In their great treatises, both Palladio and Alberti paid their homage to Vitruvius, and both of them devoted one section in their work to the topic of bridges, an object totally neglected by Vitruvius. And just like Alberti, Palladio's interest in bridges was also stimulated by practical reasons.

In October 1567, the old Bassano Bridge over the Brenta river was washed away by a flood. It was a covered bridge with a trestle structure, made entirely of wood. Palladio was in charge of the reconstruction project. He first suggested a stone bridge according to the old Roman example, but this proposal was rejected by the city council. They requested that the restoration be kept as close as possible to the original bridge. Palladio presented his final design in 1569, a restoration of the old form, with new solutions to the technical and static problems (Puppi 2000, 197–8). In his *I quattro libri dell'architettura,*

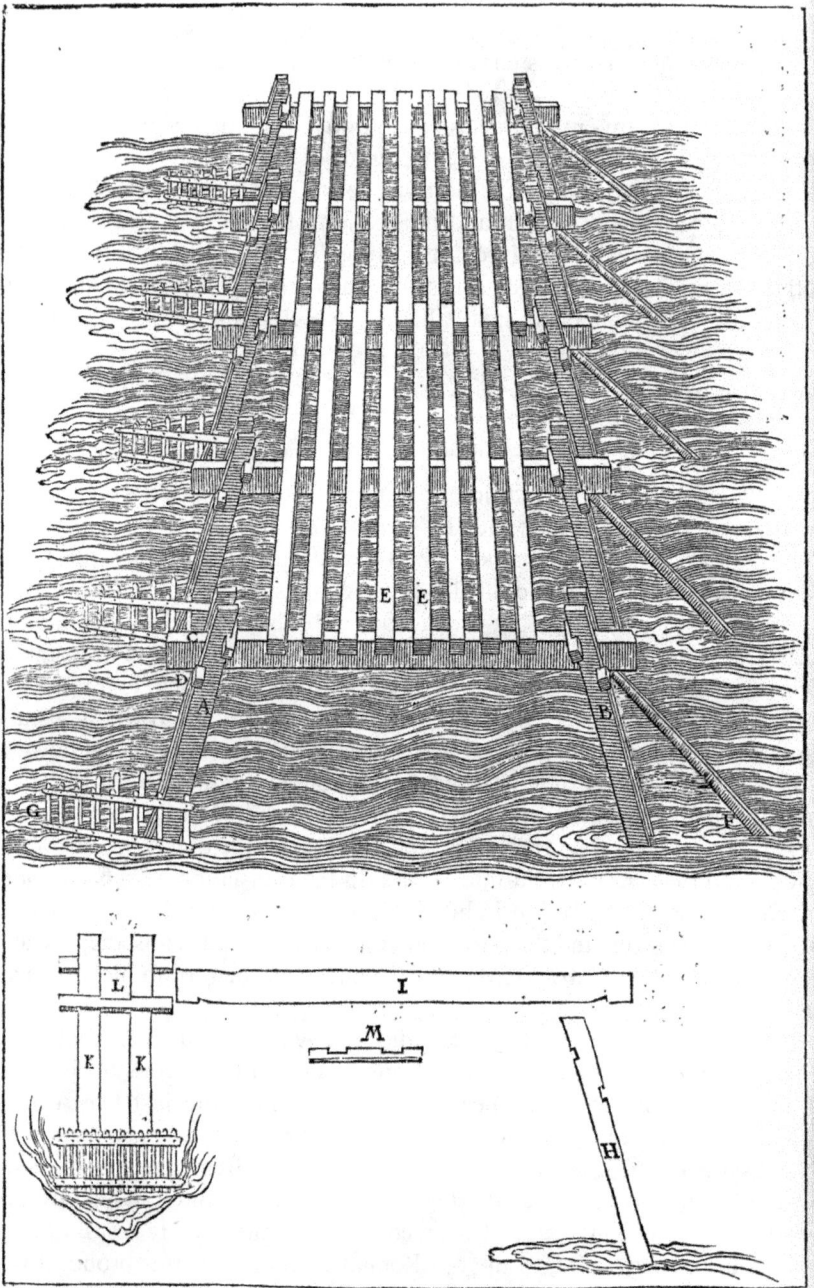

Figure 1.8 Restoration of the Rhine Bridge by Andrea Palladio (1570).
Source: Palladio 1570.

the Bassano Bridge including the plans for it, took up the last chapter of the wooden bridge section.

1.1.3.2 I Commentarii di C. Giulio Cesare

In 1575, Palladio published his edition of the Italian translation of the commentary –*Commentarii di C. Giulio Cesare*, in which the drawing of the Rhine Bridge was reproduced as in his former work, except that "*fibulis*" is rendered as "*legature*" here.

The most striking feature of Palladio's edition, aside from the usual maps, drawings on fortifications and bridges, is that it was the first edition to include numerous etchings of battles, formations, encampments, etc., from a bird's-eye view, all drawn by himself.

These etchings were Palladio's attempt to visualize Caesar's text, and his approach aimed at conveying Caesar's military and engineering techniques to his compatriots, soldiers who were uneducated country folk and who were fighting unsuccessfully to defend their country.

He dedicated this book to Captain-General Jacopo Boncompagno, and stated in his dedication that Italy had recently been "depressed" by military defeats. The ancient Roman battle skills, despite recent changes in the situation, were still worth imitating (Griffin 2009, 353).

A practical military purpose or ideal might have been shared as a common concept by all the illustrators/publishers of Caesar's work at the time, as a reflection of the anxiety of the whole society. When Giocondo published his *Commentarii*, Italy had already suffered from the Italian wars[7] for nearly two decades, and the pain of military failure had not been alleviated in the half-century that followed. That is probably the reason why the Rhine Bridge and other military facilities received much more attention in the Commentaries tradition in Italy.

Palladio's drawings started a new illustrative tradition for the *Commentarii* in the centuries that followed, while the drawings of the battle orders were gradually reduced. In some later (nineteenth-century) editions, his Rhine Bridge or an unattributed copy of his original idea are all that remain in terms of illustrations.

1.1.3.3 Feasibility of Palladio's restoration

Caesar's description of his Rhine Bridge served as the starting point of Palladio's "invention." Although comply with Caesar's description, it is technically, less feasible for Caesar's battlefield. In his "design," posts and beams are chopped into square sections; the position of the bolts (*fibulis*), and the

---

7 Or the Renaissance Wars, a series of conflicts from 1494 to 1559, that involved most of the area of modern-day Italy.

depth and angle of their notches needed to be accurately calculated and executed, otherwise either the joint would become too loose to be stable, or a large error would be introduced into the measurements related to the inclination of the posts.

Palladio's design is an exquisite device for precise and on-shore manufacture with an ideal and calm river environment. It would hardly be possible to process such sophisticated joints in the water, in situ. But according to Caesar's description, the height of the posts could only be decided according to the conditions of the river.

Palladio's design, as he said himself, was a "way of invention" inspired by Caesar: neither a true historical reconstruction nor even a real construction plan. Rather, it was an intellectual puzzle to be solved by a smart young man, as he proudly stated, and as a challenging puzzle, it was solved with the joy of intelligence and then left aside.

On subsequent pages of *I quattro libri dell'architettura*, Palladio changed his focus to a new theme: wooden truss bridges. He designed four types of truss bridges (Figure 7.15), none of which bear any relationship with the Rhine Bridge concept. As a pioneer of the truss study, Palladio's profound influence has extended right down to modern iron bridges and structural science.

However, Palladio was not the first man to solve the Caesar puzzle this way. Even before he was born, another wise man had come up with a similar but more technical answer. That brilliant mind belonged to Leonardo da Vinci (1452–1519).

1.2 Leonardo da Vinci, the genius mind

1.2.1 Leonardo da Vinci's study on Caesar's Rhine Bridge

Unlike other Renaissance architects, Leonardo da Vinci's manuscripts were not published during his lifetime or even soon after. They were kept in private ownership for a long time, during which they were rearranged in terms of the order. The preserved manuscripts have been reorganized into different books (codices) roughly sorted according to the type of content. The bridge-related pages are spread throughout different codices. Thus, it is impossible for us to reconstruct the exact original timeline. Apart from limited clues as to a general time-scope for the drawings, the logic of the timeline of the drawings has to be 'dug out' from their content to restore Da Vinci's train of thought for our study.

In his letter to the Duke of Milan (c.1481–2) applying for a job as a war engineer, Da Vinci listed all his abilities for this position. At the top of this list was his skill in bridge construction:

> I have plans for very light, strong and easily portable bridges with which to pursue and, on some occasions, flee the enemy, and others, sturdy and indestructible either by fire or in battle, easy and convenient to lift

and place in position. Also means of burning and destroying those of the enemy.

(Kemp and Walker 2001, 251)

The advantages he lists about the bridges that are within his ability to build strongly remind us of the Rhine Bridge in Caesar's narration, viz. lightness, convenience, ease of assembly and disassembly; strong, firm and stable.

Such kinds of bridges are also to be found in Da Vinci's drawings of various type of military equipment. Folio 58 of the Codex Atlanticus (Figure 1.9) contains a group of constructional details, and the inclined posts grouped in pairs, the beam (either round or square) held in between, and the *fibulis* identical in shape to those in Palladio's later design, provide proof of Da Vinci's involvement in the subject of the Rhine Bridge. The abundant varieties of the knotting experiments give us an idea of his meticulous considerations in the study.

In the two sketches at the bottom of this folio, by adding a set of linking ropes pulling the whole structure together, Da Vinci shows that he realized the peculiarity of this type of device: the structure requires a centripetal force for stability. It is provided by the pulling ropes as a "prestressing force." The load from above on the beam and the lateral thrust from the side acting on the upper part of the posts were also beneficial to the structure, whereas reversed forces pushing from inside outwards would be a threat to the structure. This structural feature recalls Caesar's description: "the stability of the structure was so great and its character such that, the greater the force and thrust of the water, the tighter were the balks held in lock."

1.2.2 The clamp bridge

Da Vinci was never a *Commentarii* illustrator, for whom a combination of faithfulness to the original text and the feasibility of the restoration was the final aim. Instead, he was first and foremost a practical military engineer. The study of Roman wars gave him support for this goal. He didn't stop with Caesar's text, but rather, the idea of such an interlocking joint inspired him to develop further onto the whole structural principle.

Folio 902 of the Codex Atlanticus (Figure 1.10) depicts two types of bridges. Both of them are longitudinal structures (i.e. the main beams run along the direction of the bridge). Each of the horizontal beams is supported by an X-shaped stand which is composed of a pair of slanting posts.

In the centre of the folio (a) is an image which is drawn in clearer and more well-defined lines, depicting a bridge-type where the main beams are each supported by a row of X-stands (although only one is depicted). Each X-stand stands and functions independently.

Although the structure on the left (b) looks similar at first glance, there is a significant difference. Its X-stands of the parallel horizontal beams are set in pairs and are tied together in the middle. This would improve the stability of the whole structure.

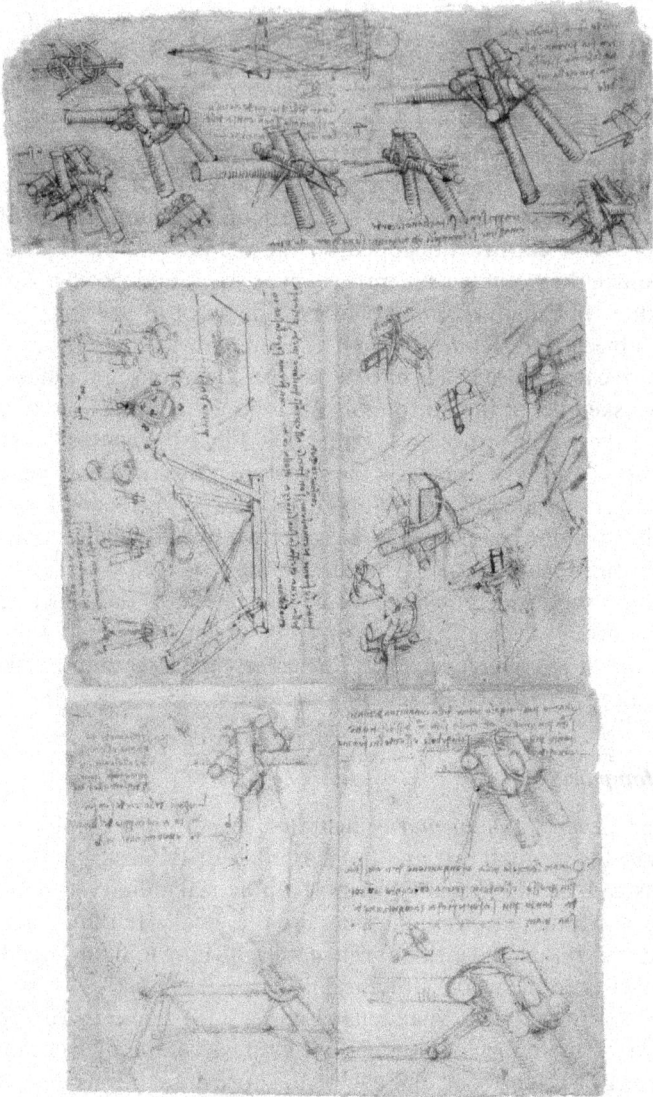

Figure 1.9 Leonardo Da Vinci, Codex Atlanticus, folio 58.
Source: ©Veneranda Biblioteca Ambrosiana/Mondadori Portfolio.

The small sketches at the top and bottom of the same folio (c) demonstrate a more interesting idea. The two posts of the X-stand now work as a clamp (bridges with such construction are called "clamp bridge" in this chapter) for holding the pair of longitudinal beams. For it to be feasible, the slope of the slanting posts must be substantially gentler.

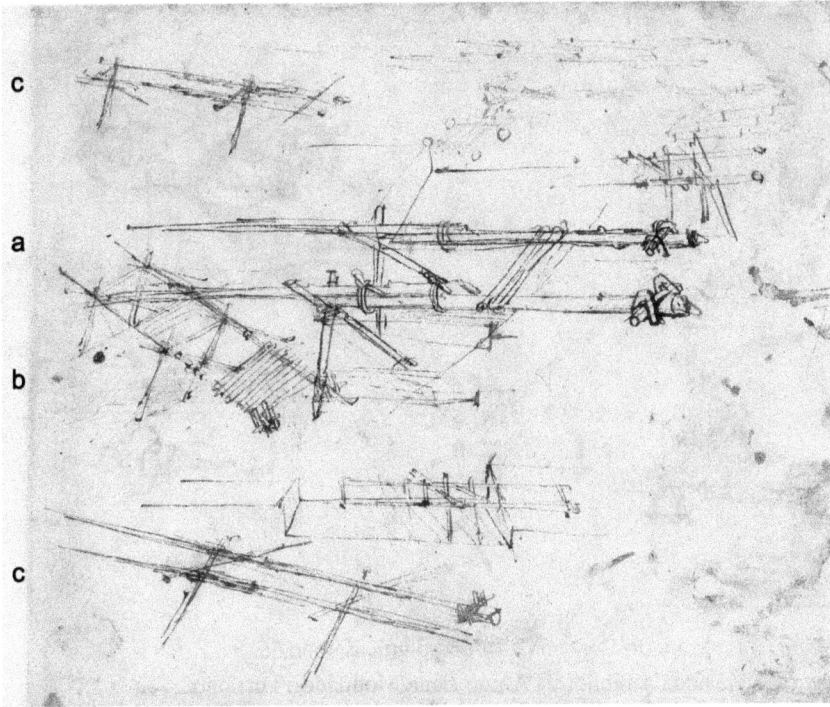

Figure 1.10 Leonardo Da Vinci, Codex Atlanticus, folio 902B.
Source: ©Veneranda Biblioteca Ambrosiana/Mondadori Portfolio.

It's not hard to tell that the clamping mechanism is a direct development from the Rhine Bridge: an exact magnification of the structure of the X-shaped joint (Figure 1.9). The construction Da Vinci studied over and over in his research on Caesar's bridge.

At first glance, the bridge drawings on folio 55 of the Codex Atlanticus (Figure 1.11) look like a more formal version of the sketches in folio 902. But in fact, there are crucial differences. The drawings in folio 902 are of different types of bridges, whereas in folio 55 they show the construction steps for a single bridge.

In the text of folio 55, written by Da Vinci in mirror-writing, he articulated the idea of processes of the construction:

Text at the top of the page:

Armadure. In che modo si debbe porre alcuno ponte con brevità, atti a fuggire o seguitare il nemico.

[Armature. How to construct a bridge rapidly, for fleeing from or pursuing the enemy.]

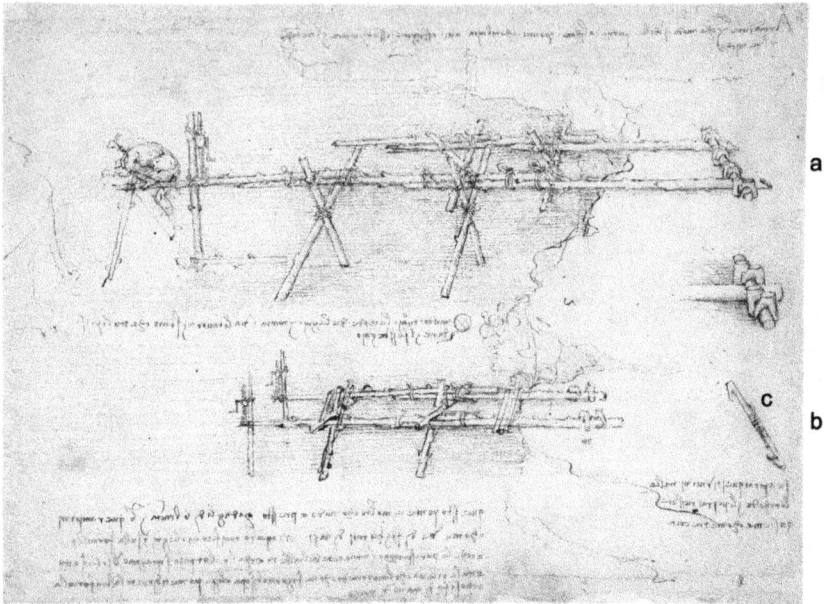

Figure 1.11 Leonardo Da Vinci, Codex Atlanticus, folio 55.
Source: ©Veneranda Biblioteca Ambrosiana/Mondadori Portfolio.

Text at the centre of the page:

Quando tu hai le code de' legni in aria, va li tanto in sommo che tu li possi dare il sostegno.

[When you have the tips of the woods (wooden posts) in the air, position them so high that it is possible for you to support them.]

Text at the bottom of the page:

Questo ponte è molto comodo e presto, ma dagli di sei braccia in sei que' rampini come vedi figurati dinanzi. Ma quando tu hai fermi i legni in sulla forcella e che le punte sportano in aria cioè de li alberi, e che pel lor peso si piegano verso l'acqua, e tu le rileva col martinetto, com'è figurato di sopra, e lega. Po' metti dirieto l'altra forcella, e così fa di mano in mano.
(Marinoni 1974, 116)

[This bridge is very convenient and quick to build, it is carried forward six arms length by six. But when you fix the woods (wooden posts) on the fork and the tips protrude in the air, and due to their weight bend toward

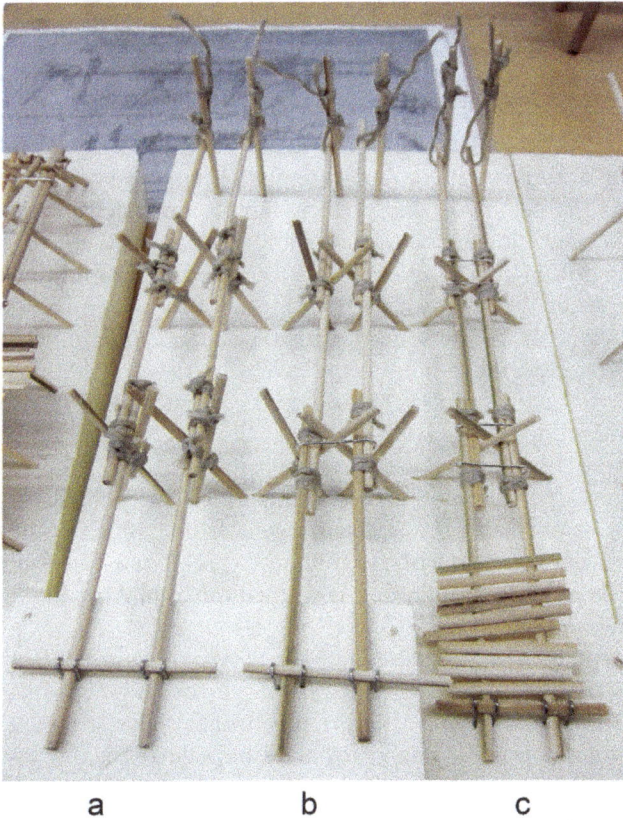

Figure 1.12 Wooden model of the image in folio 55, Codex Atlanticus.
Source: Made by the author.

the water, you lift them with a hoist, as shown above, and bind. Then position the next fork, and so on.]

The drawing above (Figure 1.11, a) shows the first step for positioning the beams in place. The X-stands underneath now function more as scaffolding (Figure 1.12, a). Afterwards, the pair of the posts of the X-stands, those with their feet at the external sides of the bridge, will be bent downwards, until they reach the beam of the opposite side. When this is done for both sides of the stands, the structure acts as a clamping device (b). Hooks (c) are then used to keep the beams in place, and those posts that are not bent and tied and are left over can be removed, and the structure is finished, as shown in Figure 1.11(b) (cf. Figure 1.12, c).

When comparing the drawing of the first step in folio 55 (Figure 1.11, a) with the central drawing in folio 902 (Figure 1.10, a and b), we notice two essential differences. Firstly, in folio 902, narrow wooden boards are used to make the deck of the bridge, defining them as finished bridge structures. Its counterpart in folio 55 is naked, without a deck, indicating that it's an unfinished structure, while in the drawing below (b) it is ready-decked as the final step of the construction.

Secondly, in the bridge shown in folio 902, the beams are resting right at the intersection of the X-stands, a natural position for a beam to sit on branching support. Whereas the beams of step one in folio 55 are fastened onto the outward posts, those would later be removed; and be positioned at a distance above the intersection of the X-stand. This is because the beams are not in their final place. After the beams have been moved into position, clamped between the inward posts, the outward posts will be untied and removed.

These details give clear indications as to Da Vinci's trains of thought with regards to the construction process when he was inventing a new type of bridge (clamp bridge).

The shortcomings of this structure are also obvious. If the beams are kept at a certain distance apart to allow an appropriate deck width for the troops to cross, the posts would finally rest at a rather gentle slope, which would dramatically reduce the height of the bridge (Figure 1.12, c). For a higher bridge, the beams must be positioned rather close to each other, making the bridge quite narrow (Figure 1.13).

To solve this problem, Da Vinci designed a new type of clamp bridge with double X-stands (Figure 1.14), each clamping a pair of narrowly-arranged longitudinal beams. By doubling the formerly troublesome structure, the large-scale bridge now has bulkily-arranged posts and a flexible width, enough for any army.

One notable characteristic of this bridge design lies in the caption beneath this drawing: Da Vinci clearly expressed the idea of "weaving" a bridge ("*tessere il ponte*") (Giorgione 2009, 122–3). This will be discussed in more detail later.

If we look at this structure in Figure 1.14 more closely, we see that Da Vinci did not simply place two X-stands together, but added an additional pair of longitudinal beams in the centre, crossing at the post feet. The longitudinal beams serve as another pair of clamps, exactly like those above. This is a means to increase the stability of the whole frame. It also leads to an interesting result: what we now have is not two woven frames standing shoulder to shoulder, but rather an "M"-shaped folding structure, with a woven joint at every three intersections.

Moreover, at the end of the bridge, a group of four posts, perpendicular to the other posts seen from the horizontal plane, are woven into the structure using the same principle and with the help of a pair of deck-beams clamped in between. This adds yet another dimension of the woven mechanism.

1.2.3 The woven bridge

Folio 57B is the left half of a larger page, while the right half appears in the same book, viz. Codex Atlanticus, in folio 69A. On this page, we can find

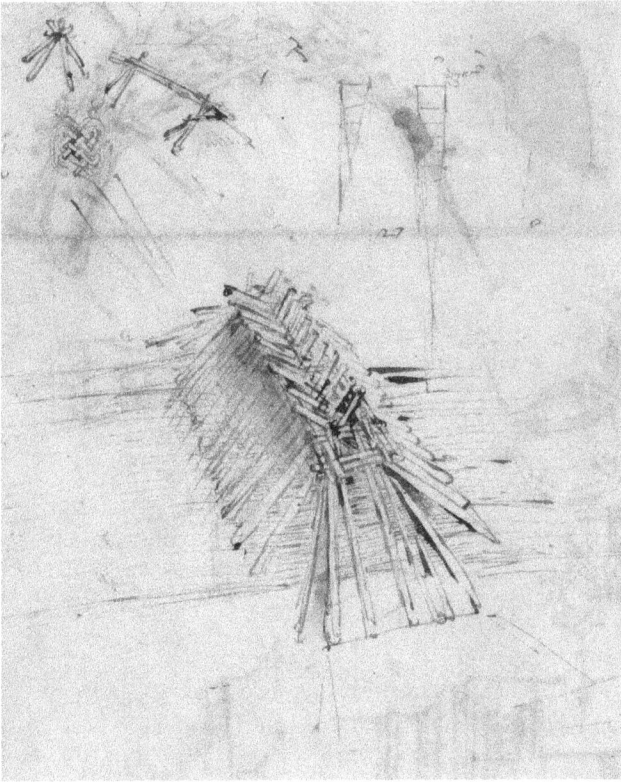

Figure 1.13 Leonardo Da Vinci, Codex Atlanticus, folio 71A.
Source: ©Veneranda Biblioteca Ambrosiana/Mondadori Portfolio.

sketches of the most attractive and well-known of Da Vinci's bridge designs, a "woven arch bridge" (Figure 1.15).

When the two folios are put together in their original position, we notice the smaller faint drawings at the connecting edges. They most probably show the respective construction steps. The one in the centre is the first step for the double X-stands bridge (b), with the longitudinal beams already in situ, the posts of the stands are grouped into fours and function as the scaffolding. From here people could push, or in Da Vinci's words, "bend" the pairs of posts into the correct position, and fix them with the longitudinal beams.

Above them (a), the group of three sketches indicates the building process for the woven arch bridge. The first step is to set the longitudinal beams one after another, then add the cross beams and tie them together (Figure 1.16). Ropes are used to ensure the feasibility of the construction.

It has been long observed by scholars that the two folios (folio 57 and folio 69) must have come from a single page (and had been separated by an earlier collector). Yet it has not yet been realized that the two main structures of

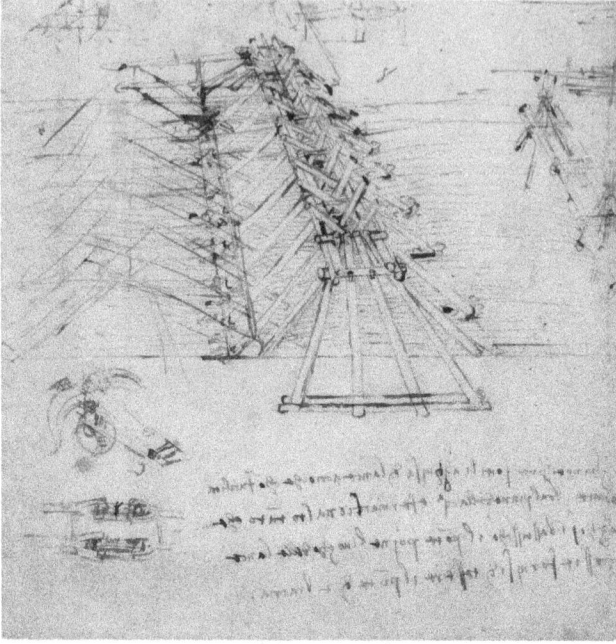

Figure 1.14 Leonardo Da Vinci, Codex Atlanticus, folio 57B.
Source: ©Veneranda Biblioteca Ambrosiana/Mondadori Portfolio.

these two halves, though different in appearance, belong to the same train of thought, a leap of thought for a peculiar invention.

How did Da Vinci get and develop the idea of a woven bridge? The sketch at the bottom of folio 183, Codex Atlanticus (Figure 1.17) could provide a clue. Here at the left side is a woven arch bridge, like the one in folio 69 (Figure 1.15), while at the right side are the details of its joints, which are also identical to the joints of Caesar's Rhine Bridge (Figure 1.9) and the X-stands of the clamp bridges (Figure 1.10, c).

From the principle of the clamp bridge, Da Vinci discovered the woven arch principle. This constructional principle freed the structure theoretically from the additional connecting elements involved in the clamp bridge (ropes and hooks). In the woven bridge, all beams interlock. The crossbeams are now clamped between the longitudinal beams, like the warp of a textile.

The idea of a woven arch bridge fascinated Da Vinci so much that he studied it over and over to explore all possible variations. In Codex Madrid folio 45 (Figure 1.18), at the background of a spiral-shaped machine, in light ink, there is a woven arch with the characteristic double crossbeams. The next folio of the same Codex (Figure 1.19) demonstrates the simplest unit of this kind of woven structure and could be seen as a transition from Caesar's Rhine Bridge

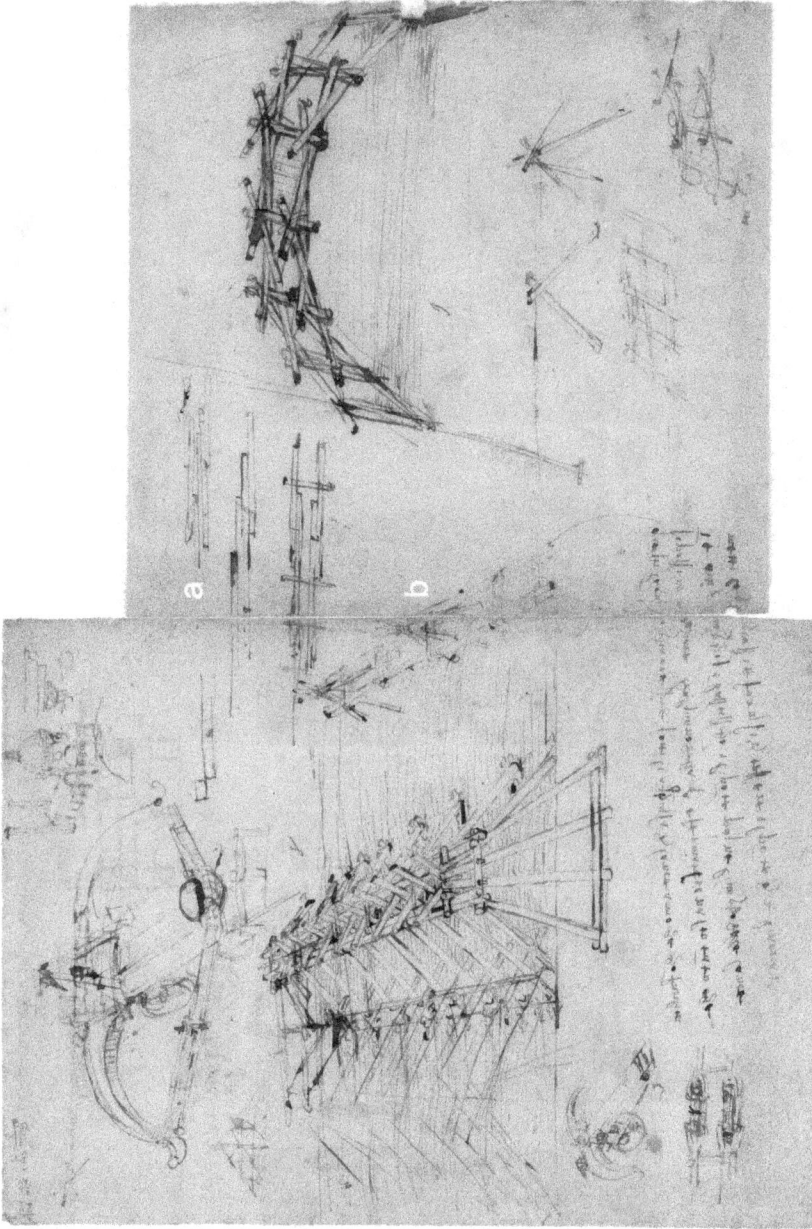

Figure 1.15 Leonardo Da Vinci, Codex Atlanticus, recombination of folios 57 and 69.
Source: ©Veneranda Biblioteca Ambrosiana/Mondadori Portfolio.

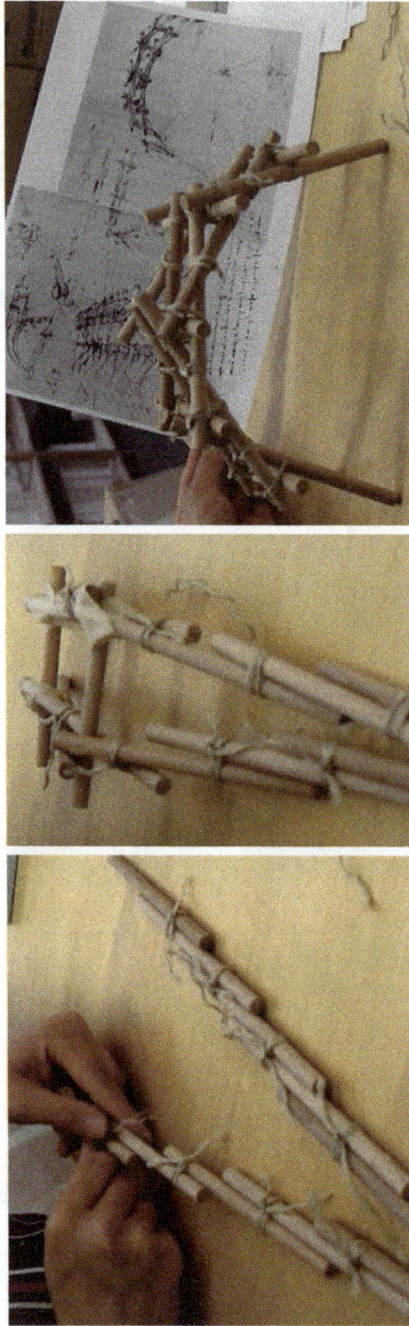

Figure 1.16 Model demonstration of the assembly process according to the Codex Atlanticus 57 and 69.

Source: Made by the author.

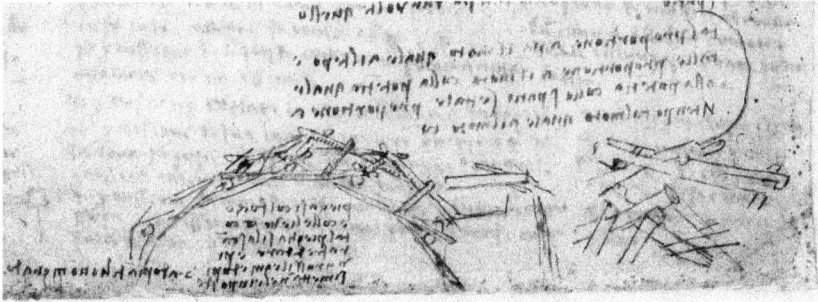

Figure 1.17 Leonardo Da Vinci, Codex Atlanticus, folio 183, the bottom of the page.
Source: ©Veneranda Biblioteca Ambrosiana/Mondadori Portfolio.

to the woven arch bridge. In folio 71 of the Codex Atlanticus (Figure 1.20), Da Vinci tried to achieve a broader bridge width by offsetting a parallel arch. This structure also comprises more woven units, and almost forms a semicircle.

There is no doubt that Da Vinci's woven arch bridges are designed to be devices of war. They have a strong character as temporary and even movable constructions. They are designed to be built on land and then laid over water, and for a group of soldiers to assemble and transport them. Consisting of unprocessed logs without woodworking joints, despite the structural feature of an interlocking mechanism, the bridges still require ropes or other fixing elements. Folio 46 (Figure 1.19) of the Codex Madrid shows how the necessary fixing is provided by the natural forks in the branches.

1.2.4 The woven vault

Elaborating on his thoughts on the woven arch, Da Vinci made a striking jump, from a two-dimensional woven structure (woven arch) to a three-dimensional woven structure (woven vault).

He took basic geometric shapes (Figure 1.21), namely regular polygons, squares and triangles/hexagons as basic units, and expanded them under the woven principle to achieve an extending structure.

Though drawn in a horizontal plane, they are certainly vault structures. The beams of the woven vault interlock. Just like the woven arch, every beam is supported at its ends by its neighbours, while supporting those same or other neighbouring beams at the middle position.

Da Vinci's woven vault structure, or the "reciprocal frame structure," as it is called by modern researchers, has been in the spotlight in recent decades.[8] The

---

8 Among the plentiful studies, there are several systematic monographs on this topic: Larsen (2008), Bertin (2012), Thönnissen (2015)

Figure 1.18 Leonardo Da Vinci, Codex Madrid, folio 45.
Source: ©Biblioteca Nacional de España.

idea of forming a larger plane area (such as a floor or a deck) using smaller pieces of wood can be traced back to the thirteenth-century master builder Villard de Honnecourt. One of his manuscripts records the method used by carpenters in the Middle Ages to span a floor or roof with shorter beams (Figure 1.22). Each member supports and is supported by its neighbours. With this "reciprocal" strategy, larger spans could be reached, theoretically without limit. This floor structure continued to be studied in the Renaissance,

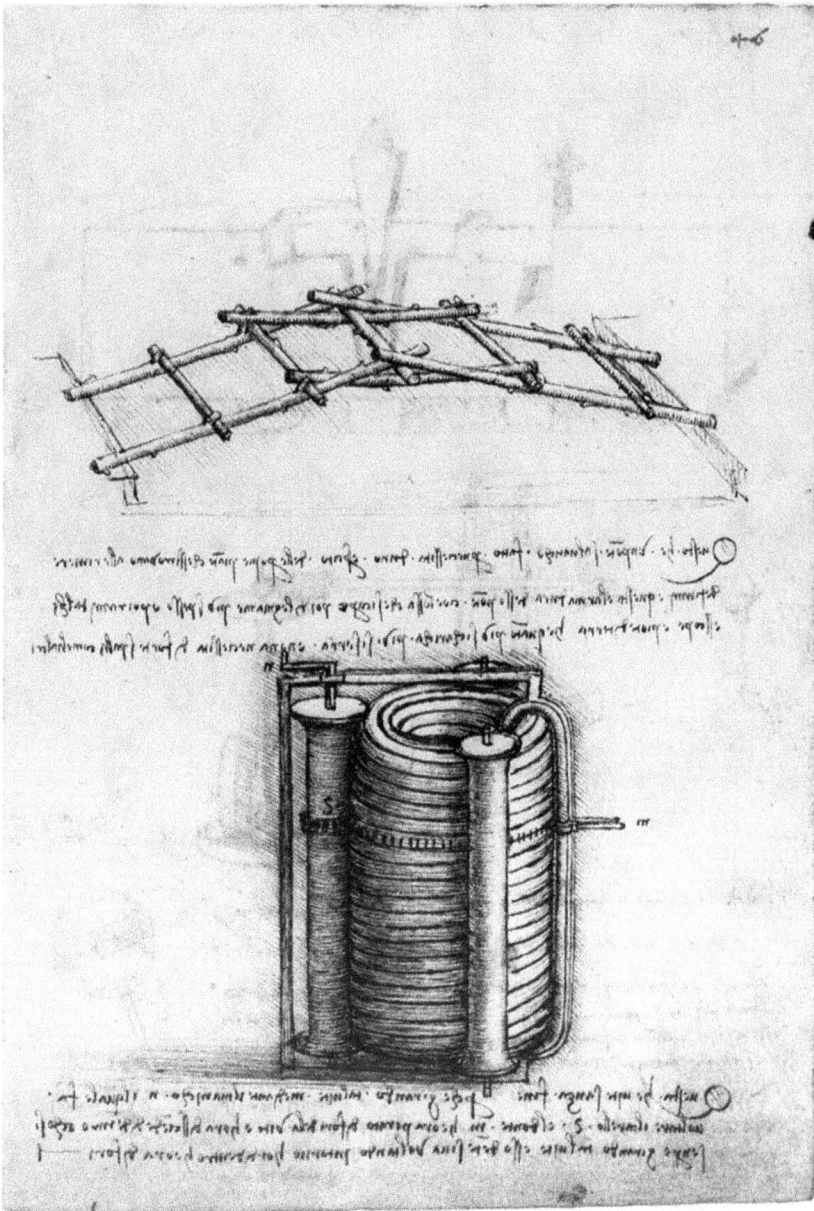

Figure 1.19 Leonardo Da Vinci, Codex Madrid, folio 46.
Source: ©Biblioteca Nacional de España,

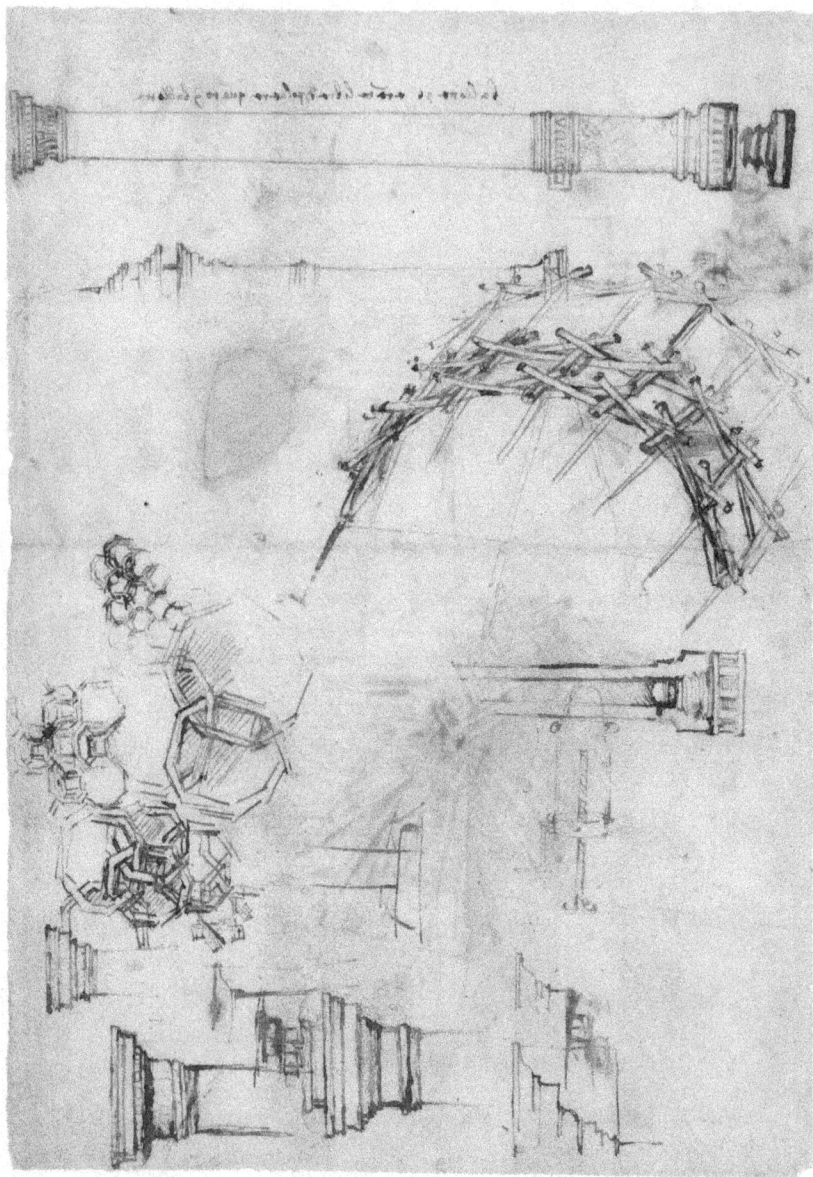

Figure 1.20 Leonardo Da Vinci, Codex Atlanticus, folio 71B.
Source: ©Veneranda Biblioteca Ambrosiana/Mondadori Portfolio.

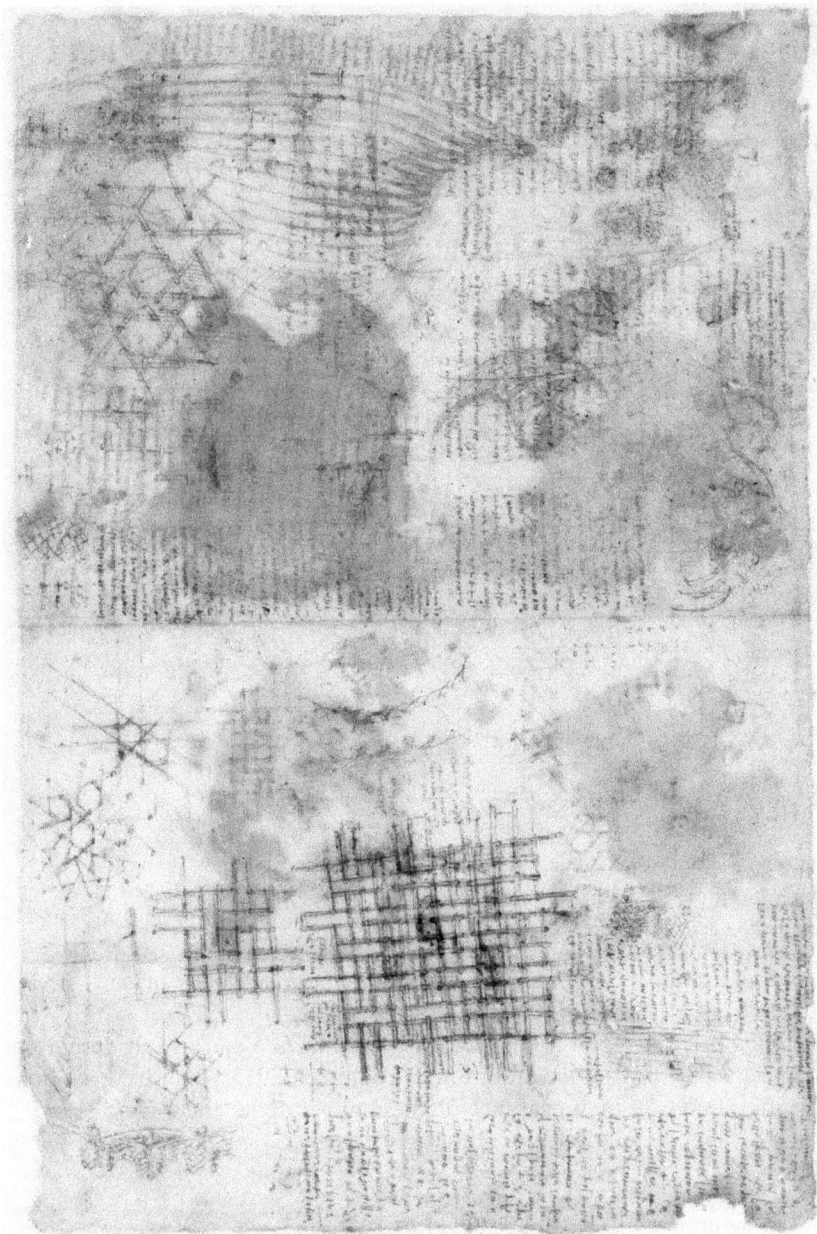

Figure 1.21 Leonardo Da Vinci, Codex Atlanticus, folio 899.
Source: ©Veneranda Biblioteca Ambrosiana/Mondadori Portfolio.

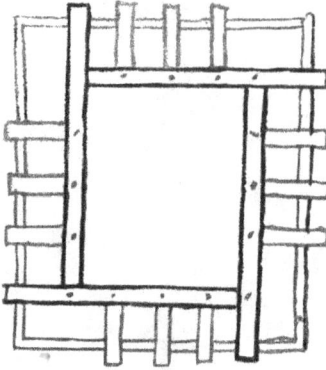

Figure 1.22 Sketch by Villard de Honnecourt.
Source: Villard 1858, table 45, fol. 23r.

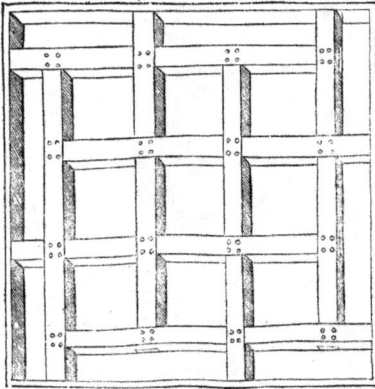

Figure 1.23 Planar system design by Sebastiano Serlio.
Source: Serlio 1584, fol 15v.

including architect Sebastiano Serlio, after whom one type of such floor is named (Figure 1.23).[9]

Some studies suggest a connection between Da Vinci's vault design and Honnecourt's construction tradition (Thönnissen 2015, 46). However, apart from the similarity in appearance of these two designs, there is no clear evidence of such a relationship. Although it is true that, in terms of planar

9 On the history of the reciprocal floor structure, see Thönnissen (2015).

(a)

(b)

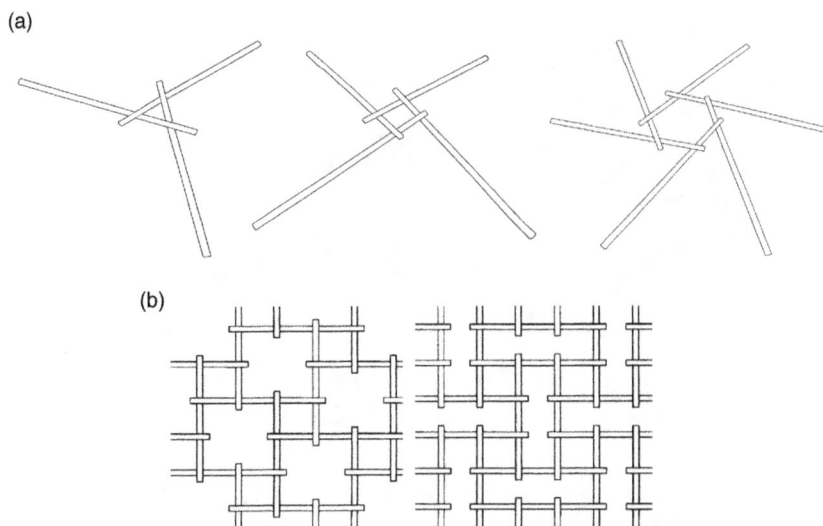

Figure 1.24 Two prototypes of reciprocal structure. (a) Radiate polygon (rotating pyramid); (b) Homogeneous grid (weaving textile).

geometry, a radiate four-beam unit could be seen as the basic or prototype form of both Da Vinci and de Honnecourt's ideas, similarity does not necessarily imply imitation.

The most important thing to bear in mind is that Da Vinci's designs are vault-shaped structures, whereas those of all his contemporaries are planar.

This difference in the shape of the structures relates to the level of practicality. The planar floor design, which is the most common use of the reciprocal structure before modern times, is absent from Da Vinci's design. Compared to the constructional problems faced by his colleagues, Da Vinci's vault structure was more of a mathematical problem than one of practical utilization – the result of a curiosity surrounding the possibilities for different forms.

From a topographical view, the reciprocal frame structure includes two basic prototypes: a rotating radiate polygon and a textile grid (Figure 1.24).[10] Most of the usual reciprocal frame structures today can be classified as one of these two types or a combination of the two.

On the other hand, Da Vinci's designs are clearly inspired by the world of textiles, which is less articulated in the design of his colleagues. In Da Vinci's grid pattern design, the idea of "weaving" is indisputably the consistent idea in both the arch and the vault study. Woven textiles were drawn multiple times on the same folio at the side of the vaults (Figure 1.25, a), and they are

10 The most thorough and theoretical typology discussion is by Bertin (2012): it starts with the "grid question," and takes the rotating unit as the basic design unit.

(a)

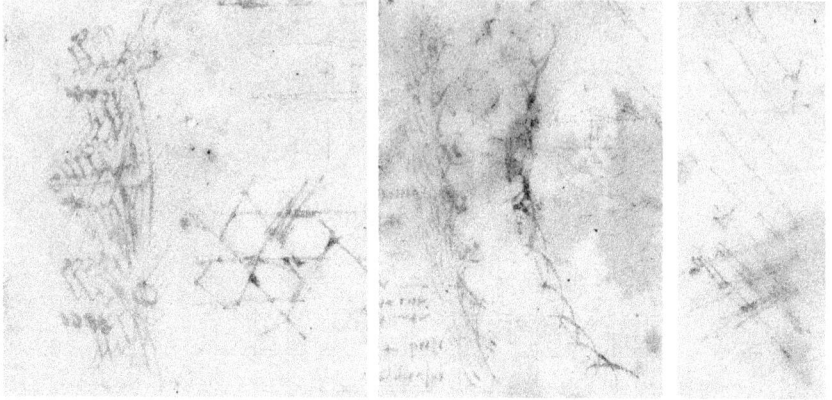

(b)

Figure 1.25 Details of Da Vinci's study on weaving construction, at the side of the woven bridges and vaults.
a. in Folio 899
b. in Folio 69

identical to the one under the arch in folio 69A (Figure 1.25, b), which, as mentioned, is next to the text where Da Vinci clearly wrote "weave the bridge" ("*tessere il ponte*").

Throughout Da Vinci's works, starting with his study of Caesar's Rhine Bridge, to his own innovations on the clamp bridges, the arch-shaped bridges and the vault-shaped structures, there is a continuous inspiration: to weave a structure. Whereas the woven vault may have remained a mental experiment before modern times, we are able to find some historical attempts to build the woven arch. During his service to Cesare Borgia as a war engineer, Da Vinci's bridge-building capabilities were well documented. In a report, his friend, the mathematician Luca Pacioli wrote:

One day Cesare Valentino, Duke of Romagna and present Lord of Piombino, found himself and his army at a river which was 24 paces wide, and could find no bridge, nor any material to make one except for a stack of wood all cut to a length of 16 paces. And from this wood, using neither iron nor rope nor any other construction, his noble engineer [Da Vinci] made a bridge sufficiently strong for the army to pass over.

(Strathern 2009, 138)

Although the description does not go into detail as to how exactly the bridge was constructed, the simplest woven arch bridge from Da Vinci's sketch (Figure 1.19) would fit the task description. However genius, Da Vinci was not the only person in history to experiment on such a woven arch bridge idea. More creations are to be found in the next chapters.

References

Alberti, Leon Battista. 1485. *De re aedificatoria*. Florenz.

Alexander, Jonathan J. G. 1992. *Medieval Illuminators and Their Methods of Work*. Haven and London: Yale University Press.

Bartoli, Cosimo, and Leon Battista Alberti. 1550. *De re aedificatoria*. Venice.

Bertin, Vito. 2012. *Leverworks: One Principle, Many Forms*. Beijing: China Architecture & Building Press.

Bundgard, J. A. 1965. "Cäsar Bridges over the Rhine." *Acta Archaeologica* 36: 87–103.

Caesar, Gaius Iulius. *Commentarii de bello gallico*. Edited by

1513. Giocondo, Giovanni. Venetiae

1517. Vrtica, Agostino, della PortaGenovese. Vitali

1519. Giocondo, Fra Giovanni. Venetia

1571 [1574]. Manutius, Aldus. Lugdunum

1575. Manuzio, Aldo. Venetiis

1575. Palladio, Andrea. Venedig

1588. Gussano. Lugdunum

1980. Reclam, Philipp. 195–8.

Caesar, Gaius Iulius. 2006. *The Gallic War*. ed. by Edwards, Henry J. Mineola: Dover Publications, 62.

Carpo, Mario, Francesco Furlan, Jean-Yves Boriaud, and Peter Hicks. 2007. *Leon Battista Alberti's Delineation of the City of Rome (Descriptio Vrbis Romæ)*. Vol. 335. Arizona Center for Medieval and Renaissance Studies (ACMRS).

Cohausen, August. 1867. *Cäsar's Rheinbrücken philologisch, militärisch und technisch untersucht, etc*. Leipzig: Teubner.

Da Vinci, Leonardo. *Codex Atlanticus* (Biblioteca Ambrosiana)

Da Vinci, Leonardo. *Codex Madrid* (Biblioteca Nacional de España). www.bne.es/en/Colecciones/Manuscritos/Leonardo/index.html.

De Hamel, Christopher. 2001. *The British Library Guide to Manuscript Illumination: History and Techniques*. Toronto: University of Toronto Press.

De Hamel, Christopher. 1992. *Scribes and Illuminators*. Toronto: University of Toronto Press.

De Honnecourt, Villard. 1858. *Album de Villard de Honnecourt; Manuscrit publié en fac-simile annoté... par JBA Lassus; ouvrage mis au jour... par Alfred Darcel.* Paris: Imprimerie impériale.

De Honnecourt, Villard. 1959. *The Sketchbook of Villard de Honnecourt.* edited by Bowie, Theodore Robert. Bloomington: Indiana University Press, 74.

Drachmann, Aage Gerhardt. 1965. *Cäsars bro over Rhinen.* Copenhagen: G.E.C.GADS Forlag.

Frank, Tenney. 1928. *Catullus and Horace.* New York: Henry Holt and Company.

Giorgione, Claudio. 2009. *Leonardo da Vinci: The Models Collection.* Edited by Museo Nazionale della Scienza e della Tecnologia Leonardo da Vinci. Milan: Museum Collections.

Griffin, Miriam. 2009. *A Companion to Julius Cäsar.* Malden, Oxford: John Wiley & Sons, 317–34, 350–8.

Gundolf, Friedrich. 1904. *Cäsar in der deutschen Literatur.* No. 33–37. Berlin: Mayer & Müller.

Kemp, Martin, and Margaret Walker. 2001. *Leonardo on Painting: An Anthology of Writings by Leonardo da Vinci, with a Selection of Documents Relating to His Career as an Artist.* New Haven, Conn.: Yale University Press.

Larsen, Olga Popovic. 2008. *Reciprocal Frame Architecture.* Oxford: Routledge.

Marinoni, Augusto. 1974. *Il codice atlantico della Biblioteca Ambrosiana di Milano.* Volume Secondo. Florenz: Giunti-Barbèra. 116.

McDermott, W. C. 1983. "Mamurra, eques formianus." *Rheinisches Museum für Philologie.* 126, no. H.3/4: 292–307.

Menge, Rudolf. 1885. "Ein beitrag zur construction von Caesars Rheinbrücke, Caes. BGall. IV, 17." *Philologus* 44, 279–90.

Moulin, François du and Albert Pigghe. 1519. *Commentaires de la Guerre Gallique.* France, Central (Paris or Blois),

Palladio, Andrea. 1570. *Quattro libri dell'architettura.* Venedig, 11–30.

Palladio, Andrea. 1965. *The Four Books of Architecture.* Mineola: Dover Publications. 62–73.

Popoleschi, Dante. 1518. *Commentarii di Iulio Cesare.* Firenze.

Puppi, Lionello. 2000. *Andrea Palladio: das Gesamtwerk.* Munich: Dt. Verl.-Anst.197–8.

Rheinhard, August. 1883. *C. Jul. Cäsar's Rhein-Brücke: eine technische-kritische Studie.* Stuttgart: Verlag von Paul Neff,

Ringmann, Matthias. 1507. *Julius der erst römisch Keyser von seinem Kriege.* Straßburg.

Ruffel, Pierre, and Jean Soubiran. 1962. *Vitruve ou Mamurra?: Pierre Ruffel et J. Soubiran.* Paris : Faculté des Lettres.

Rykwert, Joseph, Neil Leach, and Robert Tavernor. 1988. *On the Art of Building in Ten Books.* Cambridge: MIT Press.

Rykwert, Joseph. 1998. "Theory as rhetoric: Leon Battista Alberti in theory and in practice." In *Paper Palaces: The Rise of the Renaissance Architectural Treatise.* Edited by Vaughan Hart with Peter Hicks, 33–50. New Haven: Yale University Press.

Saatmann, Karl, Emil Jüngst, and Paul Thielscher. 1939. *Cäsars Rheinbrücke.* Berlin: Weidmann Verlagsbuchhandlung.

Serlio, Sebastiano, 1584. *Tutte l'opere d'architettura di Sebastiano Serlio bolognese,* Venedig.

Schleussinger, August. 1884. *Studie zu Cäsars Rheinbrücke.* München: J. Lindauer'sche Buchhandlung.

Schramm, Erwin. 1922. *Cäsars Rheinbrücke55 v. Chr.* Berlin: Gruyter.

Strathern, Paul. 2009. *The Artist, the Philosopher, and the Warrior*. New York: Random House Publishing Group, 138.

Tampone, Gennaro. 2003. "Palladio's timber bridges." In *Proceedings of the First International Congress on Construction History: Madrid, 20th–24th january 2003*. Madrid:Instituto Juan de Herrera.

Thielscher, Paul. 1961. "Vitruvius." In *Paulys Real-encyclopädie der classischen Altertumswissenschaft*. 17.2 Reihe 2, 419–89. Halddb. Stuttgart.

Thönnissen, Udo. 2015. *Hebelstabwerke: Tradition und Innovation*. Zurich: gta Verlag.

Vasari, Giorgio. 2009. *Lives of the Most Eminent Painters Sculptors and Architects: Vol. 06 (of 10) Fra Giocondo to Niccolo Soggi*. translated by Gaston du C.De Vere. Gutenberg EBook. 4–5.

Watson, Rowan. 2003. *Illuminated Manuscripts and Their Makers*. London: V&A Publishing.

Zimmerhaeckel, F. 1899. *C. Julius Cäsars Rheinbrücke. Comm. de bell. gall. IV. 17; Ein Rekonstruktionsversuch.* Leipzig: B. G. Teubner Verlag.

2 A full moon in another land

The Moon Bridge in the Japanese Garden of the Huntington Library

2.1 A Japanese landscape in California

2.1.1 The Marsh family and the earliest Japanese garden in California[1]

In the latter half of the nineteenth century, Japan went through radical social changes. The Meiji Restoration, which began in 1868, influenced cultural aspects as well as all other social aspects, including arts and crafts. Disempowered samurai had to sell their art collections, and craftsmen lost their patrons and had to search for new patronage or employment. At the same time, the whole of Japanese society was attracted to various international expositions and took them as an opportunity to sell Japanese goods and learn western technologies. The Japanese works of art that were shown at the Vienna Exposition of 1873 and the Philadelphia Centennial Exposition of 1876 generated enthusiasm for Japanese art in the western world.

In 1872, when the Marsh family moved from Australia to the United States, they had a brief stay in Japan. George Turner Marsh (1857–1932) –"G. T." for short – though still a teenager at the time, was immediately fascinated by Japanese culture and art. When his family left, he stayed on in Japan for four years, travelling and accumulating knowledge about Japanese art and culture.

In 1876, G. T. went to San Francisco to join the family business. Soon, after another trip back to Japan to purchase artwork, he opened a store called "Japanese Art Repository" in San Francisco (Figure 2.1), which was probably the first shop selling Japanese art in America.

With the background of the wave of railroad construction and silver mining in California, together with the first taste of Japanese art brought in by the 1876 Centennial Exposition in Philadelphia, G. T. Marsh's business became an immediate success. G. T.'s frequent travel to the east – between 40 and 60 trips – is an indication of his business success.

In 1893, the United States was hit by serious economic depression. The California Midwinter International Exposition, which opened in 1894 in Golden Gate Park in San Francisco, was held to revive the economy and

1 Information on the Marsh Family in this section comes from Wolf and Piercy (1998).

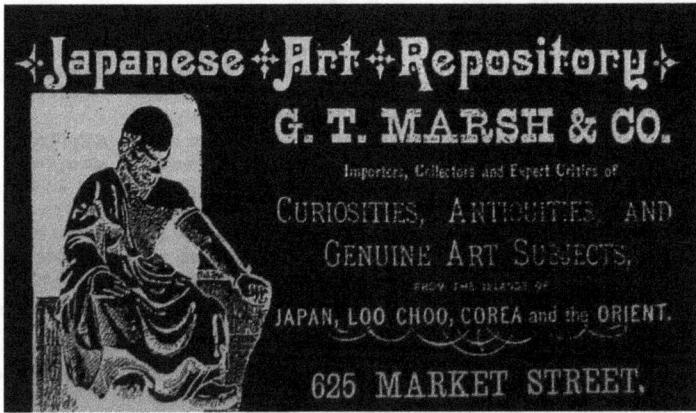

Figure 2.1 Store pamphlet of G.T. Marsh and Co. in San Francisco.
Source: Wolf and Piercy 1998.

improve the local neighbourhoods. For the Exposition, G. T. worked as the chief administrator of the fair's Japanese Village, and helped to organize the Asian exhibition. For the construction of the village, various Japanese artisans and craftsmen, together with the necessary materials and supplies, were brought in from Japan to the US. The buildings for the Exposition include a bell tower, bridges, a two-storey house, a Japanese theatre, and a *Torii* (symbolic gate for shrines). Plants and other articles were also brought from Japan.

Another important contributor to the Japanese Garden was the Japanese government commissioner Nakatani Shinshichi (中谷新七,1842–1922).[2] Among his creations for the fair, he designed and built a semi-circle Bridge (Drum Bridge).

The Japanese Village was a great success. When the fair finished, the park commissioners purchased the Japanese Garden from G. T. It was repeatedly destroyed by fire: in 1893, 1925, 1932–3, and was reconstructed several times in the century that followed.

The success of the business enabled the Marsh family to open further branches of their Asian art store in California, including branches in Pasadena, Santa Barbara, San Diego, Los Angeles, etc. (and even one in Ensenada in Mexico, although this one closed down after a short time). The

2 In order to fund the completion of the Drum Bridge and to fund the construction of the bell tower gate (*Shoro-no-mon*鍾楼の门) in the Golden Gate Park, Nakatani Shinshichi sold his family's rice field, and asked his son to remain and work in San Francisco for nearly half a century, earning money to repurchase the family field. (Information from the display board in Golden Gate Park.)

shop in Pasadena was run by G. T.'s brother, Victor Marsh. It contains a Japanese Garden, which had an original Japanese tea-house purchased from Japan and decorated with water, bridges, stone lanterns and Japanese plants. However, most of the branches were closed down or sold during the century that followed. The branches that were left were run by G. T.'s children. The branch in Pasadena was purchased by Henry E. Huntington and was included in his estate.

2.1.2 The Japanese Garden in the Huntington Library[3]

"The Huntington Library, Art Collections and Botanical Gardens," known as "Huntington Library," or "the Huntington" for short, is located in San Marino, Southern California, a small city to the northeast of Los Angeles, next to Pasadena. It covers an area of about 84 hectares (207 acres) and consists of 14 specialized botanical gardens from various cultures and geographical regions.

The creator of the Japanese garden, the railway magnate Henry Edward Huntington (1850–1927) purchased the estate in 1903 and began building on it. This ground belonged to the newly-established city of San Marino, neighbouring the city of Pasadena. In 1911, when the home was almost finished and ready to welcome his family, Huntington decided to improve a small valley at the west side of the ranch and turn it into a Japanese garden.

There were at least two sources of Huntington's knowledge about Japanese gardens. The first was the Japanese Tea Garden of the Midwinter Exposition of 1894. At that time, the Huntington family also resided in San Francisco,[4] and it is likely that they were familiar with the garden in the Golden Gate Park. The other source is a book entitled *Landscape Gardening in Japan,*[5] written by an English architect Josiah Conder, the first western scholar of Japanese architecture.

The first curator of the Huntington botanical gardens, William Hertrich, organized the construction of the Japanese garden. While preparing the site, he set out to look for Oriental plants all around California. When he visited the Marsh family's Pasadena commercial Japanese garden, he was informed by Victor Marsh that the entire property was going to be put up for sale. So Huntington brought the property, together with all its plants, buildings, fixtures, and fittings. Thanks to the many items of Marsh's he had bought, Hertrich was able to finish the establishment of the garden in an amazing

---

3 Information in this section comes from the Archive of the Japanese Garden, Huntington Library and from Hertrich (1988).

4 "Henry E. Huntington's Japanese Garden" in the Archive of the Japanese Garden, Huntington Library. Also see: Bennett (2013).

5 Conder, 1893. Cynthia Dickey (administrative assistant of the Huntington Library), email to author, 26 August 2010.

three months, getting it ready right before the arrival of the Huntington family in the winter of 1912.

After it was established, Hertrich continued to improve the garden, adding new fixtures such as stone lanterns, miniature pagodas, and stone idols. He also hired a Japanese carpenter, who played an important role in this garden (more about this later).

The construction of the Japanese garden continued in the following century and was expanded on and developed, with several sub-themes.

2.1.3 San Francisco Drum Bridge

Curved bridges made out of either wood or stone are generally known as *taiko bashi* (drum bridge太鼓橋) or *sori bashi* (reversed bridge反橋); Curved bridges with a semi-circle shape are known as *"engetsukyou"* (moon bridge, full moon bridge円月橋 or halfmoon bridge偃月橋), describing the appearance of the bridge with its reflection on the water as the moon or the round surface of a Japanese drum. Thanks to their elegance and delicacy and the romantic cultural associations they recalled, they were greatly favoured in Japanese garden design and were often depicted in eighteenth- and nineteenth-century artworks. They became a symbol of Japanese culture.

It is generally believed that the curved bridges in Japan have their roots in China. Curved wooden bridges were in fashion in the Tang (618–907) and Song (960–1279) dynasties in Chinese history, as witnessed by many landscape paintings and an abundance of Buddhist paintings. It is probable that the spread of the Buddhism, together with the architectural and garden culture, brought the concept of wooden curved bridges to Japan in the Heian (794–1185) period at the latest, as witnessed by the masterpiece "Treatise on Garden Making" (*Sakuteiki*作庭記).

The drum bridge in the Golden Gate Park in San Francisco was designed and manufactured in Japan and transported to San Francisco where the final assembly was carried out. Old photos of the Exposition (Figure 2.2) show an exaggerated semi-circle shape, which looks almost the same as it is today (Figure 2.3), while the bridge that currently stands in the garden has been reconstructed several times – unfortunately, without being documented. This bridge in a foreign land is much steeper than any bridges of this kind in Japan. In 1893, the Drum Bridge, as a symbol of Japanese aesthetic culture, was expressed in an even more exaggerated style, sent out from Japan to the western world.

In traditional Japanese carpentry construction, the curved building members, such as the undulating bargeboards (*karahafu*唐破風) are cut from larger timber. However, the bridge currently standing in the Golden Gate Park shows a structure of western construction (Figure 2.4). The curved beams under the bridge deck consist of multiple layers of thin boards attached by metal bolts (Figure 2.5). Since this construction method has no root in the Japanese building tradition, the current structure indicates that it

Figure 2.2 Drum Bridge in the Japanese Tea Garden, San Francisco, USA, 1894.
Source: Archive of the Japanese Garden, Huntington Library.

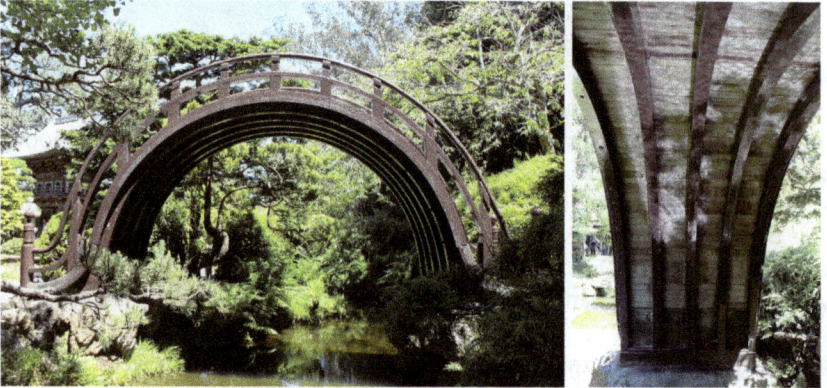

Figure 2.3 Drum Bridge in the Japanese Tea Garden, Golden Gate Park, San Francisco, USA, 2013.

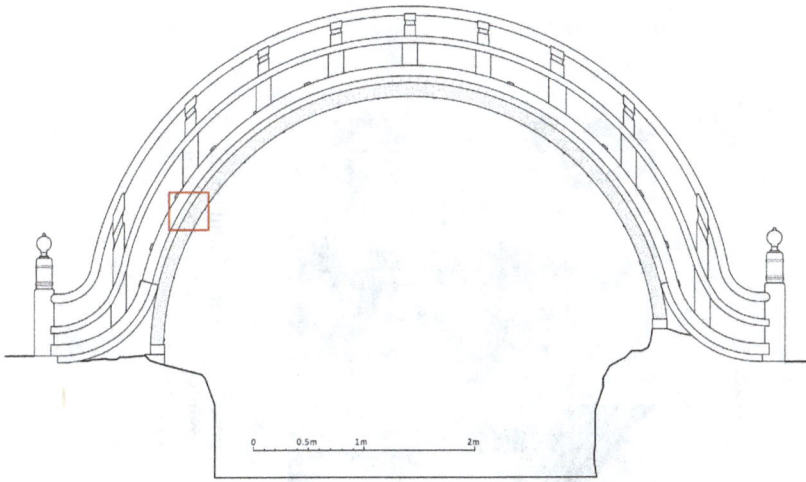

Figure 2.4 Elevation of the Drum Bridge in the Golden Gate Park.

is a later reconstruction, even though the shape of the bridge is true to the original image.

There is no doubt that the exaggerated Drum Bridge must have made a strong impression on H. E. Huntington, who happened to be living in San Francisco at the time of the California Exposition. As soon as his Japanese garden was established, he ordered a "Full Moon Bridge" (which is later referred to also as "Moon Bridge" in the Huntington archives) built in his garden by a Japanese carpenter, Toichiro Kawai.

2.2 Toichiro Kawai, the Japanese carpenter[6]

2.2.1 From Yokohama to Pasadena

Toichiro Kawai (1861–1943) was born in a rural section of Shizuoka-ken, Enshu. He lost his father at the age of 14 and was raised by his mother and sent to a temple, where he studied reading, writing, and gardening, among

6 Information in this section comes from the archive of the Japanese Garden, Huntington Library, and documents from the archive of the Oral History Collection of the Pasadena Museum of History, 1985, Interview with Toichiro Kawai's son, Mr Nobu T. Kawai:

Nobu Kawai. Kawai Family Background. 1985.01. From: Archive of the Japanese Garten, Huntington Library.
  Oral History Project. Interview with Nobu Kawai. From: Oral History Collection. Special Collections on Toichiro Kawai, unpublished manuscript. (Un-edited material). Pasadena Museum of History.
  Biographical sketch III. From: Archive of the Japanese Garden, Huntington Library.

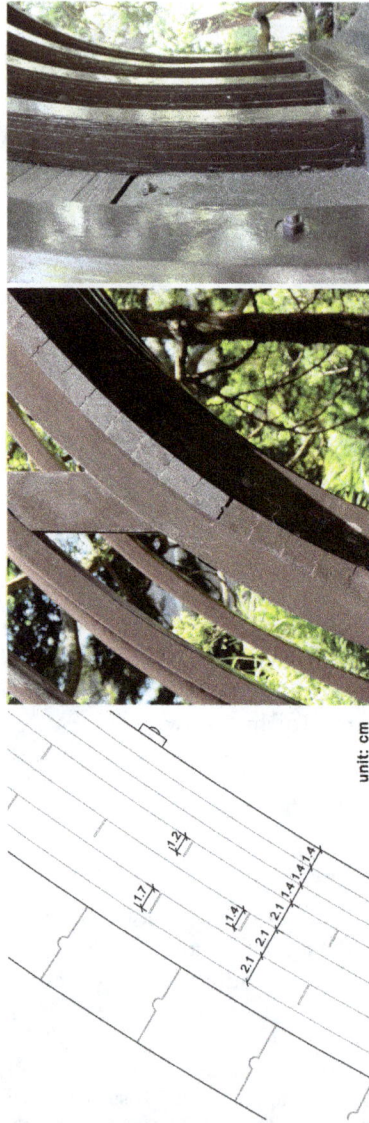

Figure 2.5 Details of the Drum Bridge in the Golden Gate Park.
The construction of the curved beam: every beam under the bridge deck is composed of 6 layers of thin boards: 3 thinner ones and 3 thicker ones. Cuts have been made in the thicker boards at regular intervals of about 10 cm along their length, for easy bending. The layers are bolted together. The bridge deck is composed of square wooden bars with semi-circle bridle joints at the ends.

other skills. With a deep love of ships and the sea from an early age, he went to Yokosuka to study shipbuilding. Afterwards, he went to Yokohama, the prominent port city of Japan, looking for a job.

In Yokohama, he lived in a hotel. The guests there were mainly customs officials and wealthy Chinese students. The daughter of the hotel owner, Hama (Hama Ishiwatari Kawai, 1873–1956) later became his wife. Hama's mother ran the hotel and her father was a civil engineer who had designed many bridges in Yokohama.

At that time, most ships were made of wood, so the ship's carpenter was an important member of the crew. Kawai had worked on many foreign ships travelling to South and Southeast Asia, among them also British ships, where he had learned some simple English.

In Yokohama, he happened to meet Victor Marsh during one of the latter's buying trips. Thanks to his knowledge of English, Kawai was hired by Victor to be his interpreter during his buying trips to Kobe, Osaka, and Kyoto.

In 1898, an American naval vessel, the "Original," needed a ship's carpenter and Kawai was selected from among the volunteers because he understood some English.

Meeting Victor Marsh had inspired in Kawai his "American dream," and he took this trip as his passport to the future. After arriving in San Francisco, he stayed there illegally. He found Victor Marsh and got a job in Marsh's art goods store in San Francisco, to repair artworks from Asia that had got damaged during the sea voyage. His job was to carve pieces from hardwood for the furniture and for other artworks to replace the missing or damaged part and refinish it in the original finish. He was adept at sketching plants and animals, and used this talent in wood and ivory carving. Kawai's familiarity with carpentry tools and his refined taste and artistic sense made him suitable to be a restorer.

After two years working for Marsh, Kawai returned to Yokohama to get married. The next year, after his first child was born, he returned to San Francisco and continued to work for Marsh. In 1902, he brought his family to San Francisco, later moved to Pasadena to work for Marsh's Pasadena store, where (as mentioned above), there was a commercial Japanese garden with items displayed for sale, and which was taken over by H. E. Huntington later on.

2.2.2 Projects in the Huntington's Japanese garden

In 1912, when Marsh decided to close down the Pasadena store and the property was taken over by H. E. Huntington for his Japanese garden on the San Marino estate, among the structures in Marsh's commercial garden there was a teahouse which was purchased from Japan and built in the authentic Japanese way, "without the use of nails," as the documents in the Huntington Archive emphasize. Kawai was hired for the removal of the house from one property to the other. He marked every piece of the structure, disassembled it and reconstructed the house successfully.

Huntington was satisfied with Kawai's work and commissioned him to build some other structures for his Japanese garden, including a *Torii* (traditional Japanese gate), a moon bridge and a bell tower, all in the traditional Japanese way.

From the correspondence between Hertrich and Huntington, we can reconstruct the timeline of the constructions: In November 1911, Hertrich mentioned hiring a Japanese person to remove and reassemble the teahouse. In March 1912, the Japanese house was finished. In April 1912, work was started on the *Torii*, and it was finished in May. In October 1912, the construction of the moon bridge was commissioned. In November 1913, Hertrich first mentioned the bell tower – by this time the moon bridge must have been finished. In December 1914, the bell tower was completed.

When Kawai studied in the temple in his youth, he became interested in architecture and collected books on it, which helped him with his constructions in the US. In addition, the knowledge about gardens that he learned in the temple also contributed to the design of the Huntington Japanese Garden.

The *Torii* gate (Figure 2.28) is made of cedarwood from Alaska. The logs were almost a metre (three feet) in diameter, and were originally intended to be made into totem poles. They were shipped to Los Angeles, destined for an Indian Village of which Huntington was a major financial supporter. After the village was abandoned, the timber was transported to the Huntington Garden, to be used for the *Torii* gate.

2.2.3 Life in Pasadena

At the time when the Kawai family arrived in Pasadena, there were few Japanese residents in this area; the Kawai family belonged to the first wave of Japanese immigrants. As the number of Japanese residents increased in subsequent years, they founded the Pasadena Japanese Association. From 1907 to 1913, the floats in the Rose parade of the Pasadena Japanese Association were designed by Kawai. Pasadena had become the Kawai family's main place of residence.

After working on the Japanese Garden, Kawai did not continue working for Huntington. He worked for several years for the next owner of the Pasadena shop after Marsh sold it, and then went into semi-retirement. He did some repair work on furniture and other items for local people. He also made ships for a fishing village. The Kawais had eight children: six sons and two daughters (Figure 2.6). They lived in Pasadena for the latter half of their lives, and throughout their lives, they had a love for both Japanese culture and their new homeland – the US.

2.3 The Moon Bridge: the art of construction

2.3.1 Architectural features of the Moon Bridge

Located in the central basin of the natural landscape, the Moon Bridge is a visual focus of the Japanese garden. Its light and elegant structure attracts visitors from all parts of the world (Figure 2.7).

Figure 2.6 The Kawai family, circa 1915.
Source: Huntington Digital Library. PhotCL 107 fld 9 (39). The Huntington Library, San Marino, California.

Figure 2.7 Moon Bridge from the west side, Huntington Library, San Marino, USA, 2013.

The present bridge now looks rather humble compared to the splendour of the previous version. From about the 1960s to the 1980s, the bridge was painted with red lacquer, "following the Japanese tradition." In the restoration project of 1988, the person who was curator at that time considered the red colour of the bridge to be "offensive" and requested that the lacquer be removed[7]. A colourless wood stain was subsequently applied to protect the wood.

In the Huntington archive, this wooden arch bridge was named the "Full Moon Bridge" / "Moon Bridge" or "Drum Bridge." For convenience and to distinguish it from the Drum Bridge in Golden Gate Park, the Huntington bridge will hereinafter be called the "Moon Bridge."

The most striking feature of the Moon Bridge is the so-called "woven arch" structure under the bridge deck. It was this special feature that drew the author's attention to the Huntington Moon Bridge during his cross-cultural research on the woven arch structures all over the world, and this inspired him to carry out this specific case study.

During the centennial of the Moon Bridge, in June 2013, the author paid a three-week visit to the Japanese Garden and produced a set of detailed measured drawings in 1:15 scale by a thorough on-site investigation (*Bauforschung*) (Figure 2.11). Two years later, in May 2015, the author paid another visit to determine the restoration issues.

The Moon Bridge is in the shape of a circular arc, with a structure height (ground to deck top) of 2.98 m (9.8 feet), a clear span of ca. 8.34 m (27.4 feet), and a total (deck) width of some 2.37 m (7.78 feet). The structure is a combination of a woven arch below with a layer of curved beams above it (Figure 2.8).

The woven arch consists of interlaced longitudinal and crosswise beams, both of which have the same square section. Two groups of longitudinal beams are laid at the ends of the crossbeams, forming two parallel curved lines at the left and right-hand sides along the direction of the bridge span. In each group, neighbouring longitudinal beams are laid alternatively on the inner and outer sides. Horizontal beams are clamped where they intersect (Figure 2.9).

The curved beams rest on the horizontal beams of the woven arch. This layer of the arch structure has three beams, resting on the sides and in the middle of the bridge.

Deck boards are placed on the curved beams. On the bridge deck are the steps, which are made from wooden boards with the same thickness as the deck. The width and height of the steps vary according to their position along the arch (Figure 2.10).

The handrail goes along the curved shape. The handrail posts are fixed onto the curved beams at their feet with round wooden dowels.

7 Source: Archive of the Japanese Garden, Huntington Library.

Figure 2.8 Conformation of the Huntington Moon Bridge.

At the feet of the arch are two concrete sills, and at their outer sides are huge stone foundations, buried deeply in the earth. They serve as the stone steps at the bottom of the bridge steps, and provide steady foundations for the bridge to offset the side thrust of the arch structure.

2.3.2 The bridge of perfection

The Full Moon is a metaphor for "perfection" in China-centred East Asian culture, the same as the number ten. In Chinese, the word for "circle" (圓) has the same meaning as the word for "full" (滿), and the word formed by combining these two characters has the exact meaning of "perfection," while the combination of the number "ten" (十) and the word for "complete" (全) means "complete perfection." This is the same in both Chinese and Japanese *kanji* (the adopted logographic Chinese characters in Japanese are 円満、十全, respectively).

The bridge's name in the contract is "Full Moon Bridge" and it is frequently referred to by this name, especially in the earlier documents by the first curator Hertrich. As mentioned above, it was a production inspired by the semi-circular-shaped Drum Bridge in San Francisco, which, together with its reflection in the water, expressed a perfect circular form. The form of the Drum Bridge, not only its exaggerated shape, but its remarkable symbolic aspect, must have made a strong impression on Huntington.

Figure 2.9 Under-deck structure of the Huntington Moon Bridge, 2013. From the north side.

The shape of the Huntington Moon Bridge, however, is not that close to a "full moon." Perhaps its designer had other measures for approaching the metaphor.

The shape of the structure was elaborately designed. The woven arch is composed of ten longitudinal beams and ten horizontal beams, the design making the structure unstable without the help of nails (Figure 2.12). Each of the beams is square in section with the width of five *sun* (Japanese inch, 3.03 cm, 0.12 inch, see below). For the circular form of the curved beam, the innermost arc line of the parallel curves has a radius angle of almost exactly 120 degrees, and a radius of exact five metres, which gives the circular arc a perfect diameter of ten metres (Figure 2.13).

When we go on to check other geometric scales, it is obvious that neither the span nor the rise of the arch is an integral number in all three measurement systems. In this sense, this could not have been the overriding factor in the design: that must have been the diameter of the circular arc, and the ten-metre diameter is too neat to be a coincidence (Table 2.1).

Figure 2.10 Wooden model of the Huntington Moon Bridge.
Source: Model and photo made by the author.

Favouring designs which use integral numbers has a tradition in Japanese bridge construction. The famous Kintai Bridge takes 60 degrees as the radius angle, making the span equal the radius (Ren 2013).

The geometry of the bridge façade highlights the artful configuration of other aspects. The bridge site has a 23 cm (9 feet) height difference at the northern (lower) and southern (higher) arch feet. If the bridge had been designed symmetrically, the centre pair of beams would be inclined. As it is, the centre beams are almost horizontal with only a 5 cm (2 inch) difference in height, which is almost invisible to the naked eye. To achieve this, the length of the lowest longitudinal beams at the arch feet must be unequal. The difference in length is 14 cm (5.5 inches).

2.3.3 Measurement system (yardstick)

When we approach the issue of scale, we are faced with the problem of having several different measurement systems. The measurement system in America is the imperial system which uses inches, feet and miles, and in the Meiji Period (1868–1912) the Japanese used the *shakkanhō* system, in which

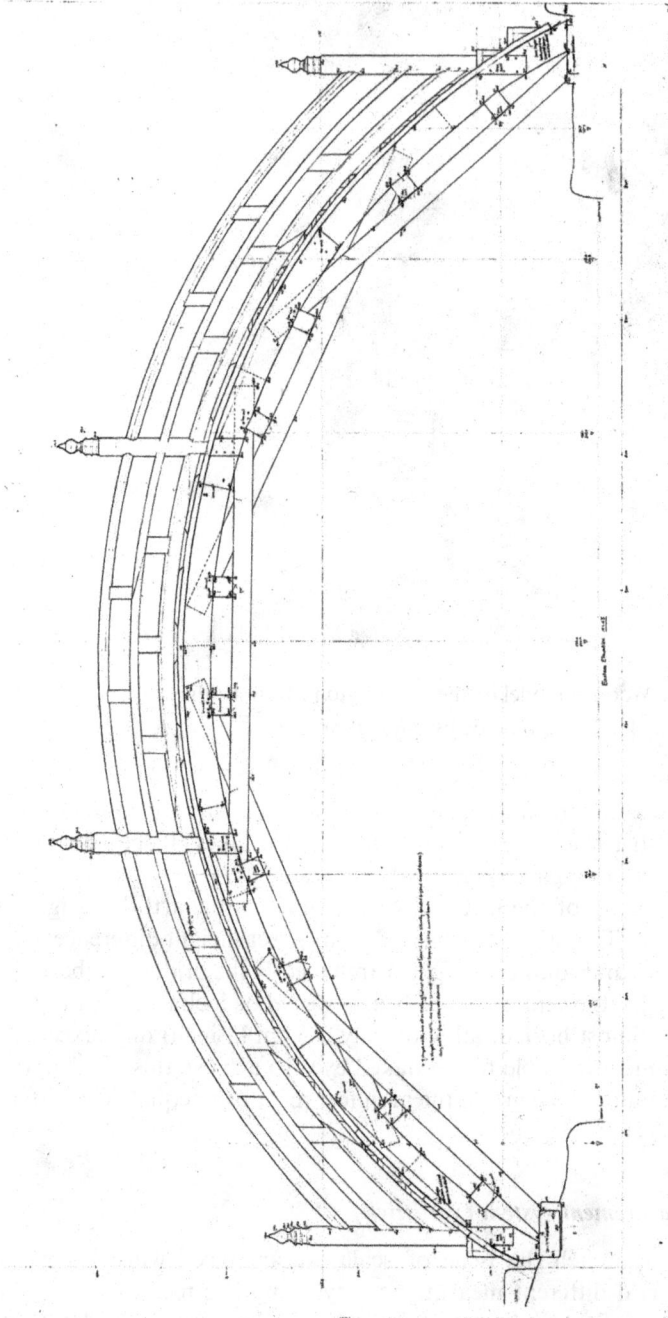

Figure 2.11 Measured drawings of the Huntington Moon Bridge, 2013. (a) East Elevation of the Moon Bridge.

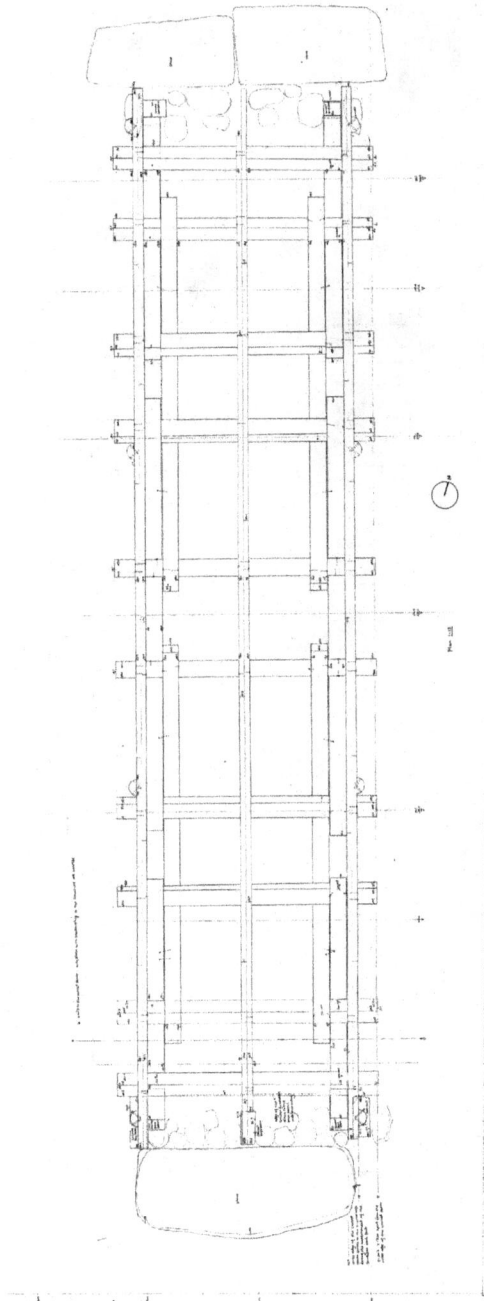

Figure 2.11 (b) Arch structure, plan view from below.

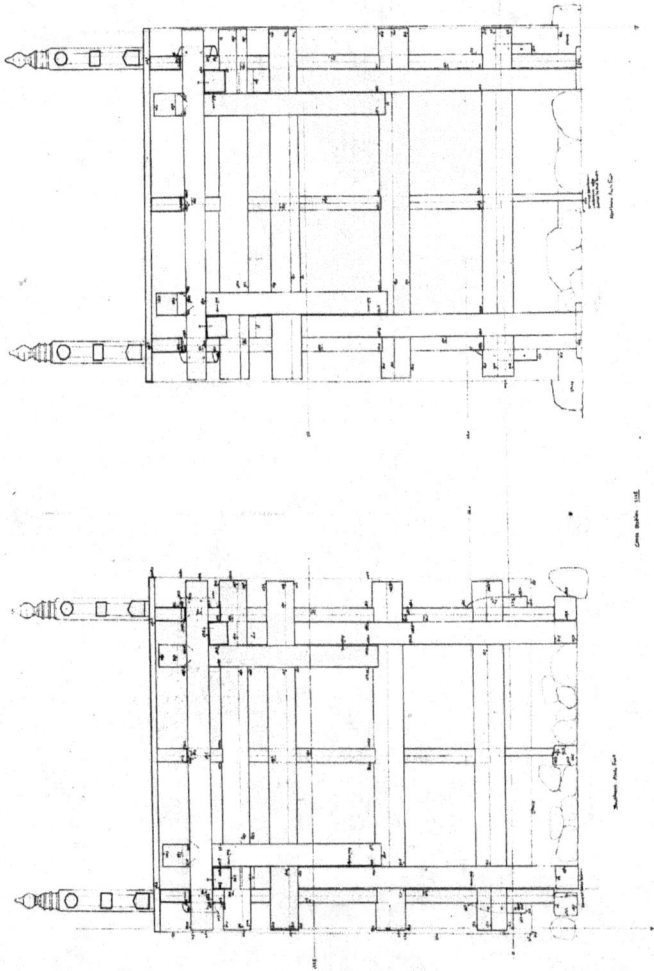

Figure 2.11 (c) Cross-section of the Moon Bridge (left: southern side; right: northern side).

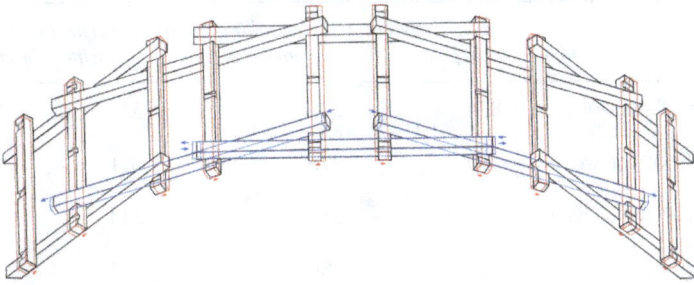

Figure 2.12 Possible movement of the beams without the nails applied at the joints.

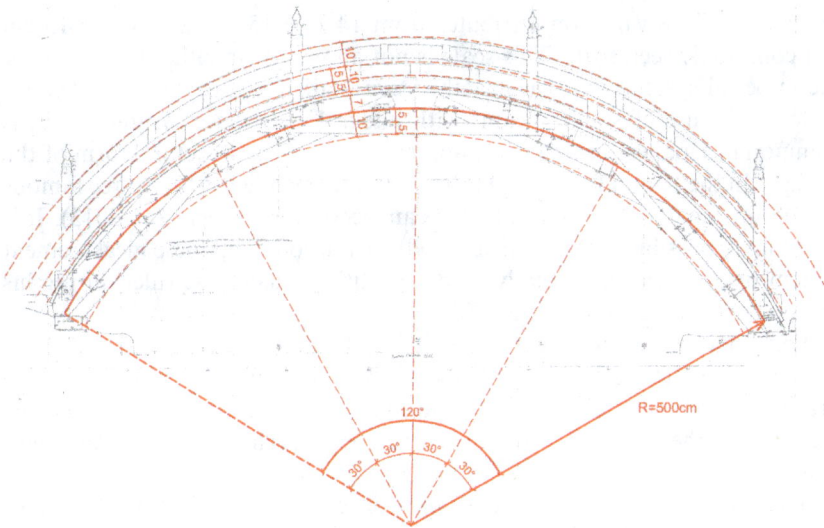

Figure 2.13 Geometric analysis of the elevation of the Moon Bridge (unit: *sun*).

one *shaku* (Japanese foot) equals 30.3 cm. The *shakkanhō* system lasted until 1924, when it was replaced by the metric system. Since Japan had entered the General Conference on Weights and Measures in 1885 and had been pro- moting the metric system since the 1890s, Kawai would have been familiar with all of these three measurement systems.

Aiming to find out the measurement system used in the design of the bridge, the author checked the scales of most of the structural elements (using metric-system rulers). The first step was to check the measurements of the most important structural elements, the squared beams. Measurements were taken of every square beam, in multiple positions. Their width of the

Table 2.1 Geometric data of the elevation of the Moon Bridge

Item measured	Metric system (cm)	Shakkanhō system (1 sun=3.03 cm)	Imperial system (1 inch=2.54 cm)
Clear span of the woven arch	834	275	328
Clear height of the woven arch	230	76	91
Clear span of the curved beam arch	875	289	344
Clear height of the curved beam arch	260	86	102
Radius of the curved beam arch	**500**	165	197

section has little variation attributes from 14.7 to 15.2 cm, while 15.15 cm is a commonly seen scale. This scale is not an integer in either the metric or the imperial system, but it is exactly 5 *sun* (half a *shaku*) in the *shakkanhō* system. The use of square beams with integral numbers measured in *sun* is common in Japanese bridge-building, regardless of the structural form of the bridges. In the Kintai Bridge, the most prominent bridge of the "three famous bridges in Japan," the length of the beam section is six *sun* (Ren 2013). It is thus highly possible that Kawai kept to the traditional Japanese measurement system or even that he actually used a traditional Japanese ruler during his bridge-building activities in the US.

For other measurements of the main building members, see Table 2.2.

From the data on other members, the following conclusions can be drawn: the specially-designed and hand-made structural members, namely the members of the woven arch and the curved beam arch, fit with the *shakkanhō* system.

The measurements of the handrail fit in only partly with the *shakkanhō* system. Interestingly, all the scales that "fit" are situated on the façade (Figure 2.14). The measurements on the cross-section drawings, namely the thickness of the members, do not fit in with the *shakkanhō* system. This indicates a façade-directed design process.

2.4 The Moon Bridge: construction secrets

2.4.1 *Construction of the wooden structure*

Curved beam arch:

The curved beam arch consists of three parallel curved beams (Figures 2.15 and 2.16), two at the sides of the bridge and one in the middle. Each curved beam is laminated with two layers of boards with nails, whereby each board layer is composed of four to six short arc boards, placed head to head in a

Table 2.2 Measurements of the main building members

Elements	metric system (cm)	shakkanhōsystem (1 sun=3.03 cm)	imperial system (1 inch=2.54 cm)
Section of the square beams	14.7—15.2	**4.85–5.02**	1.85–5.98
Length of longitudinal beams	373	123.10	146.85
	383	126.40	150.79
Length of cross beams	233	76.90	91.70
Thickness of the curved beams	9.1–9.3	**3.00–3.07**	3.58–3.66
Height of the curved beams	20.7–21.3	**6.83–7.03**	8.15–8.39
Diameter of the handrail posts	13.8	4.55	5.43
Diameter of the top beams of the handrail	9.1	**3.00**	3.58
Height of the middle beams of the handrail	11.0	3.63	4.33
Thickness of the middle beams of thehandrail	7.0	2.31	2.76
Height of the ground beams of the handrail	11.2	3.70	4.41
Thickness of the ground beams of the handrail	11.1	3.66	4.37
Width of the upper posts** of the handrail	9.2	**3.04**	3.62
Thickness of the upper posts of the handrail	6.9	2.28	2.72
Width of the lower posts** of the handrail	12.2	**4.03**	4.80
Thickness of the lower posts of the handrail	8.4	2.77	3.31
Length of the deck boards (width of the bridge)	236.4	78.0	93.1
Width of the deck boards	14.1–14.5	4.65–4.78	5.55–5.70
Thickness of the deck boards	4.1	1.35	1.61
Thickness of the stair boards	4.1	1.35	1.61

Note: figures in **bold** fit only with the *shakkanhō*system, thus we assume a Japanese ruler was used.
** "Posts" are the short wooden blocks between the handrail beams.

row. The division of the boards into two layers is alternate so as to avoid juxtaposed seams.

The division of the boards of the side beams is symmetrically spaced along the bridge. They both have a "keystone" at the top centre of the exterior layer of boards, and a centre seam in the interior layer which is invisible from the façade, but locates almost exactly in the middle of the elevation.

The curved shape of the boards is achieved by shaping larger pieces of wood by cutting. This means that boards which are much larger than the actual width of the curved boards are used. Where the material is not wide enough, a circular segment is attached along the arc chord using glue (Figure 2.16).

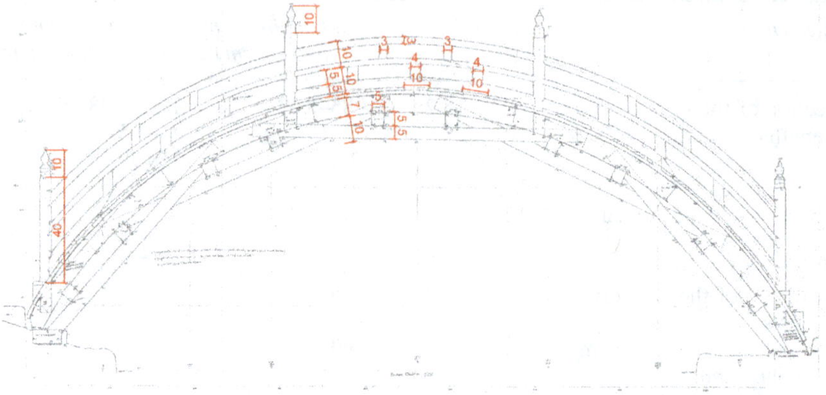

Figure 2.14 Analysis of the measurements of the building elements of the Moon Bridge (unit: *sun*).

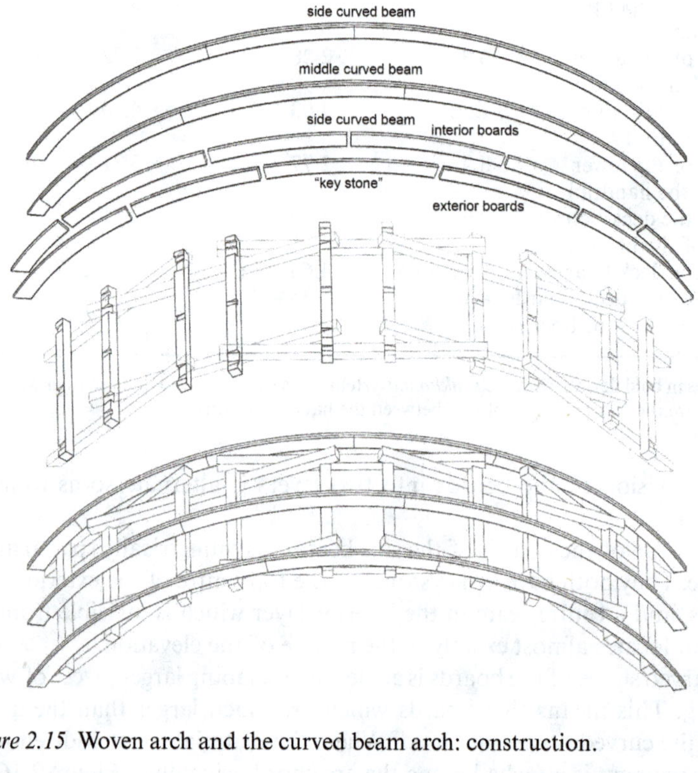

Figure 2.15 Woven arch and the curved beam arch: construction.

Woven arch:

In the arch structure, the longitudinal beams are complete square pieces that do not undergo any cutting. All the joint-notches are cut on the crossbeams, including the notches for setting the longitudinal beams and those for accepting the curved beams.

The notches are triangular in section, following the form of the element set inside. This kind of joint by itself would not be able to prevent the two pieces of jointing moving. Without additional fixing elements, the structure would rely only on frictional force to keep structural integrity, which is obviously far from enough. As a result, plenty of nails are inserted into the joint.

2.4.2 Nails

The iron nails used in the bridge are galvanized iron nails.[8] The diameter of the nail-heads is between 8 and 10 mm. They are either perfectly round and are industrially produced, or somehow irregular in shape, which means they could have been made by hand.

The nails in the under-deck structure of the Moon Bridge can be classified into three types, according to their function:

The first type is used to join the two layers of boards of the curved beams (Figure 2.17, a, right). These nails are comparatively small, with an industrial, round nail-head, and are knocked in horizontally.

The second type is applied at the intersection of the woven beams, between the longitudinal and crossbeams (Figure 2.17, a, left and bottom). They are applied singly or in pairs, are inclined downwards. The projecting nail-heads of the inclined nails are hammered down, back inside the wood surface. They are probably the longest nails in the structure, judging from the length required in that position. They also have the biggest nail-head, the form of which indicates that they are probably made by hand. The position of this group of nails is marked on the plan of the arch structure viewed from below (Figure 2.11, b) and the cross-sections (Figure 2.11, c) of the Moon Bridge.

The third type is used to connect the curved beams and the crossbeams (Figure 2.17, c). Like the second type, they are also nailed in inclined downwards. They are used on the exterior surface of the two layers of bound boards, and also, between them. Their location and direction are marked in Figure 2.18.

It is extremely difficult to locate the nails between the boards during the investigation, but thanks to the tiny gaps between structural members, some of them could be discovered with the help of a strong flashlight or by sticking

---

8 According to Andrew Mitchell, the architect in charge of maintaining the bridge, conversation with the author on 26 June 2013.

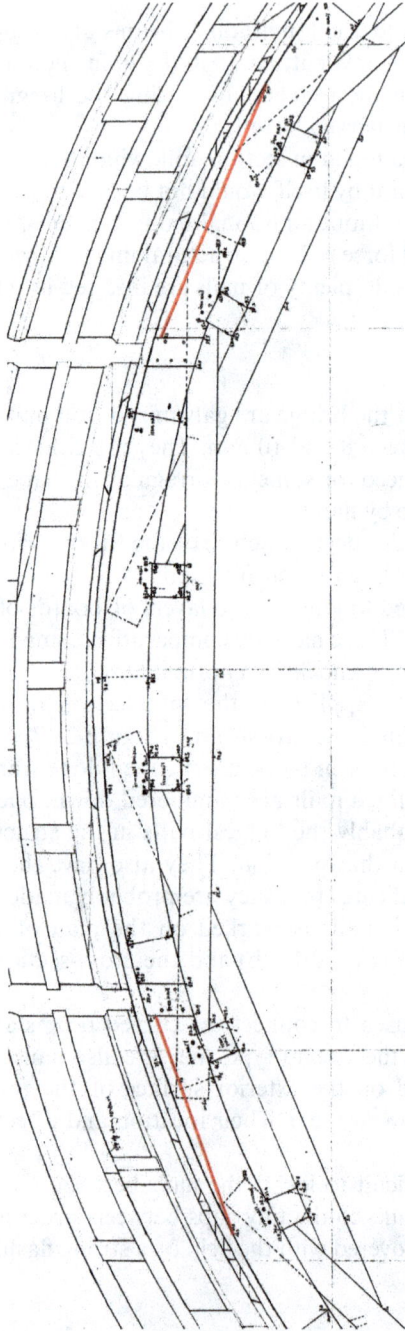

Figure 2.16 The seam of the attachment of the board (shown by red lines).

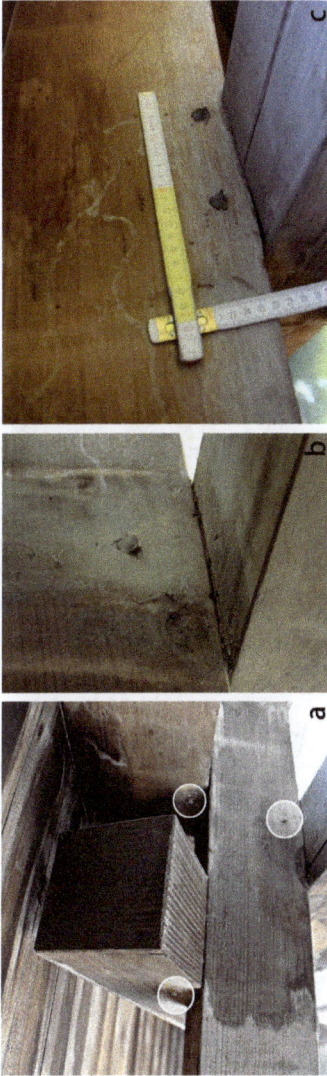

Figure 2.17 Nails used between intersecting members: (a) Nails at the intersections of the members. (b) Nail at the intersections of the members: detail. (c) The connection between the curved beam and the crossbeam, exterior surface. The nails have been inserted inclined downwards inside the board of the curved beam, and go in a further half-centimetre. The nail holes have been filled with lime. The lime has been darkened by the wood-protection substance applied during the repair activities in the 1980s.

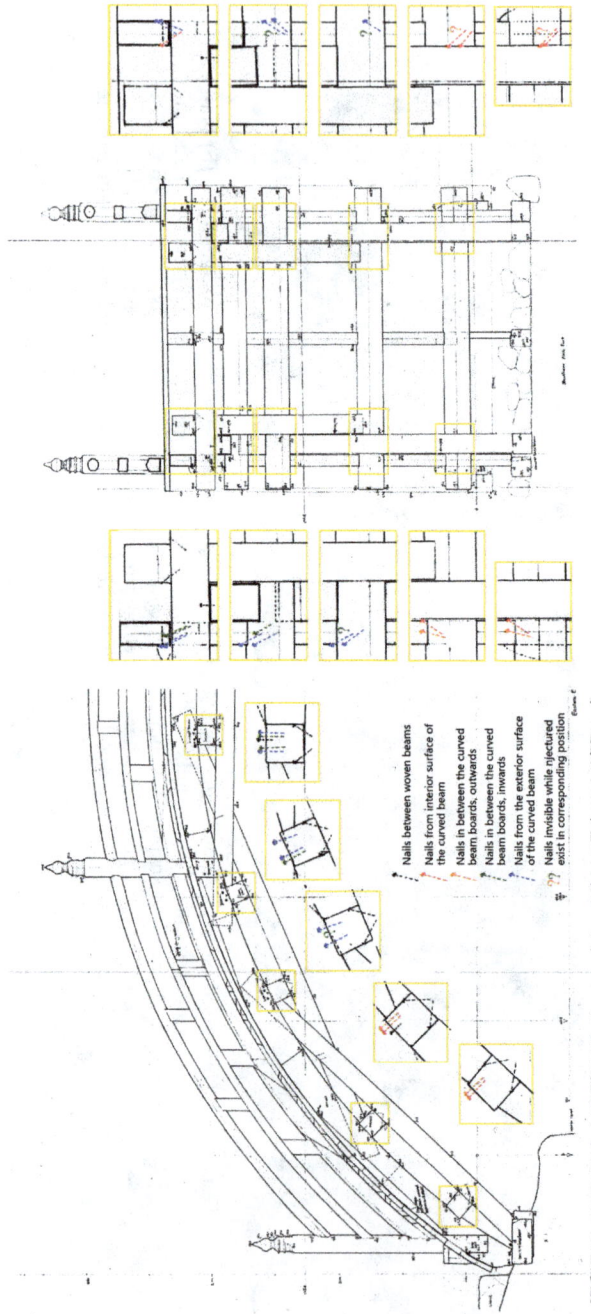

Figure 2.18 Position of the nails between curved beams and crossbeams. (a) Nails on the southern part of the arch: east elevation and section of the southern side

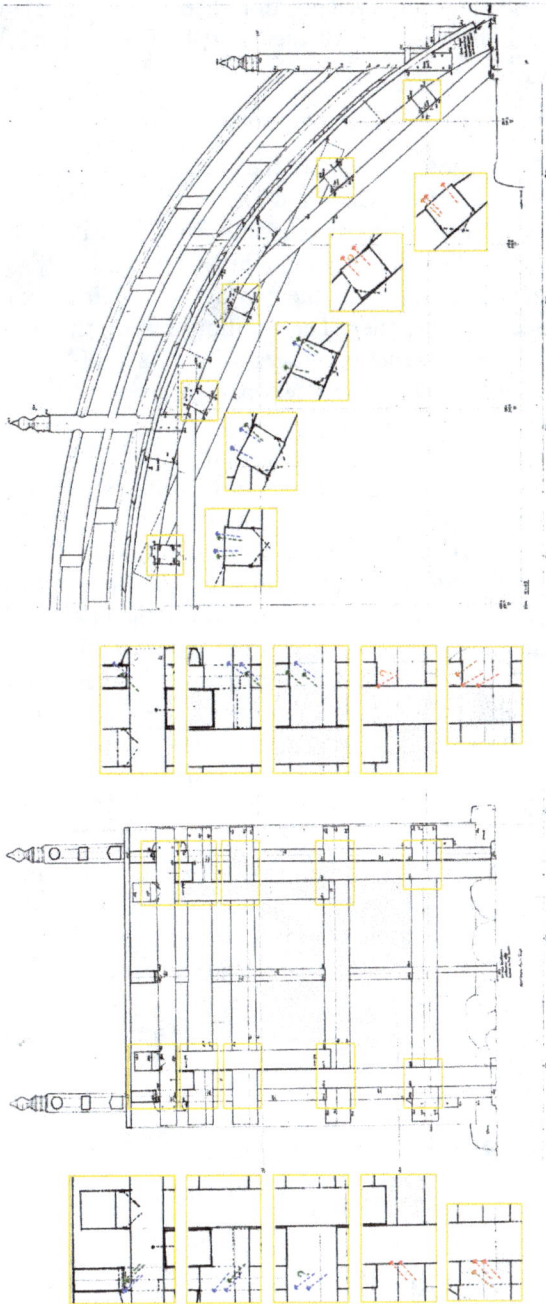

Figure 2.18 (b) Nails on the northern part of the arch: East elevation and section of the northern side.

in a thin knife. Furthermore, although a couple of joints are too tight to be inspected, the symmetrical form of the bridge means that it can be assumed that there will be nails at the mirrored position. The number of nails discovered is enough to show us the regularity of their distribution (Figure 2.18).

2.4.3 Construction process of the curved beams

With regards to the location of the nails, the most interesting aspect is related to those connecting the curved beams and the crossbeams, but found between the two layers that make up the curved beams. They indicate that the curved beams are not assembled on the ground and then attached to the woven arch as complete elements, but rather that the two layers of boards are nailed onto the crossbeams one after another. The position and direction of the nails enable us to retrace the construction process (Figure 2.19).

Generally speaking, on the lowest two pair of crossbeams, the nails are hammered in separately onto the inner surfaces of the two layers of boards, from the inside outwards. This indicates that the exterior boards must have been attached first, with the interior boards being attached afterwards (Figure 2.19, step 1–2).

On the upper three pair of crossbeams, the nails have been hammered in separately and inwards onto the exterior surfaces of the two layers of boards. This tells us that, on these corresponding positions, the interior boards were positioned and attached before the exterior ones (steps 3–6).

It seems that although the designer tried to avoid using nails on the visible sides wherever possible, on the third and fourth crossbeams (counted from the arch feet), it is impossible to avoid having nails on the façade. At these positions, the interior boards must be fixed directly onto the longitudinal beams. Accordingly, the nails on the exterior boards can be seen on the visible exterior surfaces.

Since each piece of boards (with exception of the inner pair at the arch feet), covers two neighbouring crossbeams, and the exterior and interior boards are placed in an alternating pattern, according to the distribution of the nails, there was only one possible construction process to be followed.

Now, having traced most of the building process, from design to construction, we're left with only one problem – this structure is not the bridge from 1913.

2.5 Reconstruction of an unwritten history

2.5.1 The Moon Bridge in the 1920s

In June 2013, when the author arrived at the Japanese Garden of the Huntington Library, he couldn't wait to reproduce an exact photograph of the oldest preserved image of the Moon Bridge (dated between 1913 and 1923). Despite his many attempts, he was not successful. The new photos from 2013

The curved beam is composed of five exterior boards and six interior boards.

Step 1. The exterior lower boards are attached first, fixed from the inside.

Step 2. The interior lowest boards are fixed from the inside.

Step 3. The subsequent interior boards, fixed from the inside at the second crossbeam, and from outside at the third crossbeam.

Step 4 (false). If the next boards are exterior, the inner nails at the fourth crossbeam would be in the wrong direction, and there would be no space to fix the corresponding interior boards.

Step 4. Therefore, the upper pair of interior boards are positioned during this step, attached by nails from either outside or inside.

Step 5. To complete the exterior layer of boards, the top board is probably positioned first, to assure its correct position in the centre.

Step 6. The exterior board sequence is completed using nails from the exterior surface (and nail holes are filled with lime).

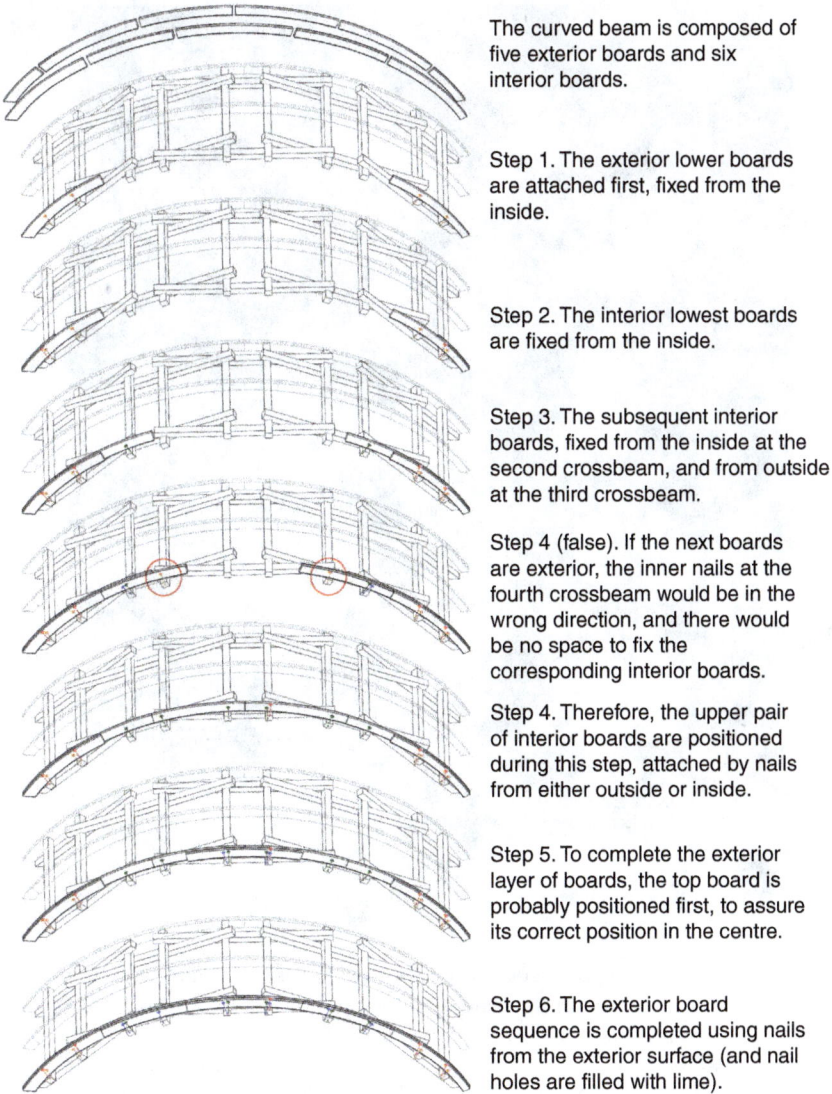

Figure 2.19 Building process of the curved beam, analysed from the nails. Eastern beam.

seem to get really close to the old photo, but there were always some proportional errors here and there (Figure 2.20). The author blamed the errors on an inaccurate standing point or image deformation from the optical principle at the edge of a photograph.

A little later, further photos from the Huntington digital archive revealed the reason for this inaccuracy: the bridge in the old photo was not the one

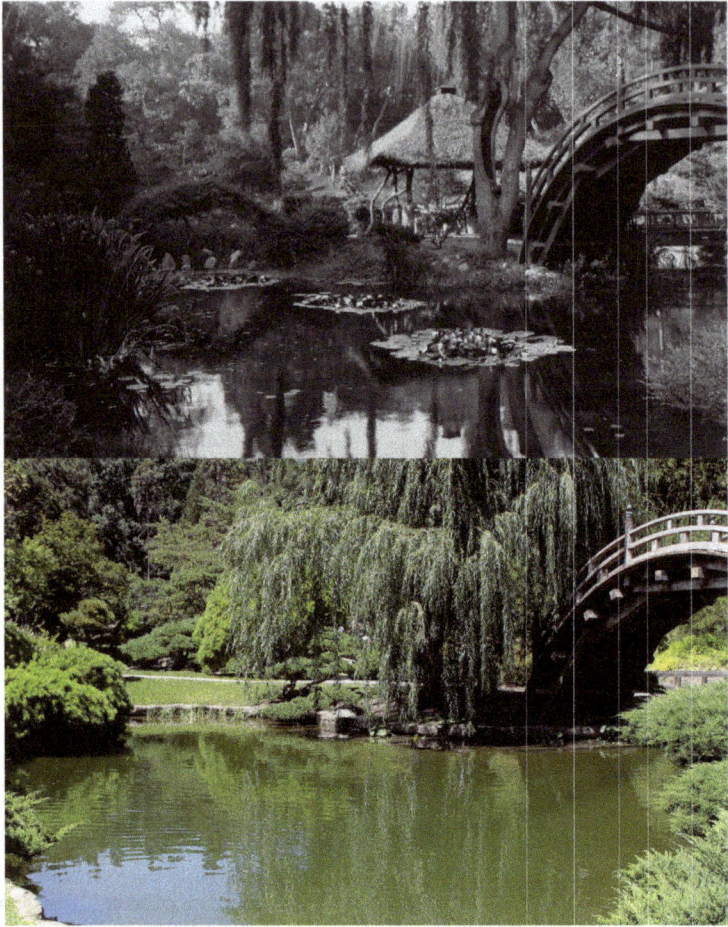

Figure 2.20 The Huntington Moon Bridge: a century lies between the two images.

The old photo was dated 1913 by Hertrich (1988). (Various copies of the same photo in the Huntington Digital Library are dated differently).*

* PhotCL 107 vol3 box 2 pg78, "Drum bridge in the Japanese garden, circa 1918. The Huntington Library, San Marino, California. PhotCL 107 fld 9 (19), "Drum bridge in the Japanese garden, circa 1920". The Huntington Library, San Marino, California.

When checking to find the correct date of this photo, the author received the reply from the library that the photo could have been taken any time between 1913 and 1923. (From correspondence with Ms. Devonne Tice, Imaging Services, Huntington Library, received by the author by email in December 2015.)

standing in the Japanese garden today. A group of old photos dating from around the 1920s show different proportions for the woven arch of the earlier structure. The distances between the crossbeams are more uniform in the old bridge, while in the present one, they lie pairwise closer.

The most helpful images which helped to restore the early proportions are shown in Figure 2.21.

The latest dated image with the same proportions is from January 1949 (Figure 2.21, f), while the earliest dated photo with the proportion that concords with the bridge today is dated 1962.[9] This puts the reconstruction of the bridge between 1949 and 1962.

With the help of these old photos, the author reconstructed the earlier proportions of the bridge (Figure 2.22). Using the reconstructed proportions, the design method, namely the geometrical principle was worked out. In Japanese culture, the traditional method to define an angle is the *kou-bai* (勾配), namely using the tangent value to define the scale of an angle. The locations of the crossbeams in the woven arch are designed according to the *kou-bai* principle, the basic angle is 2.2 *sun* in *kou-bai*. Because of the height difference between the two arch feet, the angles are not equal throughout the entire woven arch. The quota of adjustment is 0.1 *sun* in *kou-bai* (Figure 2.23). By using this method, the Moon Bridge achieves steady and well-balanced proportions in its appearance.

2.5.2   The reconstruction of the bridge between 1949 and 1962

Details from higher resolution images reveal further differences other than the proportion between the earlier structure and the bridge that stands there today.

2.5.2.1   Arch structure

Firstly, the entire structure of the woven arch and the curved beam arch from the earlier version are replaced during the reconstruction.

Two photos in Figure 2.21 (a. and b), both dated around 1925 and taken from the western side, reveal a wave-shaped texture on the middle longitudinal beam (Figure 2.24). The corresponding members of the current bridge do not exhibit this texture.

From many old photos, it is clear that the notches on the crossbeams for setting the curved beams are deeper than those in the present members (Figure 2.25). This indicates that all the crossbeams were replaced by new beams in the reconstruction.

Given the fact that the position of the crossbeams of the woven arch has been changed, and that on the crossbeams are places where the woven arch joins the curved beams, if old members of the crossbeams were reused,

9 PhotCL 107 fld 9 (20), "Drum bridge and wisteria pergola in the Japanese garden, 1962." The Huntington Library, San Marino, California.

Figure 2.21 Photos of the Moon Bridge from the 1920s and the 1940s show an image of a woven structure different to the Moon Bridge of today.

(a) "Drum bridge in the Japanese garden, circa 1925" (from western side).[i]
(b) "Drum bridge in the Japanese garden, circa 1925" (from western side).[ii]
(c) "Drum bridge and house in the Japanese garden, circa 1925" (from eastern side).[iii]
(d) "Drum bridge in the Japanese garden" (from eastern side).[iv]
(e) "Children and adults on Japanese garden drum bridge" (from eastern side).[v]
(f) "Japanese garden after snowfall, January 11, 1949" (from eastern side).[vi]

[i]   PhotCL 107 fl d 9 (17). The Huntington Library, San Marino, California.
[ii] PhotCL 107 fl d 9 (18). The Huntington Library, San Marino, California.
[iii] PhotCL 107 fl d 9 (14). The Huntington Library, San Marino, California.
[iv] PhotCL 107 fl d 9 (13). The Huntington Library, San Marino, California.
[v]   PhotCL 107 vol13 pg14 (78). The Huntington Library, San Marino, California.
[vi] PhotCL 107 fl dr19 (27). The Huntington Library, San Marino, California.

Source: Huntington Digital Library.

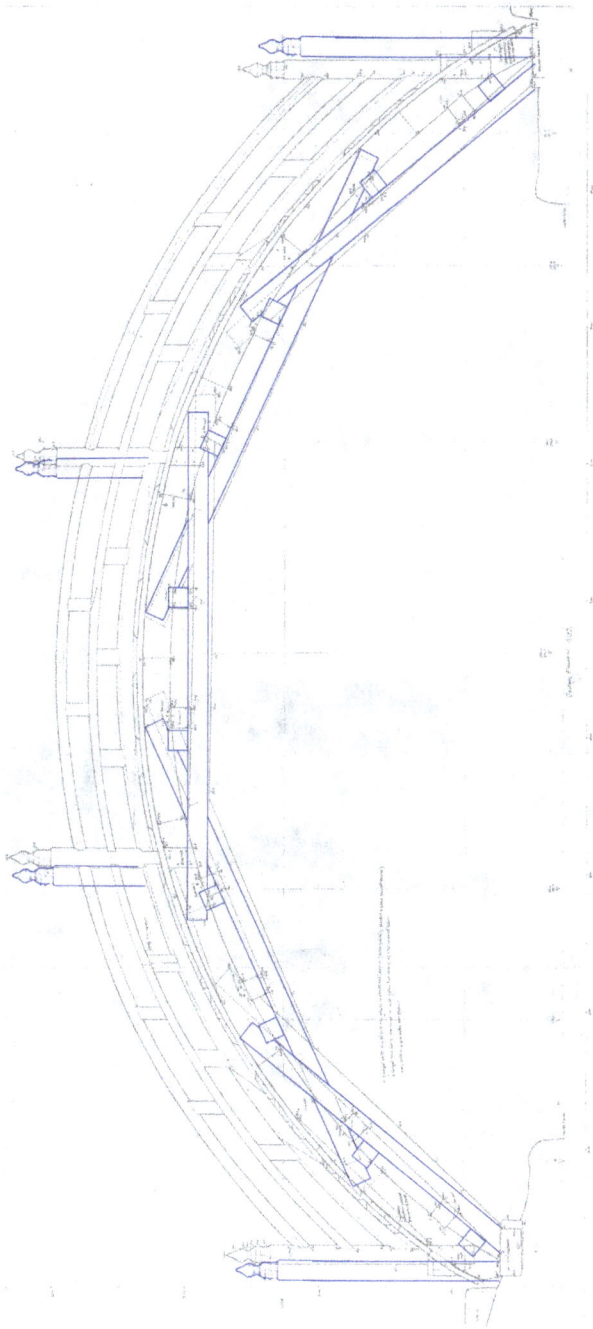

Figure 2.22 Moon Bridge, proportions of the woven arch, before 1949.

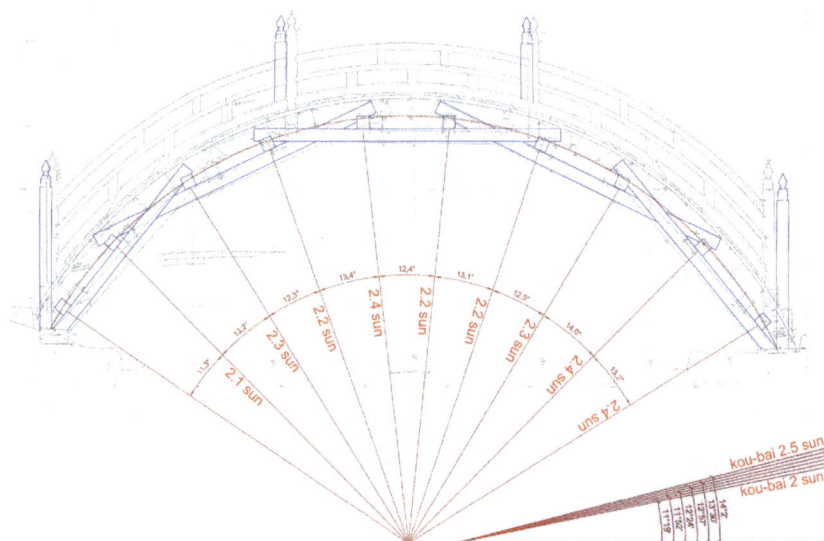

Figure 2.23 Moon Bridge, design geometry of the woven arch before 1949.

Figure 2.24 Moon Bridge, details from Figure 2.21 a (above) and b (below).
Two photos were taken from the western side, circa 1925, showing grain of the middle
longitudinal beam.

there should be traces left on the curved beams. However, the current curved beams show no evidence of old joints or other indication of reuse (e.g. old nail holes, joint scratch, or marks caused simply by the sunshine and weathering between the inner and outer sides of the joints), so we can safely conclude that the curved beams were also replaced during this reconstruction.

Members of the woven arch and the curved beams can easily be replaced since they do not need large timber like that for the handrail, which is discussed below.

After realizing that the notches in the curved beams were much deeper in the 1920s than they are today, the author looked back at the top pair of crossbeams (Figure 2.24). Between the curved beam and the crossbeams there were obvious gaps. This indicates an inefficient internal force distribution of the curved beam arch and the woven arch. Judging from these gaps, and from other aspects (more about this later), the curved beams and the crossbeams were probably not connected using nails at the time. The inefficient force distribution and the probable lacking of effective connection between the curved and the crossbeams could have contributed to the structural failure of the bridge, and resulted in the reconstruction around 1950.

2.5.2.2 Handrail

The photos a and b in Figure 2.21, which date from ca. 1925, tell us about the principle of the composition of the handrail members. Compared with the comparatively fragmentary pieces of the handrail in the current bridge, the former structure was rather neatly organized:

In the three sections divided by the posts, the middle section is the shortest. Its top and middle beams are complete pieces cut from single pieces of timber; only the bottom beam is composed of multiple pieces of wood. The horizontal seam line (Figure 2.26, a, red lines) indicates that its main part was cut from a piece of board.

The side sections are longer than the middle. The top beams are also complete pieces, as they are today. In the workshop of the Japanese Garden, Andrew Mitchell, the architect who was in charge of the bridge repair in 1988, showed the author a huge piece of redwood (Figure 2.27), which was probably left over from the earliest construction. From this wood, Mitchell cut out a replacement piece for a top handrail-beam on the southwest side of the bridge.

In the image from the 1920s, the middle beam of the side section consists of two pieces halved together (Figure 2.26, a). This kind of lap joint is not used in the current bridge.

The bottom beam of the side section also consists of two pieces, simply joined with a butt joint.

Figure 2.25 Moon Bridge, the second crossbeam from the arch foot, northeastern side of the arch (a) Detail from Figure 2.21 e. (b) The present (2015) bridge for comparison.
The curved beam went much deeper into the crossbeam in the 1920s than it does today.

All three sections in the 1920s photos are longer than in the current version of the bridge (Figure 2.22). This fact, together with the overlapping of the seams of the original and the present structure (shown by the red and black division lines in Figure 2.26, a) indicate a maximized re-use of the original building elements during the reconstruction carried out around 1950.

Details from the old photos also show that the handrail posts at the sides of the bridge used to rest on the base (Figure 2.26, b, detail from Figure 2.20). In the current version of the structure, they have been cut short.

2.5.3 Nobu T. Kawai's account of the collapse of the bridge

Toichiro Kawai's son, Nobu T. Kawai, mentioned a very interesting event which does not appear in the Huntington construction report. Apparently, sometime after they were built, both the Moon Bridge and the *Torii* (Figure 2.28) collapsed. The Moon Bridge was subsequently reconstructed, while the *Torii* was not.

The collapse of the Moon Bridge was mentioned by Nobu on three different occasions,[10,11,12] two of which are dated, but they contradict with

10 Oral History Project in Pasadena History Museum: Long, Long ago oral history project. Answers to Questions by Nobu T. Kawai. Question No.4.
11 Oral History Project in Pasadena History Museum: Interview with Nobu Kawai, 55 Harkness Street, Pasadena, Thu, 27 September 1984, pp. 6–7.
12 Kawai family background. Compiled by Nobu. T. Kawai. January 1985, p. 4. From: Archive of Japanese Garden, Huntington Library.

(a)

(b)

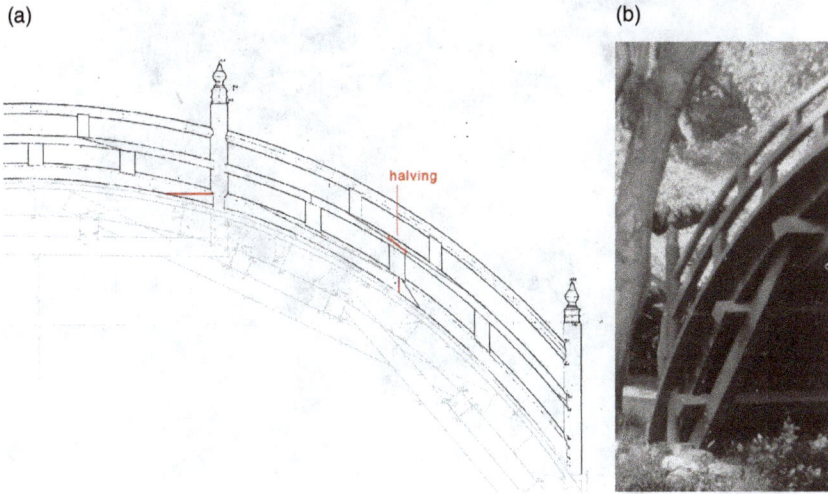

halving

Figure 2.26 Moon Bridge, details of the handrail. (a) Sketch of the component elements of the handrail (southwest side). The 1920s seam lines are shown by
red lines. (b) Handrail post (northwest side, from Figure 2.20).

each other. In one interview in 1984, Nobu stated that the collapse happened
"in about 1914," and that the bridge standing in its original location was an
"exact duplicate" made to his father's original design:

> He built the *Taiko-bashi*,[13] which is a drum bridge that is still there and
> also the bell tower and helped in designing other structures or replace-
> ment of stone lanterns and statues throughout the Japanese garden. Now
> that [was] back in about 1914 and in the meantime, the original bridge had
> rotted through termites and dry rot and so forth and collapsed. So they
> have rebuilt that bridge in the exact design that it was built by my father.
> So the bridge that is standing today is not the one that my father built
> but it is an exact duplicate of the one that he built at the same location.[11]

However, in another material compiled by Nobu in 1985, he referred to the
time of the collapse as being "several years ago." In the same paragraph, he
talked about how his father studied the bridge at home using a model:

> He made a model of the bridge which was around the house until it got
> lost during our wartime evacuation. It was built with interlaced timbers
> without the use of nails. The more weight placed upon the bridge caused

13 This word is mistakenly spelt "tyco bushee," indicating that the interview was transcribed
from an audio recording.

Figure 2.27 A huge piece of redwood in Andrew Mitchell's workshop area. Roughly 3.6 m long, 9.5 cm thick. Japanese Garden of the Huntington Library. 2015.

the joints to become tighter and stronger. Age, termites and dry rot made it necessary to rebuild the bridge several years ago. However, it was rebuilt in the same design on the same spot as the original.[12]

On all three occasions, Nobu attributed the collapse to "age, termites and dry rot" (the same reasons he gave for the collapse of the *Torii*).

Toichiro Kawai built three buildings for the Huntington Japanese Garden: the Moon Bridge, the *Torii*, and the bell tower. Nobu has mentioned how proud his father was of the bell tower (Figure 2.29). It was his favourite work, and still stands in good condition in the garden today. In the interview of 1985, Nobu described it as follows:

Dad often quoted Japanese proverbs. One was, 'When a bull dies, it leaves its horns; when a tiger dies, it leaves its hide; when a man dies he leaves his name. 'I recall his staying up night after night drawing pages of blueprints. He wanted this to be one with which he would be proud to have his name identified. All his talents were directed to make each joint fit perfectly. That he was satisfied with his work was indicated when he told us he had tacked a plaque, inscribed with his name as the builder and the date of construction, [hidden] in the attic of the building.[12]

Figure 2.28 The *Torii* of the Japanese garden, circa 1925.*
*PhotCL 107 fld 9 (50). The Huntington Library, San Marino, California.

2.5.4 A collapse in 1914? A thought experiment

In conversations with the author on the findings related to the reconstruction history, Andrew Mitchell mentioned that he had heard from the locals that the bridge had probably been rebuilt at least twice.[14]

Nobu T. Kawai did not mention a second reconstruction. Both the times he mentioned (in 1914 and some time before 1984) are not reliable. The "several

---

14 Conversation between the author and Andrew Mitchell, Huntington Japanese Garden, 21 May 2015.

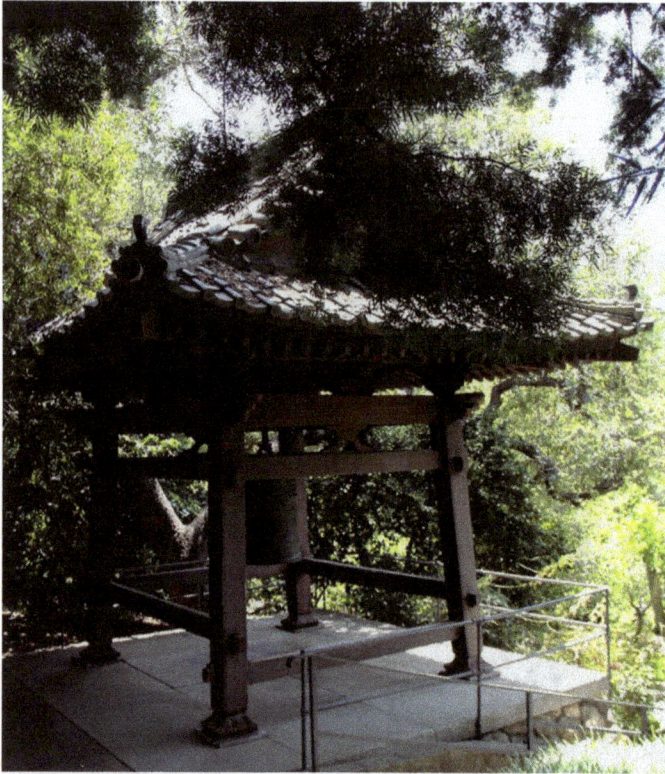

Figure 2.29 The bell tower.
Japanese Garden of the Huntington Library. 2013.

years ago" from the 1984 report might be reluctantly referring to the reconstruction between 1949 and 1962. In any case, the photo archives confirmed that since 1962, the bridge was never actually rebuilt – only repaired. The date of "1914" is even more confusing, because in 1914, Toichiro Kawai was still working on the Huntington Japanese Garden bell tower, which was only finished that December. Was this simply a slip of the tongue during the interview? Or was there a first collapse? Why did Toichiro Kawai, a craftsman with professional pride, never show his children how proud he was of the Moon Bridge, which was indeed a marvellous design? Therefore, it's possible that there was an early accident, one which occurred long before Toichiro Kawai's death (1943).

At the same time, since the photographs of the bridge between the 1920s and 1949 exhibit identical proportions, that would indicate that if the bridge was indeed reconstructed twice, the first time must have been very early on.

There are not enough clues on this issue for us here, neither in the historical records nor the already much-altered structure. What we can do is to use our imagination to explore and to get a better understanding of the structural features:

Is it possible that the Moon Bridge stood for several months, then 'collapsed', and was subsequently reassembled within a short time – either towards the end of 1914, as per Nobu's description, or before the first photos taken in the early 1920s – without damaging the handrails above the deck (which were in fairly good condition in the images taken in the 1920s)?

The answer is "Yes."

The reason is that, as mentioned before (Figure 2.12), without nails or similar fixing mechanisms, the woven arch has two degrees of freedom of movement: the crossbeams could slide along the longitudinal beams under them, and the longitudinal beams could slide along their axis. Thus, the friction force would not be enough to keep the structure stable.

Toichiro Kawai must have understood this structural problem. The model he made at home, described by his son, is definitely a stable woven structure even without the use of nails. However, by aiming for the metaphor of the number ten, perfection, thus by insisting on having ten members and an arc of 10 metres in diameter, the design that finally appeared in front of us had sacrificed its structural stability.

Kawai must have had some means to keep the joints fixed, but probably without the use of nails, as his son proudly recalled. Furthermore, it is often mentioned in the early archives of the Japanese Garden that the Japanese carpenters do not use nails, and that was precisely the reason that Kawai got his first job with the Huntington. Although this belief is still popular in America and even in Asia today, in fact, the concept of "construction without nails" refers only to the *main structure* of Japanese (and Chinese) wooden architecture; in the joinery, including windows, doors, ceilings and decking, plenty of iron nails are used since at the latest a millennium ago. In Japan, nails and other iron elements, for example iron cramp and iron band, are also common in traditional wooden bridges. It is thus possible that Kawai, a proud Japanese carpenter, might have been "hijacked" by the legend of his career tradition, and he might have avoided the use of iron elements in the main structure, and used other types of joints instead.

There are many ways to restrict the motion between the beams. For example placing wooden dowels between the intersecting beam members, or hidden dovetails at the position of the present notches – this kind of wood joint in that position would be functional enough for a while, but it would also be weak enough to be damaged over a relatively short period of time (Figure 2.30).

According to this assumption, the first "non-nail" Moon Bridge collapsed shortly after its completion (in 1914 or later), because of the failure of the

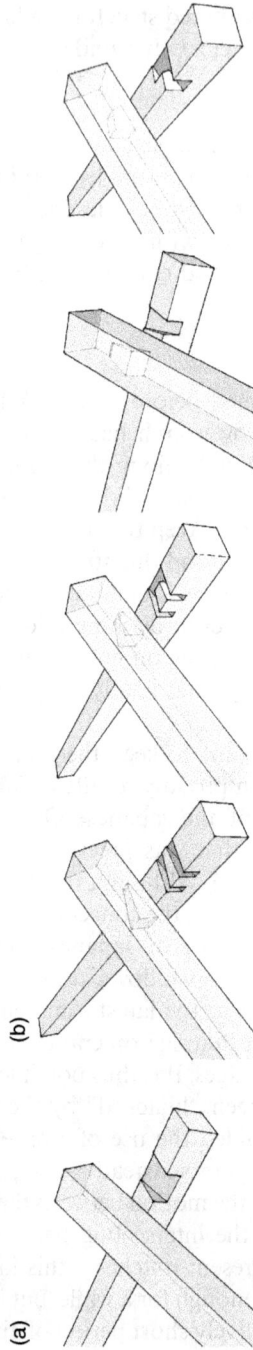

Figure 2.30 Joint form analysis. (a) Current joint form of the longitudinal and crossbeams. (b) Possible joint forms for the longitudinal and crossbeams.

wooden joints of the woven arch. In this case, the arch-beams would remain standing, since they are rigid structures by themselves, and the handrail would also not be affected. The builders would have put the fallen beams back in place by simply nailing them together, even without Kawai's instruction or help. The broken joints were kept inside and unseen, and thus did not need to be replaced. In 1950s, after several decades of use, the damage to the old material was assessed as serious, and the bridge underwent a reconstruction. The entire structure was dismantled: the collection of the woven arch and arch-beams were completely replaced. New members assembled according to the process described in Figure 2.19, while the handrail was reused as much as possible, and the ends of its rotten members were cut short and repaired. The bridge was also painted red, and the red lacquer stayed in place until 1988 when it was removed on the instructions of a new curator – and in that same year, part of the handrail was replaced or repaired. Those were the last notable changes made to the Moon Bridge to date.

2.6 Conclusions: curriculum vitae of the Huntington Moon Bridge

At the end of our journey tracing the century-long legendary history of the Moon Bridge and its creators, we can now build up a complete biography of the mysterious Moon Bridge in the Huntington Japanese Garden.

The Moon Bridge was a result of the trend for Japonisme in the western world – here, specifically in the US – which went from the later part of the nineteenth century to the beginning of the twentieth century, accompanied by the wave of world expositions.

The Drum Bridge in the Japanese Village of the 1894 California Midwinter International Exposition was the direct inspiration for the Huntington Moon Bridge, even serving as a model for it when Henry E. Huntington commissioned a "full moon bridge" to be erected in his garden in 1912.

The Marsh family, pioneer Far Eastern art dealer in the west coast of the US, played a key role in the story of the Moon Bridge, as it was they who established the California Japanese Garden, which in turn stimulated Huntington's desire to have his own Japanese garden and even his own "full moon bridge." In addition, the Marsh family introduced Huntington to the Japanese carpenter Toichiro Kawai.

Kawai, the creator of the Moon Bridge, a Japanese immigrant, was an open-minded craftsman. He was well-trained in traditional Japanese carpentry, he was flexible, and he was eager to make changes in his life and his career.

The Moon Bridge was constructed partly following the Japanese drum-bridge tradition, and partly introducing the innovative adaptation of a woven arch system which was (and is) not common in either Japan or America.

We have not mentioned the possible origins of Toichiro Kawai's woven-arch-structure idea – they will be left open until the concluding chapter of this book. Here we are ready simply to state that the woven arch structure built by Kawai was an immature attempt and it had structural weaknesses.

Bearing the name "full moon," which also stands for "perfection" in East Asian culture, the bridge utilized multiple means to fulfil this metaphor. However, it does that not by being built in the form of a semi-circle (as its model, the Golden Gate Park Drum Bridge was), but rather by incorporating the number ten in its design, including its composition of ten longitudinal beams and ten horizontal beams and the 10 metre diameter of the arc form, etc.

A further analysis based on a set of measured drawings showed that the bridge was designed in a façade-directed process, using the traditional Japanese method and yardstick system.

As a direct result of these structural weaknesses, the Moon Bridge underwent reconstruction at least once, if not twice. The first reconstruction, if indeed it took place, would have to have been soon after it was erected in 1913, and would probably have been due to the weak joints of the original beams. After the collapse, beams members would have been nailed back by Huntington Garden staff, without Kawai's involvement.

Between 1949 and 1962, the bridge was definitely reconstructed (again). This time, members of the woven arch and the curved beams were completely replaced. The handrails were repaired and re-used. Industrial nails were largely applied. The reconstruction caused a change in the proportion, which could be confirmed through the images from the photo archive.

References

Bennett, Shelley M. 2013. *The Art of Wealth: The Huntingtons in the Gilded Age*. San Marino, Calif.: Huntington Library Press.

Conder, Josiah. 1893. *Landscape Gardening in Japan*. Shanghai: Kelly and Walsh.

Hertrich, William. 1988. *The Huntington Botanical Gardens 1905–1949: Personal Recollections of William Hertrich.* San Marino, Calif.: Huntington Library.

Ren, Congcong
任叢叢. 2013. "岩国錦帯橋技術に関する調査(その1設計過程)."日本建築学会大会学術講演梗概集・建築デザイン発表梗概集.巻2013ページ：ROMBUNNO.22254.

Wolf, Barry, and Annabelle Piercy. 1998. "George Turner Marsh and Japanese art in America." *Orientations* 29. 4: 47–57.

3 Searching for the Chinese wooden arch bridges

3.1 Discovery of the Chinese wooden arch bridges: from the Rainbow Bridge to the Min-Zhe area bridges

3.1.1 Discovery of the Rainbow Bridges

In 1953, four years after the founding of the People's Republic of China, the Palace Museum in Beijing, the former Forbidden City of the imperial court in the last two dynasties, first exhibited to the public the Qingming Scroll (*Qingming Shanghe Tu* 清明上河图). It is now the most famous Chinese scroll-painting in the world.

It was at this time that Tang Huancheng[1] (唐寰澄, 1926–2014), a young bridge engineer who had graduated just five years previously, had his first glimpse of the painting. A fleeting glance, in dim light, this encounter did not leave a strong impression on him.

Later that the same year, Tang had a second opportunity to see the Qingming Scroll, when the journal *New Observation* published an introductory article about the scroll, including a reproduction of areas of the painting. This time, a "light and beautiful" bridge, quite different from the common bridge scenes, caught his attention.

The scroll, 24.8 cm wide and 528.7 cm long, depicts a view of the city of Bianliang (today Kaifeng, Henan Province), then the capital city of the Northern Song Dynasty (960–1127), from the suburbs to the inner city, showing the everyday life of that prosperous metropolis. From right to the left, it depicts, in sequence, the landscape of the suburbs, the Bian River with its many boats, the city gate and the prosperous city street with various shops. The bridge that is of interest to us (Figure 3.1), namely the Rainbow Bridge, as it is now known, in the exact middle of the scroll, stands outside the city gate. It is composed of many wooden beams interwoven together to form an arch.

---

1 In this book, all Chinese names are written with the surname before the given name, following the Chinese naming rules. For more details on this see Chapter 6, Sec. 6.1.

Figure 3.1 Rainbow Bridge, in *Qingming Shanghe Tu.*
Source: ©Palace Museum, Beijing, China.

The structure is so unusual, that Tang believed it was the only example of its kind in the world. Even after the discovery of the wooden arch bridges in Southeast China in 1980 (more on this later), which he believed to be the result of a surviving construction technique for the Rainbow Bridge, he was still convinced that this kind of bridge existed only in China. This theory, however, has already been refuted through the European and Japanese examples in Chapters 1 and 2, and as in the previous chapters, this kind of interlocked wooden arch structure is referred to as "woven arch" structure in this book.

Tang raised many questions about the Rainbow Bridge:

> Who was the painter and what was the historical background of the appearance of this painting?
> Was the bridge a real structure or was it purely a fantasy?
> How did the bridge come into being?
> How was the bridge constructed?
> Have any other bridges in this form survived in China?

(2010, 22–3)

Half a century later, in his last book *Chinese Wooden Arch Bridges* (2010), which was published four years before he passed away, Tang claimed that he had found answers to all these questions. This chapter reviews and examines his approaches and answers.

3.1.2 Historical background of the Rainbow Bridge

Of the many historical scroll-paintings bearing the same name "*Qingming Shanghe Tu*," the one kept in the imperial collection *Shiqu Baoji* (石渠宝笈) in the Forbidden City is recognized as being the original. This is borne out not only by the many comments inscribed on the painting during the following dynasties, but also by the sophisticated painting technique and the city scenes' realistic depiction of life during the Northern Song Dynasty.

Our knowledge about the painter comes mainly from a postscript made in the Jin Dynasty, added only half a century after the end of the Song Dynasty, which means it could therefore be reliable. From the text, we know that the painter, Zhang Zeduan (张择端), was an official painter at the court of the Emperor Huizong of Song (1101–1125). He came from Shandong Province and had studied in the capital city. He was famous for painting type *jiehua* (界画, lit. "ruler painting," a special type of traditional Chinese painting depicting architecture, vessels and other subjects, and drawn with a ruler) and was especially skilful at drawing boats, vehicles, cities, bridges, and streets.

During later dynasties, the Qingming Scroll was mainly kept in the imperial collections or powerful private collections, and thus was kept out of sight from the outside world. Nevertheless, it enjoyed prestige and inspired countless imitations. The influence of the Qingming Scroll reached as far as Japan (e.g. the *Kidai Shoran*熙代勝覧 scroll of 1805, kept in the Museum of Asian Art, Berlin). However, since most artists had not seen the original work, they painted according to textual descriptions of the famous scroll or simply presented their own experiences and ideas. A large group of these images were modelled on the city of Suzhou of the Ming and Qing Dynasties rather than the city of Bianliang of Song (including two famous versions kept in National Palace Museum in Taipei). It is also noteworthy that all the later versions of the Qingming Scroll represent the Rainbow Bridge as a masonry arch bridge. The wooden arch bridge is only found in the original Song painting.

Tang (1986; 2010) found some clues about the Rainbow Bridge over the Bian River in the literature of the Song Dynasty. The Bian River, which was actually part of the Grand Canal, connected the capital city in northern China with the fertile Lower Yangtze River area in southern China, and served as an important waterway, a lifeline for the empire. Thousands of boats carrying grain and other goods plied the river day and night.

The name of the bridge appears in a memoir entitled *Dongjing Meng Hua Lu* (东京梦华录 lit. "Dreams of the Splendour of the Eastern Capital"), a collection of essays about city life and city views under the Song Dynasty, written by a contemporary scholar. It mentioned many bridges inside and outside the capital city:

> The bridge, seven miles outside the East Water Gate, is called Rainbow Bridge. This bridge has no pillars. It is constructed entirely of large wooden

beams, painted with vermilion paint, and it arches like a flying rainbow. (东水门外七里曰虹桥。其桥无柱，皆以巨木虚架，饰以丹雘，宛如飞虹)

In addition to this bridge, the memoir mentions two other bridges, the Upper- and Lower-Earth Bridge, which are similar in structure (其上下土桥亦如之).

From the historical descriptions and from the pictorial representation in the scroll, Tang concluded that the bridge was 18 to 20 m long and 8 to 9 m wide.

The compendium (of governmental statutes) *Song Huiyao* (宋会要), which is today part of the Yongle Encyclopaedia from the fifteenth century, recorded the partial construction of a bridge as follows: The Bian River, fast-flowing and forceful, caused much damage to boats and to the previous bridge piers which stood in the water. In the year 1017, the first year of the Tianxi reign, a new bridge form, "a bridge without feet" (无脚桥) was proposed to solve the problem. This kind of bridge is "made of *woven* wood and connected with nails" (编木为之, 钉贯其中). However, the project was stopped and then can- celled due to excessive expenditure.

Despite this failure, two examples of the "flying bridge without pillars" were realized in the neighbouring provinces a few decades later. Wang Pizhi (王辟之) reported in his Work *Shengshui Yan Tan Lu* (渑水燕谈录 1099) that the people in Qingzhou (now Qingzhou, Shandong) had a problem with summer floods coming from the mountains, and damaging bridge piers standing in the Yang River. This problem was resolved in the Mingdao reign (1032–1033), by following a clever idea proposed by a lower servant in the prison (牢城废卒). He strengthened the riverbanks with large stones and built a flying bridge without piers with dozens of large wooden beams (取大木数十相贯，架为飞桥，无柱).[2] At the time when the above book was written, the bridge had already been standing for over 50 years and was still in good condition. Later, in the Qingli reign (1041–1048), the people of Suzhou (now Suzhou, Anhui), under the administration of the district magistrate Chen Xiliang (陈希亮), faced a similar problem. Chen ordered a flying bridge to be built according to the Qingzhou model. After this project, flying bridges became common in this area over the Fen River and the Bian River, and they became known as "rainbow bridges." A record of the flying bridge of Cheng Xiliang also appears in the *Song Shi* (宋史), the official history of the Song Dynasty.[3]

Although historical records show that bridges of this kind were often built during the North Song Dynasty, this form of bridge is missing in testimonies from later dynasties. In the 1950s, the rainbow bridge in China was considered to be a lost bridge type in academic circles.

2 According to this literature, Tang named the structure form of the Rainbow Bridge as "*Guan mu gong*" (贯木拱, lit. wooden arch composed of beams head to head following one another) in his books since 1980s.

3 On the Rainbow Bridges in the Song Dynasty, see Tang (2010) and Fang (1995).

3.1.3 Discovery of wooden arch bridges in the MZ area

In 1980, during the Conference on Bridge History in Hangzhou, Zhejiang, scholars from the Zhejiang Transportation Department mentioned a special type of "strut bridge" (bridge with deck beams and slant struts), in the mountains of this province. After careful examination of the photos, Tang raised many questions on the structure. Together with colleagues, he travelled 435 km to investigate one of these bridges. After arrival, he confirmed that this kind of bridge belonged to a bridge-type closely related to the Rainbow Bridge in the Qingming scroll (Tang 2010, 51).

This was the discovery of the Min-Zhe wooden arch bridges (henceforth "MZ bridges").[4] They are found only in the boundary areas between the provinces Fujian (Min) and Zhejiang (Zhe) (henceforth "MZ area/mountains"). This type of bridge, Tang was convinced, was the result of a long and creative development rooted in the rainbow bridges of the Song Dynasty, the building technique has spread from the north to the south and survived in this remote mountainous region of southeast China for nearly a thousand years.

It is speculated that there were some 200 such bridges in the MZ mountains until the 1980s (Fang 1995). In the past few decades, many of them have been destroyed by fire or flood or dismantled and gave way to modern structures. Today, about 100 historical bridges have survived.

Tang explained the structural principle of these bridges, using the Meichong Bridge (梅崇桥, Figure 3.2) in Jingning, Zhejiang as an example.

According to Tang's explanation, the main structure of the Meichong Bridge consists of two supporting systems: the first being a three-sided arch, and the second, a five-sided arch. Each system consists of longitudinal beams and transverse beams (crossbeams), connected by wooden joints, as opposed to the Rainbow Bridge, where the beams are connected using nails.

Aside from the main arch (the woven arch) structure, the MZ bridges also have a set of additional horizontal beams at either side of the woven arch, providing a flatter deck for the convenience of the pedestrians. Between the woven arch and the deck-beams, there are also various strut structures (X-shaped struts and middle struts) serving to stabilize the entire bridge.

3.1.4 The traditional game of "chopstick-bridge"

After the discovery of the MZ bridges in the MZ mountains, local cultural workers joined in the study of these bridges in and around their home towns. A local folk game attracted people's attention, the "chopstick-bridge." In this game, people build a mini arch bridge using thin pieces of wood, often chopsticks. The sticks are interlocked, and the simplest form is composed of

4 In fact, Tang was not the first bridge historian who "discovered" the MZ bridges. For example Luo Ying (罗英) decripbed some of these "wooden arch bridges" in 1959 (63–7), but his research didn't receive enough attention at the time.

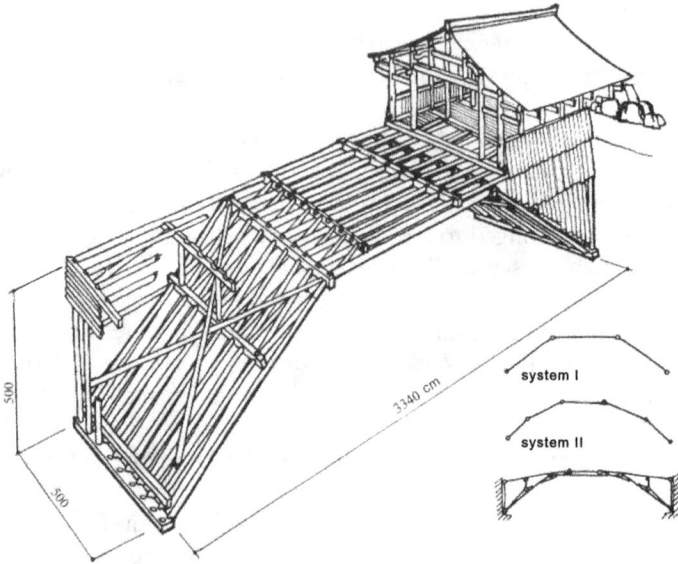

Figure 3.2 Tang's illustration of the structure of the Meichong Bridge, Jingning, Zhejiang, China. This bridge was destroyed by fire in 2005.

Source: Tang 2000, Fig. 5-45.

six sticks. It is possible to use more sticks to expand the form according to the same principle (Figure 3.3).

With appropriate notches cut on the beams, the idea of a chopstick-bridge can be expanded into a circle (Figure 3.4).

This game is particularly well known in the MZ mountains. In a local dialect in northern Fujian (e.g. in Songxi) the wooden arch bridges are commonly known as "chopstick-bridges" (Zhou et al. 2011, 174). A wooden arch bridge built in the 1960s in Ankou Village, Jianyang, Fujian, was literally named "Chopstick Bridge" (Kuaizi Bridge, Figure 6.63).

Although the chopstick-bridge is frequently referred to among local cultural workers in their study of the MZ bridges, this game did not receive equal attention in scientific circles. Strangely, although scholars used the game-like demonstration to explain the structural principle – Tang making a model of the rainbow bridge out of matches in 1950 and Lu Bingjie, professor at the Tongji University in Shanghai, using chopsticks to demonstrate a similar concept during a talk in Japan in the 1980s – neither of them mentioned a possible relationship between the traditional game and the construction history of the bridge.

During his research, the author took every opportunity to collect information about this game in his travels and from his social media and lecture

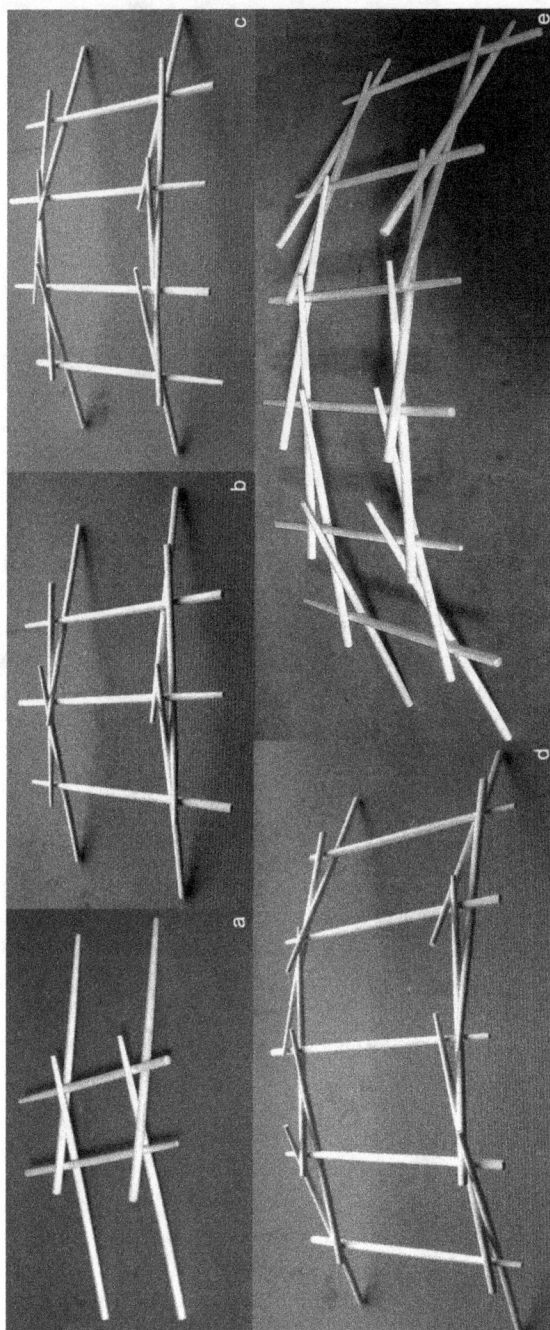

Figure 3.3 Expanding the chopstick-bridge-game.
Source: Made by the author.

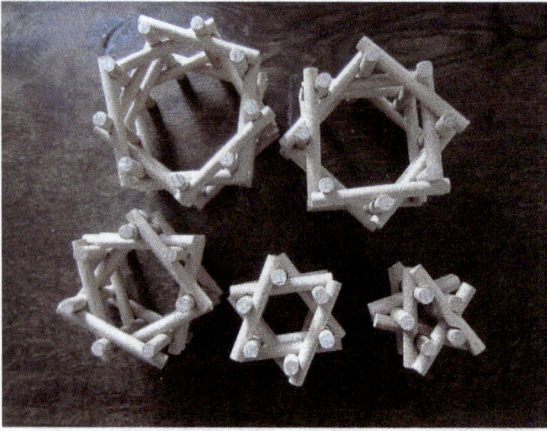

Figure 3.4 Woven circles.
Source: Made by the author.

audience. Information collected from older people shows that until the mid-twentieth century (before the advent of television in China), this game was at least common in Harbin (northeast China), Hebei (north China) and the MZ Bridge area (southeast China). Since the majority of reports come from the younger generation, the game must have been even more widespread before the appearance of the Internet (before 1994). Respondents who answered affirmatively said that they had been familiar with this game since childhood. It is noteworthy that this was also the case in the Hubei and Chongqing area, where there are other types of traditional bridges with a woven structure (Figure 3.11, Figure 3.12).

Although widespread, the chopstick-bridge-game is only popular in limited areas (well known in the MZ area; more common in northeast China and north China compared with other areas). In most parts of China, this game was relatively unknown. If it appeared, it was more or less a magic show to fascinate and astonish the audience. One interviewee[5] told the author that he witnessed this game at the beginning of this century, in Maoming, Guangdong, where a man selling ointments played this game at a market. It was a kind of advertisement for his business as if he had special knowledge. This indicates that in Guangdong, the southernmost corner of China, the game was not well-known at that time.

Unfortunately, there are no written records of this game. However, there is reason to speculate that there is a possible relationship between the game and the construction history of the bridge, not only because of the similarity between the chopstick-bridge and the rainbow bridge, but also the dominant

5 ID "邱浔" from douban.com. December 2015.

role of chopsticks in Chinese food culture. Before embarking on the construction of a real bridge, the playful chopstick game could serve as a most convenient model to assist analysis and demonstrate the design. Therefore, if, for example there really was a lower prison-servant who submitted a bridge design to his superior officer (or any similar situation), it was highly probable that the clever idea was demonstrated using chopsticks, the most commonly-seen material on hand in every house.

3.1.5 Discovery of the building tradition of the MZ bridges

As a result of China's modernization, which eventually reached the remote mountains, wooden arch bridges lost their market and gave way to domination by concrete bridges in the middle of the last century. After the last projects in the 1970s, the construction tradition of woven arch bridges disappeared in the MZ mountains. In the 1980s, when scholars went into the mountains looking for the historical bridges, they were also curious to know if any carpenters still knew how to build bridges of this kind: this question remained unanswered during the two decades that followed.

At the beginning of the twenty-first century, scholars from schools of architecture became the strongest force in research on wooden arch bridges. At the same time, research groups from Nanjing and Shanghai went into the MZ mountains. Their trips began in Taishun County in Zhejiang and went west and southward into Fujian Province. They sought surviving bridges as well as bridge carpenters.

The first scholar who focused his attention on the bridge carpenters was, again, Tang Huancheng. During his investigations, he noticed that bridge carpenters' names were written in ink on bridge beams, and he recorded the list of names on the Meichong Bridge (1986a, 108) and the Xuezhai Bridge in Taishun (2000, 481) in his books.

Using Tang's information as a reference, Liu Jie (刘杰), a PhD student at the Tongji University in Shanghai at the time, visited Shouning County in Fujian in the spring of 2001. With the help of a local cultural worker, Gong Difa (龚迪发), who had been researching bridges in the past years, they visited the bridge carpenter Zheng Duojin (郑多金) in Kengdi, Shouning.[6]

In the same summer, and with the help of the same cultural worker Gong Difa, Zhao Chen (赵辰), a professor at Nanjing University arrived at the door of the same carpenter family. The information that led him there was the inscription from another bridge Dachikeng Bridge (大赤坑桥) in Jingning, Zhejiang.[7]

Later that same year, as part of its TV series *Research & Discovery* (探索 • 发现), the Chinese Central Television (CCTV) made a documentary

---

6 Information given to the author by Liu Jie, 2011.
7 Information given to the author by Zhao Chen, 2011.

film called "Looking for the Rainbow Bridge" (虹桥寻踪). For this film, master carpenter Zheng Duojin built a small wooden arch bridge for demonstration. The building process, construction method and construction progress were recorded on film. Unfortunately, the structure was dismantled immediately after the shooting of the film.[8]

Since then the wooden arch bridges and the bridge technology have received great attention from local governments as well as researchers. More bridge carpenter families were found by local cultural workers, including the Zhang family from Xiajian area, Zhouning County, Fujian. This family, known locally and in this book as Xiajian Masters, have long been the leading experts in this profession, and have the longest tradition (over two centuries) in this career. The family have also secured the largest number of bridge contracts (between the master carpenter and the project directors), which serve as one of the most important bodies of literature for the study of this bridge tradition.

In 2009, a product of cooperation between the local authorities and scholars, the *"Traditional design and practices for building Chinese wooden arch bridges"* was inscribed into the UNESCO's List of Intangible Cultural Heritage in Need of Urgent Safeguarding. A documentary film entitled "Homeland of the Wooden Arch Bridges" (木拱廊桥之乡) was part of the documentation supporting the application. For this film, a small wooden arch bridge – the Shijin Bridge (十锦桥) – was built by bridge master Huang Chuncai (黄春财) in Pingnan County, Fujian. The construction method and the construction process were documented.

After it was inscribed into the UNESCO heritage list, the wooden arch bridges aroused the interest of the government and the public, and many new bridges were built and continue to be built. New projects also provided the author with opportunities to explore and document bridge technology. In 2015, the first MZ bridge built overseas was erected in the Nepal Himalaya Park in Regensburg, Germany, as part of a cooperation between a descendant of the Zhang family (Xiajian Masters) and the author (see Chapter 4).

3.2 Rainbow Bridge re-observed through the lens of the MZ bridges

3.2.1 *Tang's change of mind with regard to the Rainbow Bridge*

After the discovery of the MZ bridges, Tang Huancheng's comments regarding the Rainbow Bridge changed significantly. Before he knew about the MZ bridges, Tang Huacheng saw the Rainbow Bridge as an arch that:

8 Because of the poverty of this family and that of the county government, even the timber used for this bridge was borrowed, so the structure was dismantled afterwards to give the timber back. Information given to the author by Zheng Duojin's younger brother Zheng Duoxiong, 2013.

is composed of five arch-beams. Each arch-beam is placed on two crossbeams, which lie in the middle of two other arch-beams. A single series of arch-beams is not self-sufficient, at least two rows of by crossbeams connected arches are necessary.

(1957, 28)

Since the 1980s, after the discovery of the MZ bridges, whose under-deck structure appears to consist of two polygonal arches, Tang changed his description. In 1986(a), he described the Rainbow Bridge as a combination of two systems, as follows:

The first system forms the exterior group (seen from the front); it is composed of two long and two short arch-beams. The second system is the inner group; it is formed of three equal arch-beams.

(105)

A year later, Tang again changed his description, this time saying that the arch was still composed of two systems – but reversing their order. Now, the first system was named after the one with three long beams, and the second system is composed of two long and two short beams (1987, 74). In 2000, he designated the two systems number I, and number II. The order/definition of the two systems remained unchanged thereafter.

3.2.2 Reconstruction of the Rainbow Bridge in the 1950s and the 1990s

As a bridge engineer and bridge historian, Tang Huancheng twice had the opportunity to build replicas of the Rainbow Bridge.

The first time was in 1958, shortly after he had "discovered" the Rainbow Bridge in the Qingming Scroll, and he was working as a designer for the Wuhan Yangtze Great Bridge in Hanyang, Wuhan, the first modern bridge over the Yangtze River. As part of the project, Tang had the opportunity to build a small-scale Rainbow Bridge (hereinafter as the "Hanyang Bridge" for short). Unfortunately, this bridge was dismantled in the 1970s during the Cultural Revolution.

His second opportunity came in 1998, when the American WGBH station's TV Nova series made a documentary film about the Rainbow Bridge. Then, as part of the international team consisting of Chinese and American scientists and engineers, Tang was involved in building a small replica of the Rainbow Bridge in Jinze ("Jinze Bridge" for short), a town near Shanghai with plenty of waterways.

3.2.2.1 Examination on the structure form

Both of the two projects were small-scale reproductions. The "Hanyang Bridge" had a span of 12 m and was built with 12 cm thick beams. The "Jinze

Bridge" had a 13.2 m span and was built using 18 cm thick beams. Though aimed at reconstructing the Rainbow Bridge in the Qingming Scroll in miniature, both bridges diverged from the original image in their construction. In Tang's projects, the arch members were round beams and the longitudinal beams were set in pairs, whereas in the original drawing – as modern high-resolution imaging allows us to see clearly – all longitudinal beams are square pieces of wood and they are set singly.

Tang's error about the Rainbow Bridge was probably due to the unsatisfactory image sources available at the very beginning. In his comments from 1986, he still describes the longitudinal beams as being of "round wood," but "their upper and lower sides were flattened with saw or adze." However, a perspective drawing in the same book shows them as square beams. This indicates that by the 1980s at the latest, Tang had obtained better sources of images, which had provided more precise information on the construction. Nevertheless, he still used the in-pairs arranged round woods for the "Jinze Bridge."

In the "Hanyang Bridge," Tang had encountered a structural problem – the lack of lateral stability. Therefore, in the "Jinze bridge," he inserted a layer of X-shaped beams (2000, 467; 2010, 71–2, 113; Figure 3.5). This idea, he wrote, came not only from the experience in Hanyang but also from the MZ bridges, which have a pair of X-shaped struts between the woven arch and the deck. It is reasonable for a civil engineer to use the X-shaped struts to assist with stability, but this is not actually in line with the Chinese traditional carpentry system, under which structural stability is achieved through different means of construction. In fact, in the real Rainbow Bridge, there is no problem with stability. The woven arch composed of squared beams is stiff and stable in itself and functions almost as a rigid hull (Figure 3.6), thus there was no need

Figure 3.5 "Jinze Bridge," Jinze, Shanghai, China. X-shaped struts in the structure.
Source: Tang 2010, Fig. 6-21.

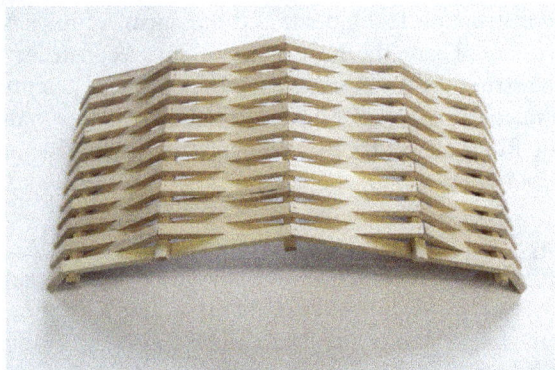

Figure 3.6 Model of the Rainbow Bridge with square beams.
Source: Made by the author.

for the X-shaped beams. Even in the MZ bridges, which have X-shaped struts, they do not serve to assist the woven arch, but other parts of the structure (more on this see Sec.4.6.1.6).

3.2.2.2 Examination on the construction method

Tang's understanding of the structure of the woven arch bridge changed over time from 1957 to 2010, due to the discovery of the MZ bridges and his project experience. This is revealed by his technical approach to the projects and the related texts.

The first text was written in 1986, almost 30 years after the bridge project, Tang mentioned the Hanyang Bridge in a book entitled *Technique History of Historical Bridges in China*. In the section "Erection of wooden arch bridges," he discussed the construction method for the Rainbow bridges and compared it briefly with that of the MZ bridges. Notably, at that time, not knowing about the surviving bridge carpenters and the techniques they used, the description of the construction methods for the MZ bridges is a product of Tang's own speculation.

Then, in 1998, the Jinze Bridge was built by Chinese and American engineers: this was after the discovery of the MZ bridges, but before the tacit knowledge of the bridge carpenters was revealed. The construction was built using common traditional techniques. The construction method and the construction process are described in detail in Tang's last book (2010).

Overall, Tang's writing – the 1986 text on the 1958 Hanyang Bridge project; the 1986 text on the MZ bridges, and the 2010 text on the 1998 Jinze Bridge project – comprised two reports on actual construction projects and a hypothetical narration. From the historical study standpoint, all three texts are reconstructions of historical techniques in the sense of experimental archaeology. Their importance is twofold: the contents are the record and/ or explanation of the construction techniques, and the narrative reflects the author's own understanding. From the narrative, we can reconstruct how Tang's perception of the structures changed during the course of his professional life, as he was influenced by new discoveries and experiences: namely, his understanding of the Rainbow bridge was influenced by new information from the MZ bridges:

(1) The Hanyang Bridge was built (1958) before the discovery of the MZ bridges (1980), but reported on (1986b) after the discovery of the MZ bridges.
(2) The hypothetical narrative about the MZ bridges (1986b) coincided with the discussion of the Hanyang Bridge.
(3) The Jinze Bridge was constructed (1998) after the discovery of the MZ bridges (1980) and before the discovery of the bridge-building families and traditions behind them (2001), but was reported on (2010) after the discovery of those building traditions.

A HANYANG BRIDGE (PROJECT 1958; REPORT 1986)

On the construction process, Tang writes:

> At the intersection of the arch-beams of the System I, we erected two provisional post frames to support the System I and to place the crossbeams of the System I. Thereby, System I became a stable structure. On this stable foundation, we placed the beams of System II and fixed them. Then we removed the temporary posts. Only now was the construction really independent. On this we put the deck system.
>
> (1986b, 206–7) (Figure 3.7)

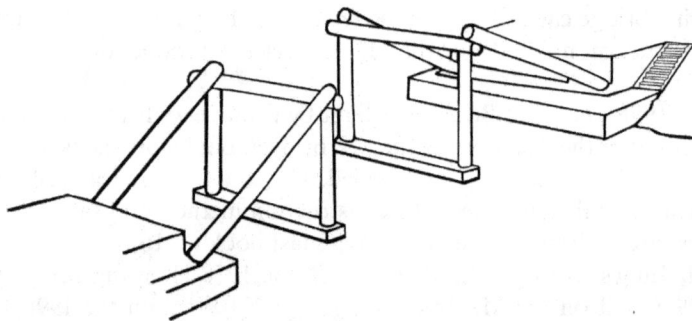

Figure 3.7 Tang's depiction of the building process of the Hanyang Bridge.
Source: Tang 1986, Fig. 6-32.

B MZ BRIDGE (HYPOTHETICAL NARRATION 1986)

In 1986(b), before the discovery of the surviving building traditions, Tang suggested that the construction process of the MZ bridges was as follows:

> The construction begins with the slanted beams of System I. First, the slanted beams of System I are assembled with the crossbeams at the top and at the bottom to form vertical frames on the riverbank. Then the top ends of the frames are tied. Then the cords are loosened and the frames are inclined towards the centre of the river, into the correct position. The horizontal beams have a dovetail at both ends: these are lifted from the bottom of the frames to the top and inserted into the sockets of the crossbeams. Thanks to the dovetail joint form, System I is pressed tight during this step. The joints can stand some bending movement. After System I come the beams of System II, then the deck system.
>
> (207) (Figure 3.8)

C JINZE RAINBOW BRIDGE (PROJECT 1998; REPORT 2010)

For the construction of the Jinze Bridge, American engineers originally suggested the use of modern machines, but Tang rejected this idea and used the means and methods that corresponded to the technical capabilities and conditions of the Song era.

Figure 3.8 Tang's hypothetical reconstruction of the MZ bridges.
Source: Tang 1986, Fig. 6-33.

All the elements were precisely manufactured in advance. Before the real construction over the water commenced, the bridge was assembled on land to assure project success.

The four steps of the final construction Tang describes as follows:

(1) Preparation: Two boats were anchored in the river. Boards and scaffolding were set on the boats. Traditional cranes were built on land consisting of A-shaped wooden frames with strings running from them to hold the arch-beams in position (Figure 3.9, a).
(2) Assembly the lateral beams of System I: Using the traditional cranes and strings, two slant beams were placed on either side. Two crossbeams were then fixed onto the beams at their upper middle part and at the upper end under the beams (Figure 3.9, b). Then the horizontal beams were added in the middle (Figure 3.9, c).
(3) Assembly of System II: Lastly, the beams of System II were pushed from the centre to the sides: the long beams in the middle span came first, followed by the short lateral beams (Figure 3.9, d). After all parts had been positioned in the right places, the longitudinal and transverse beams were bound with bamboo strips and fixed with long iron nails.
(4) Assembly of the rest of the bridge: Thanks to the standing two lateral rows of beams, the structure is now stable, enabling the rest of the arch members and other bridge elements to be installed and finished.

Figure 3.9 Construction process of the Jinze Bridge, Jinze, China.
Source: Tang 2010, Fig. 6-12,3,4,6.

Tang was an experienced bridge engineer and historian of technology. His building projects show his in-depth insight into the practical construction of the woven arch bridges. Although the differences between his (reconstructed) Rainbow Bridges and his (theoretically reconstructed) MZ bridges – both in terms of their physical structure and their intangible construction methods – it is obvious to us today, that Tang was less sensitive to express the differences in structural form and construction methods between these two types of woven arch bridges. Ironically, although he had realized and reproduced the differences in practice, the written description was rather vague and confusing.

One of the greatest differences between the Rainbow Bridge and the MZ bridges is the uniform or the bipartite section of the woven structure. In the MZ bridges, the woven arch consists of two different structural systems, a primary and a secondary arch; in the Rainbow Bridge, this division is not so clear-cut: although the longitudinal beams are set in two different groups according to their position and number, members of these two groups do not differ significantly in shape and function. The Chopsticks Bridge is an extreme example of the rainbow bridge type, with all the beams equivalent in form and function: the longitudinal beams are arranged in different rows, but the difference in position does not identify them with different systems. Strangely, this difference, though obvious and simple, was ignored by Tang.

Let us re-examine the three construction projects, leaving Tang's interpretation to one side:

a   Hanyang Bridge: Assembled using fixed scaffolding; The three-sided arch ("System I" in Tang's description) was built first, and the beams of "System II" were put into position afterwards.
b   MZ bridges: according to Tang's hypothetical narrative
    Work was carried out using suspended support; The three-sided arch (System I) was built first, and System II was positioned afterwards.
c   Jinze Bridge:
    Work was carried out using suspended support. The exterior arch frames (two lateral rows of woven beams) were constructed first, then the interior beams, irrespective of whether they belong to "System I" or "System II" according to Tang's definition.

In Hanyang project, construction was processed under the condition with fixed and stable scaffolding; in both the MZ bridges hypothesis and the Jinze Bridge, suspended supports are used, which is a less stable base for the on-going construction. Since the woven arch structure is (by its geometrical principle) unstable until the completion of the woven frame, both latter cases have to erect a stiff frame as the first step of construction as quickly as possible, and the difference in their measurements expresses the real essential distinction between these two types of woven arch bridges, namely, the idea of the woven arch structure composed of two systems (MZ bridges) and the "uniformly composed woven arch structure" (Rainbow Bridge and chopstick-bridge).

In the MZ bridges the stiff frame is achieved through dovetail jointed System I, and in the Jinze Bridge is the lateral part of the woven arch, which is by itself a complete constitution of a woven frame. This stiff frame, as a part of the bridge structure, then serves as a kind of scaffolding for the later construction steps, and frees the construction from the external supports (here, in both cases, the suspended support).

When we compare the two projects in which an (entire) three-sided arch was built first, namely, the Hanyang bridge and the MZ bridges hypothesis, another essential difference of structural feature becomes distinct: thanks to the wooden (dovetail) joints, the three-sided arch of the MZ bridges is almost a stiff frame thus it could be simply supported by a swaying suspended system. On the contrary, in the Hanyang Bridge, the connections are tied or nailed together, and are therefore rotatable, so its "System I" is not an independent system by itself. To support the unstable system I, solid scaffolding was used. So in this project, "System I" was first only temporarily fixed and strongly relied on the scaffolding. Nevertheless, Tang writes for the Hanyang Bridge "thereby, System I became a stable structure. Onto this stable foundation, we placed the beams of System II and fixed them," a description which fits more with the case of the MZ bridges.

The same happened also onto the Jinze Bridge, where although in the real process he had carried out, the lateral stiff woven frame came first, followed by the internal beams, he described it in a sequence as if the System I followed by System II. To put it more directly, in both cases of his Rainbow Bridge projects, his understanding of the structure and the construction methods is heavily affected by the knowledge of the MZ bridges – not in the way that he carried out the construction, but in the way that described the construction in later years.

Despite all his confusions regarding the Rainbow Bridge and the MZ bridges, Tang's experimental projects are leading works and they inspired later scholars.

3.2.3 Tang's typology of the wooden arch bridges

In 1986, when he first discussed woven arch bridges in the MZ area (he called them "arch bridges of Rainbow Bridge type"), Tang also introduced another example with a similar structure: the Baling Bridge (灞陵桥) in Weiyuan, Gansu Province. Actually, Tang had already mentioned this bridge in his 1957 book together with the Rainbow Bridge, but at the time, he had defined it as overlapping cantilever bridge (bridge composed of layers of cantilever beams which project/corbel over one another to reduce the bridge span). Then, in 1986, he described it as another example of the "Rainbow Bridge type," following his recent discovery that at the uppermost part of the cantilever beams, a woven structure connects the two sides of the bridge (Figure 3.10).

Afterwards, Tang found even more examples with similar structural principles in other parts of China, and together with the Baling Bridge, he defined them

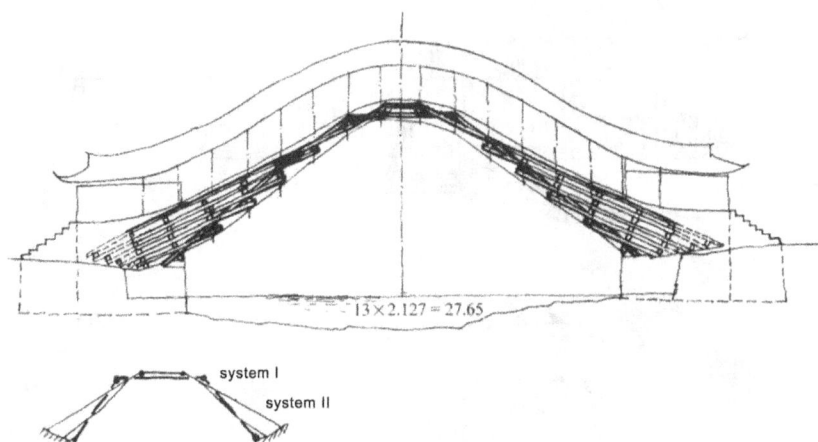

Figure 3.10 Tang's illustration of the Baling Bridge (Wo Bridge) in Weiyuan, Gansu, Northwest China.

Source: Tang 2000, Fig. 5-83.

as "the variations of the wooden arch bridge," which "may not derive directly from the Rainbow Bridge, but have more or less a relationship with it."[9]

Similar to the MZ bridges and the Rainbow Bridge, he understood their construction as a combination of two systems. However, this division, like his attribution of this to the structure of the Rainbow Bridge, is problematic. As mentioned above, in the bipartite section of the woven arch structure (MZ bridges), there are two systems, a main one and a secondary one. Neither of the two systems is a stable structure by itself: it is through being interwoven that they support and stiffen each other. On the contrary, in the Rainbow Bridge/chopstick-bridge, the effort to differentiate the two systems is incorrect, since the woven structure is uniformly composed.

In the Qunce Bridge (群策桥, Figure 3.11, Figure 7.8), which is actually a Chopstick-Bridge-like woven structure with doubled beams, Tang divided the two layers into two identical systems, whereas in fact, each system is a Chopstick-Bridge-like woven arch structure in itself.

Tang's description on the Youyang Bridge (酉阳桥) is much more problematic. His diagram (Figure 3.12) which again attempted to differentiate the two systems, makes the structure unreadable. According to the author's on-site investigation in 2016 (Figure 3.13), this complicated structure is not a woven structure at all, but rather, a combination of multiple layers of struts.

9 Tang classified the following two bridges first into the group "bridge with multiple struts" (Tang 2000, 457–61), but later changed their classfication as "variations of the wooden arch bridge" (Tang 2010, 82–4).

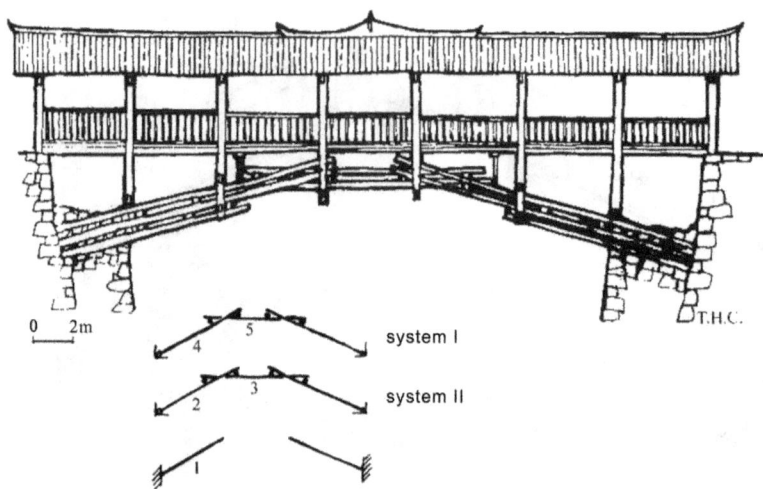

Figure 3.11 Tang's illustration of the Qunce Bridge in the border area between Hubei and Chongqing, Mid-South China.

Source: Tang 2010, Fig. 4-59.

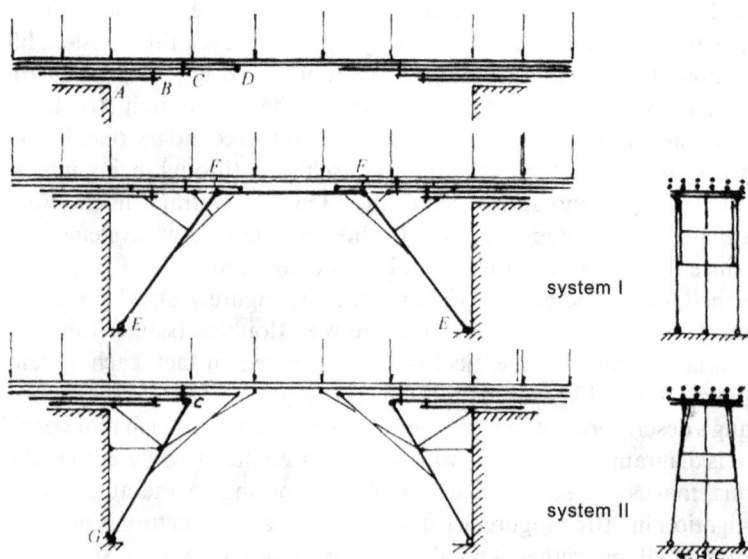

Figure 3.12 Tang's illustration of the Youyang Bridge in Chongqing, Mid-South (or Southwest) China.

Source: Tang 2010, Fig. 4-63.

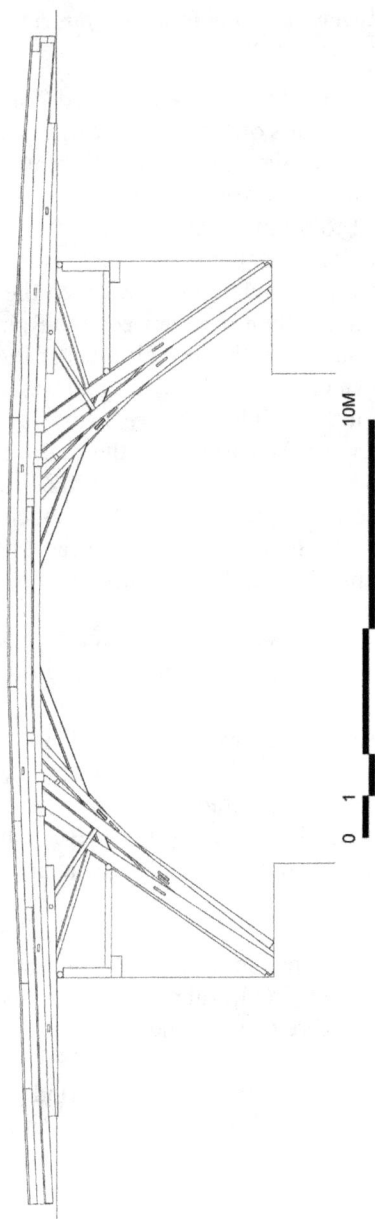

Figure 3.13 Elevation of the under-deck structure of the Youyang Bridge.

Source: Drawn by the undergraduate students Liang Xiaorui, Du Mengzeshan, Lin Yu, and Li Xueqi in Nanjing University under the instruction of the author.

3.3 The origin of the MZ bridges

3.3.1 Tang Huancheng's theory: surviving technique for the "lost" Rainbow Bridge

Tang was convinced that, unlike the above-mentioned "variations," the MZ bridges were the direct descendants of the Rainbow Bridge that had survived overtime down through the dynasties to the twentieth century. In his opinion, only the MZ bridges, and not the other woven structures, originated directly from the Rainbow Bridge. From our point of view today, Tang's conclusion resembles an obsession, and is neither consistent nor reasonable, incompatible even with his own knowledge. The relationship between the Rainbow Bridge and the MZ bridges is no closer than that between the Rainbow Bridge and the "variations." Tang's insistence on the special status of the MZ bridges can be explained only by his own emotions. Since his goal was always to find the lost traditional technique for the Rainbow Bridge, to his way of thinking, only the MZ bridges bear witness to the survival of the tradition and that it has been handed down until today. Thus, in his eyes, the MZ bridges are the clear descendants of the Rainbow Bridge tradition. This conclusion was deeply imprinted in his mind from his first encounter with the MZ bridges – when he celebrated that a technique thought lost had been found – but it was never reflected upon or re-examined.

This enthusiasm is clearly demonstrated in his last book (2010). He described his feelings after the discovery of the MZ bridges as follows:

> People can easily understand how exultant a scholar on the bridge history would be, to discover examples of a lost bridge type which enable him to repair the broken chain of technique evolution, and carry on further research! The Rainbow Bridge is not lost! However, why did it flower in the Central Plains of China, but bear fruits only in Southeast China?
>
> (51)

In order to find out how the technique spread from the Rainbow Bridge, Tang looked for answers from the cultural and historical background in the Song Dynasty. He picked out three possibilities:

Possibility 1. At the end of the Northern Song period (960–1127), shortly before the Jin Dynasty took over in north China (1115–1234), the government of the Song Dynasty fled to the south and started the Southern Song Dynasty (1127–1279). Scholars, artists and craftsmen, including bridge builders, migrated southwards together with the government.

Possibility 2. On the Bian River which belonged to the Grand Canal, many boats came from southern China. Boats from the MZ provinces are explicitly mentioned in the literature (e.g. in the

Biandu Fu 汴都赋 by Zhou Bangyan 周邦彦). It is therefore possible that people on deck, especially the boat carpenters, saw the Rainbow Bridge with their own eyes and brought the construction idea back with them.

Possibility 3. It was possible that the construction technology was brought south by government officials. For example the above-mentioned Zhou Bangyan, who was a famous scholar of the Song Dynasty, served in both the capital city and in the MZ area. (2010, 51–67)

In any case, Tang was convinced that the MZ area, rich in forests and close to the sea, was a fertile ground for the development of bridge technology. As soon as the technique spread into the mountainous area in Southeast China under the Song Dynasty, it took root and survived until today.

3.3.2 Zhao Chen's theory: local origin of MZ bridges

Contrary to Tang's theory of origin, another idea emerged as soon as the MZ bridges became a topic of study among more researchers, and this was that the MZ bridges originated from the local area instead of from the Central Plains of China. Not surprisingly, this idea was first raised by local cultural workers (Zhang 2001; Zhou et al. 2011,15-8) based on an emotion of localism. They attempted to identify the origin of the MZ bridges in their hometown by tracing the history of the technique on-site to as far back as possible. Local records showing the existence of early bridges and old tiles marked with Tang or Song dynasty inscriptions are used as proof of the technique's origins and history. However, none of these is reliable evidence of the existence of woven arch bridges in early times. The early bridges described in the literature might be bridges with other structural forms (cantilever bridge, strut bridge, or simple beam bridge). Indeed the earliest preserved woven arch bridge in the MZ area has been dated to 1625 (Rulong Bridge, see Chapter 5), and the earliest literature on a confirmed woven arch bridge found by the author is dated in the sixteenth century (see Sec.6.5.3).

The first scientific work to suggest a local origin for the MZ bridges is the typology study carried out by Zhao Chen (2001). Based on his extensive field observation in the MZ area, Zhao suggested that the construction technique for the MZ woven arch bridges is closely related to the local bridge-building tradition. He put forward a typological theory that there was a continuous evolution in wooden bridges in the MZ area from the beam bridge to the woven arch bridge (Figure 3.14):

(1) Beam bridge
(2) Beam bridge with wooden posts
(3) Beam bridge with slanted struts
(4) Beam bridge with one three-sided arch/strut bridge
(5) Beam bridge with multiple types of struts and/or supports

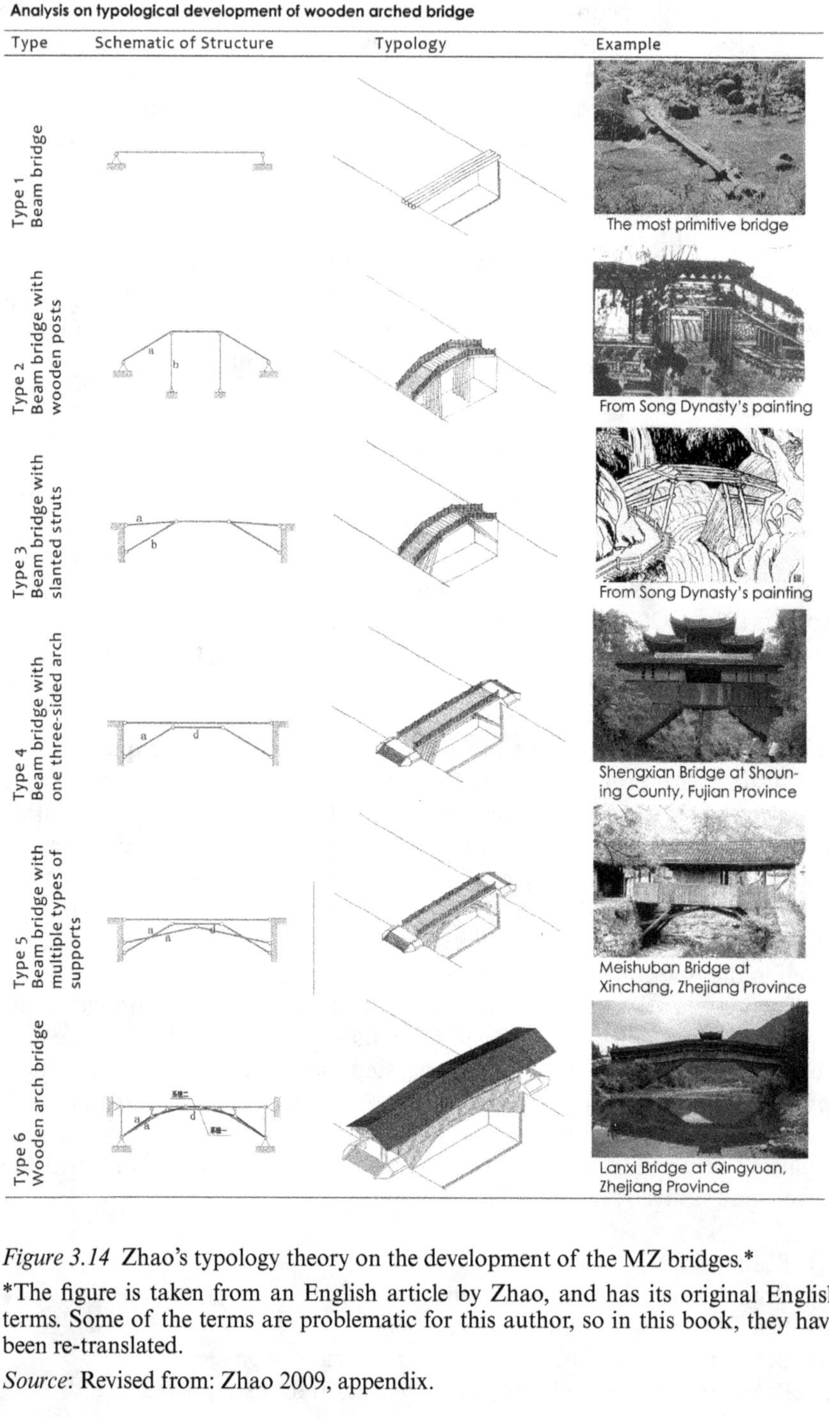

Analysis on typological development of wooden arched bridge

Type	Schematic of Structure	Typology	Example
Type 1 Beam bridge			The most primitive bridge
Type 2 Beam bridge with wooden posts			From Song Dynasty's painting
Type 3 Beam bridge with slanted struts			From Song Dynasty's painting
Type 4 Beam bridge with one three-sided arch			Shengxian Bridge at Shouning County, Fujian Province
Type 5 Beam bridge with multiple types of supports			Meishuban Bridge at Xinchang, Zhejiang Province
Type 6 Wooden arch bridge			Lanxi Bridge at Qingyuan, Zhejiang Province

Figure 3.14 Zhao's typology theory on the development of the MZ bridges.*

*The figure is taken from an English article by Zhao, and has its original English terms. Some of the terms are problematic for this author, so in this book, they have been re-translated.

Source: Revised from: Zhao 2009, appendix.

(6) Wooden arch bridge with woven arch (of three-sided and five-sided arches).[10]

Tang's and Zhao's opposing theories regarding the origin of the building technique for the MZ bridges triggered an academic debate between the two scholars.[11] Tang's argument was looked to classical literature as a basis for the appearance of woven arch bridges and the possibility of the technique spread. Zhao, on the other hand, refused to consider that any legendary event played a role in the origin of this type of vernacular architecture ("cannot have been invented by [a] specific person or persons at a specific time") (Zhao 2011), and is convinced that it emerged as a response to local requirements by local craftsmen and is the fruit of the totality of construction techniques of the region.

The arguments regarding the origin of the wooden arch bridges reflect two basic attitudes in a nationwide discussion taking place in the field of historical and archaeological studies: the origin of Chinese civilization. Since the 1980s, the old theory that the Central Plains of China were the cradle of Chinese culture has been giving way to the new theory that the Chinese culture has multiple cultural origins from remotest antiquity, and that among these multiple origins, the Yangtze River area was one of the most developed areas, on a par with the Yellow River area. Since 1990, South China has also received new attention in the field of architecture, in which Zhao Chen also played a role. The focus of architectural research is shifting from the monumental and governmental buildings in Northern China to vernacular architecture, and the wealthy and advanced region of Southeast China is one of the most discussed areas.

Against this background, Tang's theory is rooted in a cultural history concept centred on the Central China Plains, whereas Zhao's theory corresponds to the trend of emphasis on the cultural status of Southeast China and the study of vernacular architecture.

However, Zhao's theory has some weaknesses.

Firstly, of the examples he provides as proof of his six-link evolutionary chain, three links do not come from the MZ area. Two examples come from historical paintings: Link No. 2 "Beam bridge with wooden posts" from the *Shuidian Zhaoliang Tu* (水殿招凉图) by Li Song, South Song Dynasty and Link No. 3 "Beam bridge with slanted struts" comes from the *Qiulin Feipu Tu* (秋林飞瀑图) by Fan Kuan, North Song Dynasty. And one – link No. 5 – "Beam bridge with multiple types of struts and/or supports") comes from the middle of Zhejiang province, 200 km north of the MZ area.

Secondly, in his evolutionary chain, there is a huge jump between the penultimate step and the final step. The overlapped struts in Link No. 5 do not consequently lead to the woven arch structure. When facing larger structural

---

10 The terms for the structural forms have been re-translated by the author and are thus different from the original English terms from Zhao's article.

11 Letters between Zhao and Tang. Copies of the letters were provided to the author by Zhao Chen.

challenges, the overlapping of multiple kinds of supports is a natural solution in every building culture, but the concept of a woven mechanism is a fundamentally distinct structural idea. The technical origin, namely the actual constructional principle of the woven structure, is skipped in Zhao's theory of the evolutionary chain.

In 2009, local cultural workers found a group of "wooden arch bridges with two three-sided arches" (3+3 bridge) (Su and Liu 2010) on the southern edge of the MZ area. Compared with the common woven arch bridges (3+5 bridge), the 3+3 bridges are simpler in form and smaller in scale. They are now considered to represent a step before the mature woven arch bridges in Zhao's typological chain. On this basis, the author of this book, a student of Zhao's and in the early stages of his doctoral study, completed Zhao's evolutionary chain by inserting this link between the penultimate and last links (Su and Liu 2010; Liu 2011). Soon after, local cultural workers discovered even more examples of 3+3 bridges and 3+4 bridges. Seen as intermediate steps in the technological revolution, they seemed to provide even more solid proof of Zhao's evolution theory.

3.3.3 Evaluation of both theories

From today's viewpoint, both Tang's and Zhao's theory are problematic.

The main problem with Tang's theory lies in his assumption that the MZ bridges are evidence of the survival of the construction technique of the "lost" Rainbow Bridge. As a result, he over-emphasizes their similarities and overlooks the differences between them. This mistake is so unexpected,especially considering his status as an experienced bridge engineer who twice built smaller replicas of the Rainbow Bridge, but can ultimately be explained by his personal emotions and standpoint.

The main problem with Zhao's theory is his assumption that the MZ bridges are a product of local building culture and local conditions. Zhao began his exploration of the MZ bridges with the clear intention to prove his tectonics theories about the Chinese building culture. The MZ bridges are one of the two branches he deliberately selected as part of his study framework. The goal of his research was a foregone conclusion: to reveal and emphasize a relationship between building technology and local backgrounds (Zhao 2001).

Before we formulate a new theory on the origin of the MZ bridges, which will only appear in the final chapter of this book, we have to understand the real question that needs to be asked in this context. Regarding the possible spread of bridge technology, if there was such a technique, and it was transferred (from the Central China Plains to the MZ mountains or from any one place to another), there are two levels of aspects to consider:

(1) The woven arch principle, namely, the concept of forming an arch-shaped structure by interlocking beams together. No craftsmanship or specialists are necessary for the spread of this idea. The chopstick game or a simple

sketch is ideal media to illustrate or transfer this idea simply and straight-forwardly. If the idea of the woven arch bridge is spread by these means and practised in different places, carpenters in each place, having had no direct contact with the real bridges, may realize this concept according to their own imagination and construction experience. Therefore, the bridge forms and construction details might be quite different from one place to another.

(2) The technical knowledge and its practical application. At this level of technique spread and/or development, people with specialist knowledge – in this case bridge carpenters – are required. If the bridge technique is transferred from one place to another and from one generation to the next, although the structures might not be exactly the same, some major construction features must be practically identical.

Tang's discussion focuses mostly on the second aspect (since he is looking for historical possibilities for the migration of carpenters to transfer the tech-nology), but his theory accepts the first aspect as well. In contrast, Zhao rejects the second possibility outright. He focuses on the tectonics of the MZ bridges, accepting only the possibility of the local environment (origin and development on-site). At the same time, his attitude to the first aspect is ambivalent. Although he expresses an openness towards accepting the idea of multiple routes for technique spread (Zhao 2001), the evolution route he suggests for the structural form leaves no space for the first possibility. To put it another way, in order to prove his typology theory, he has to refute both possibilities at the same time: the direct technique spread, and the pos-sible role of the chopstick-bridge in the spread and development of the bridge technique as well.

By comparing the woven arch traditions in the Central China Plains and in the MZ area, the second possibility has already been self-evidently refuted by the huge differences in structural forms and construction features. To refute the first possibility, however, is almost impossible. The chopstick-bridge-game is too common a folk game in the MZ area, and its relationship with their woven arch bridges is a firmly-held belief by the carpenters and by the general public. Moreover, even examples from early modern times show that the chopstick game was used to introduce the bridge technique to carpenters outside the traditional MZ Bridge area (more on this see Sec.6.4.5). The game, with its convenient material and playful form, has a great vitality for surviving and spreading naturally. There is no visible his-torical evidence to prove its role in the variations of woven arch bridges throughout China, but it is equally impossible to refute its potential role in bridge construction history.

The chopstick-bridge-game in the MZ area could have been introduced from the north or somewhere else, or it could have emerged locally. No matter the origin of the game, as soon as we accept the possibility that the game could have served as a medium for the transmission and development of the bridge technique, we can get a new viewpoint for discussing and observing the history of the bridge.

Firstly, The chopstick-bridge-game can be played in a great variety of forms, making the idea of a linear evolution chain, from less to more, from simple to complicated, untenable. In other words, the development of the bridge at the level of actual built structures does not lie in the change of shape and form (e.g. the number of struts and beams), as Zhao Chen suggested, but in other aspects of construction.

Secondly, even though the game might have provided the idea of weaving a large span, the MZ bridges are far different from the composition of the game. Carpenters still face great challenges regarding the structural details and construction methods. The real history of the technique is the evolution which solves all these technical problems, and this is one of the foci of Part Two of this book.

3.4 Wooden arch bridges: the dedication of two generations of Chinese scholars

In his last book *Chinese Wooden Arch Bridges* (2010), Tang wrote about his 50-year search for Chinese wooden arch bridges. The decisive impulse, the one which shaped his entire professional career as a bridge engineer and bridge historian, came as follows:

> In 1948, when I graduated from the Shanghai National Jiaotong University, as a young engineer, I worked on the design of the Wuhan Yangtze Great Bridge, the first bridge over the longest river in China. The design principle, as dictated by the government was 'socialist content and Chinese form'. By 'socialist content', I understood 'everything for the people'... but what 'Chinese form' meant, I had no idea. To answer this question, I looked for material related to this subject whenever and wherever possible. Ultimately, I concentrated on the history and aesthetics of bridges, although I was just a bridge engineer.
>
> (22)

The memoir also explains his later obsession with searching for the cultural roots of the Rainbow Bridge: national pride. In the 1950s, the first decade after the founding of the People's Republic, China's national attitude to its own technical history is clearly reflected in the speech of Prime Minister Zhou Enlai in 1953: "Despite our many historical accomplishments, our achievements in the field of architecture, generally speaking, are less well developed. We have to move forward gradually."[12]

---

12 Zhou Enlai's speech on 10 June 1953.

As a bridge engineer, Tang had clearly recognized the rather discouraging situation of Chinese construction technology. Therefore, we can well understand Tang's emotional reaction to the Rainbow Bridge, with its complicated and marvellous construction, a structure thought to be the "only bridge of its kind in the world" and a great achievement of the Chinese bridge-building history, at a time when China longed for advanced building technology. As a Chinese bridge historian, he devoted his life to the exploration. Tang's pursuit of the Chinese wooden arch bridges is thus also the pursuit of the highest value of Chinese construction technology and the epitome of Chinese building culture.

Unlike most other scholars working on the bridge-building history, Zhao Chen chose MZ bridges for his own specific theoretical reasons. From the standpoint of an architectural theorist, Zhao (2009) used Chinese vernacular architecture as a door to criticism of the classical tradition of Chinese architectural research:

> Compared to Western architectural history, Chinese architectural history is obviously not as rich or colourful…More specifically, according to the western architectural theory after World War II, the traditional Chinese design and construction system could be catalogued under 'anonymous architecture', or 'architecture without architects'. The 'Chinese Classical', as a counterpart to western classical in the sense of classical Western theory and methodology, as defined by the first generation of Chinese architectural scholars, does not necessarily exist in traditional Chinese culture. … All forms and styles of Chinese architecture are a direct reflection of local construction systems, customs and methods. That is why I have always doubted whether Chinese architectural history can be explained and classified according to Western models… My experience with vernacular architecture gave me faith that the value of traditional Chinese building culture and building art is inherent in construction.
>
> (162–3)

Zhao accepted the theory of tectonics during his doctoral studies at the ETH Zurich (Swiss Federal Institute of Technology in Zurich, Switzerland). This theory, which originated in the German-speaking area, gave him a new approach for looking at Chinese architectural tradition and allowed him to fight the Eurocentrism in Chinese architectural research:

> Eurocentrism, which has aroused the discontent of scholars and theorists outside of Europe, is based directly on the style theory of art history. As far as style is concerned, evaluation is always based on inferiority or superiority (167)…However, any study focusing on construction will avoid this evaluation… From the tectonics point of view, the essential cultural value of human construction is seen as 'the stable and durable expressive force which originates from the construction form', and has a huge significance in aesthetics. Since the tectonics theory is a good approach to the German

"*Baukunst*", it should be of value in understanding Chinese construction and design.

(172)

Zhao's tectonics theory, based on the architectural theory of the German-speaking area, focuses on the Chinese tradition of construction in wooden. Zhao divides his tectonic theme into three aspects: material properties of the wood, construction of the joints, and structure as a whole. As an architectural theorist, he is most interested in the design form, which is determined by the span and height of the timber construction. He chose the MZ bridges for the discussion on the problem of span, and chose the drum tower of the Dong people in Southwest China for the problem of height. In these studies, he poses the same questions: on the one hand focusing on the construction principle ("How are they built?"), and on the other hand, the developmental typology ("How did they come into being?") (2007,106).

Zhao's interpretation of the Chinese wooden building culture from a tectonic perspective and his criticism of the classical Chinese research tradition are important milestones in Chinese architectural history and historiography. As far as the MZ bridges are concerned, Zhao was the first to look at these bridges and their architectural history from the viewpoint of architectural theory, thus raising the discussion of vernacular architecture to an academic level.

Although many scholars have made their contribution to the study of the woven arch bridges, this chapter focuses on Tang Huancheng and Zhao Chen as representatives of the roads that the Chinese scholars worked out on the topic of Chinese wooden arch bridges. Not only because these two authors had the strongest influence on the author of this book, but most importantly, they represent two generations of modern Chinese scholars who have been driven by different, yet similar motivations, and are both branded with a strong mark of their identity and of the times. As Chinese scholars, their background is inextricably linked with and profoundly underpins the objects of their research: to identify not only aspects of Chinese culture and technology, but to contextualize themselves in the history of human cultures and between the western and eastern civilizations, to relieve their academic anxiety which is rooted in a strong notion of national identity. With this legacy, their studies of Chinese wooden arch bridges were also imbued with deeper emotions and profound expectations. For the first generation, represented by Tang Huancheng, the Chinese wooden arch bridges are a source of national pride and the responsibility for finding the advanced building technology that was thought to be lost. For the second generation, represented by Zhao Chen, the MZ bridges are evidence of an outstanding example of how to solve the methodological problem of carrying out a comparative study between cultures.

At the same time, a legacy could become a burden, and an expectation could become a blinder. In the second part of this book, readers will explore new visions to become reacquainted with wooden arch bridges in China.

Searching for Chinese wooden arch bridges 119

References

Fang, Yong
方拥. 1995. 虹桥考. 建筑学报, (11). 55–60.
Liu, Yan
刘妍. 2011.浙闽木拱桥类型学研究——以桥板苗系统为视角. 东南大学学报(自然科学版) (03): 430–6.
Luo, Ying
罗英. 1959. 中国桥梁史料（初稿）.上海：上海科学技术出版社.63–7.
Su, Xudong (and Liu Yan)
苏旭东,刘妍.2010.“双三节苗”木拱桥——木拱桥发展体系中的重要形式. 华中建筑 (10)，39–42
Tang, Huancheng
唐寰澄. 1957. 中国古代桥梁. 北京：文物出版社.: 27–9.
唐寰澄, 张尚杰. 1986a. 竹木拱桥. in: 茅以升. 中国古桥技术史. 北京：北京出版社: 100–9
唐寰澄. 1986b. 木拱施工. in: 茅以升. 中国古桥技术史. 北京：北京出版社: 206–207
唐寰澄. 1987. 中国古代桥梁. 北京：文物出版社: 64–77
唐寰澄. 2000.中国科学技术史·桥梁卷. 北京：科学出版社: 461–92.
唐寰澄. 2010.中国木拱桥. 北京：中国建筑工业出版社.
Zhang, Jun
张俊. 2001. 泰顺木拱廊桥发展历史探讨. 小城镇建设 (09). 51–4.
Zhao, Chen
赵辰, 冯金龙, 冷天, 毕胜.2001. 木拱桥作为山地人居文化遗产的重新评价. 出自:山地人居与生态环境可持续发展国际学术研讨会论文集. 北京：中国建筑工业出版社.406–12.
赵辰. 2007. 对中国木构传统的重新诠释. 出自:赵辰. 立面的误会. 96–117 .北京：生活·读书·新知三联书店.
赵辰. 2009.“建构热”后话建构. 出自:丁沃沃，胡恒主编. 建筑文化研究（第一辑）. 159–82.北京：中央编译出版社.
Zhao, Chen and Feng Jinlong, Bi Sheng. 2009. Tectonic Studies on Wooden Arched Bridge. As the Case of Span in Chinese Wooden Construction Tradition. *Proceedings of the Third International Congress on Construction History*, Cottbus. 1547–55. Berlin: Neunplus1.
Zhou, Fengfang, Lu Zeqi and Su Xudong
周芬芳,陆则起,苏旭东. 2011. 中国木拱桥传统营造技艺. 杭州：浙江人民出版社.

Part II

Woven arch bridges in Southeast China

Introduction to Part II

The woven arch bridges (or wooden arch bridges)[1] in Southeast China (hereinafter "MZ bridges") are located in a small and rather isolated mountain area on the border between Zhejiang (Zhe) and Fujian (Min) provinces ("MZ area") (Figure Intro.1). The remoteness of the mountains, an undeveloped road system and an under-developed economy meant that the MZ bridges and their bridge-building tradition, isolated from and unknown to the outside world for centuries, were protected from the influence of the rapidly-changing modern world. Today, there are some 100 historical wooden arch bridges (Figure Intro.2) in existence, and traditional bridge carpenters[2] active in this career. A database of known historical woven arch bridges together with an interactive map is available (in Chinese) on the website *http://w-bridge.wiki* established by the author.

The second part (Chapters 4, 5, 6) of this book focuses on the traditional building techniques for the MZ bridges, as practised today, and their technical evolution among the bridge carpenter families/groups over the centuries. The beginning provides a brief overview and a general understanding of this kind of bridge with the help of a wooden model of a typical example (Figure Intro.3), the Jielong Bridge.

The Jielong Bridge in Jingning, Zhejiang, is a large bridge over a deep gorge, with a clear span of 30 m. According to the ink inscriptions written on the bridge corridor beams, the bridge was finished in 1917, and the carpenter masters who were in charge of the under-deck structure were of the famous Zhang family from Ningde County (Xiajian Masters, see Ch. 4 and 6), whereas the bridge corridor was built by local carpenters.

The wooden model, 1:20 in scale, was constructed by the author together with Professor Philip Caston of the Neubrandenburg University of Applied Sciences in 2012, in Neubrandenburg, Germany. The construction of the

---

1 As mentioned in Chapter 3, the woven arch bridge in Southeast China is listed as the "Chinese Wooden Arch Bridge" in the context of the UNESCO's List of Intangible Cultural Heritage in Need of Urgent Safeguarding.
2 In this book, the term "bridge carpenter" denotes carpenters with the specific knowledge, skills, and experience required to build MZ woven arch bridges in the traditional way.

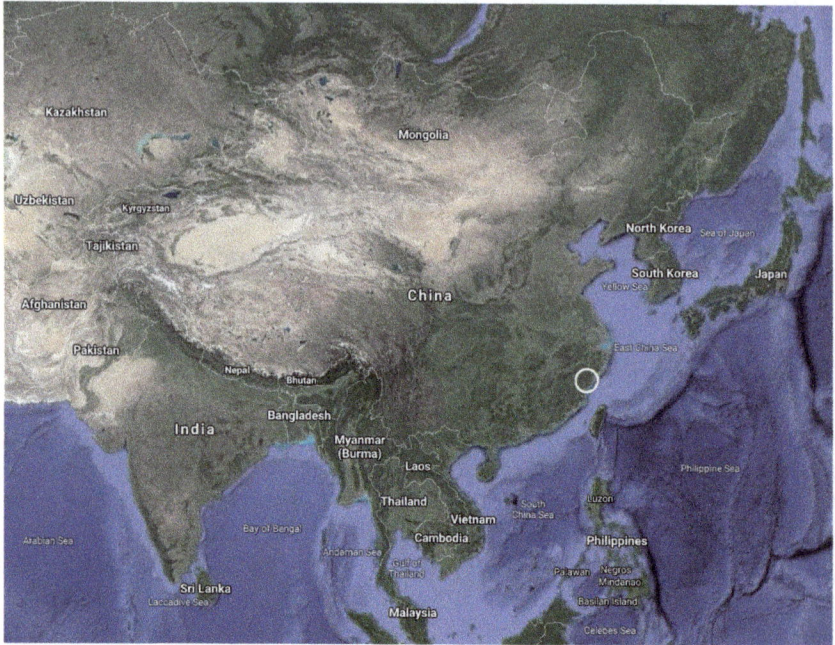

Figure Intro.II.1 MZ area on the geographical map of China.
Source: Based on Google Maps.

model is true to scale and in joint forms to the measured drawings of the bridge, which are based on the survey made by the author.

As mentioned in earlier chapters, the MZ bridges are a type of woven arch bridge with a woven arch structure under-deck. A typical MZ bridge woven arch has two systems, a three-sided arch and a five-sided arch. Each system is composed of longitudinal beams (slanted and horizontal beams), and crosswise beams (crossbeams and foot beams). The longitudinal beams are natural shaped logs, connected to the squared crosswise beams by means of wooden joints. This fundamental structural principle is demonstrated as follows with the exploded drawings (Figure Intro.4, Figure Intro.5) and the model (Figure Intro.6).

II.1 Rules of terminology and abbreviation of the structural members

Names of the structural members and their abbreviated written form are also found in Figure Intro.4 and Figure Intro.5. Members of the first and second arch system are indicated by the suffix [1] and [2] for short, respectively. The slanted beam is shortened to "S-Beam," the horizontal beam to "H-Beam," the crossbeam to "C-Beam" and the foot beam to "F-Beam."

Figure Intro.II.2 Topographic map of the wooden arch bridges in MZ area, Southeast China.

Note: Among some 100 bridges marked on the map, 71 have been investigated by the author in situ with measured drawings. Counties with the most woven arch bridges include: Zhejiang Province: Jingning, Qingyuan, Taishun. Fujian Province: Shouning, Zhouning, Pingnan, Zhenghe, Gutian, Minhou.

Source: Based on Google Maps.

(a)   (b)

Figure Intro.II.3 Jielong Bridge (接龙桥), Jingning, Zhejiang.

Members of the first system do not have a prefix, while those of the second system need a prefix to indicate their position. In the second system, the upper (slanted and cross-) beams are prefixed with "U-," while the lower beams are prefixed with "L-." For example "U-S-Beam[2]" refers to the upper slanted beam of the second system, the "L-C-Beam[2]" refers to the lower crossbeam of the second system.

In a typical 3+5 bridge as described above, where the woven arch is composed of a three-sided arch and a five-sided arch, most sections of the longitudinal beams are parallel in the two systems (i.e. the L-S-Beams[2] are parallel to the S-Beams[1], and the H-Beams[2] are parallel to the H-Beams[1]), except for the upper slanted beams of the second system (U-S-Beams[2]). The U-S-Beams[2] take on the role of weaving inside the woven structure, and are therefore referred to as "weaving-beams" from now on, in the context of the MZ bridges.

The lateral horizontal beam, which connects the woven arch and the abutment and serves as the deck beam, is labelled as "LH-Beam."

II.2 General building process of a typical MZ bridge

II.3 Basic information of the research

In order to achieve a deeper understanding of the craftsmanship and the traditional building technique used in the MZ bridges, the author used multiple study approaches. By making an extensive survey of the existing historical

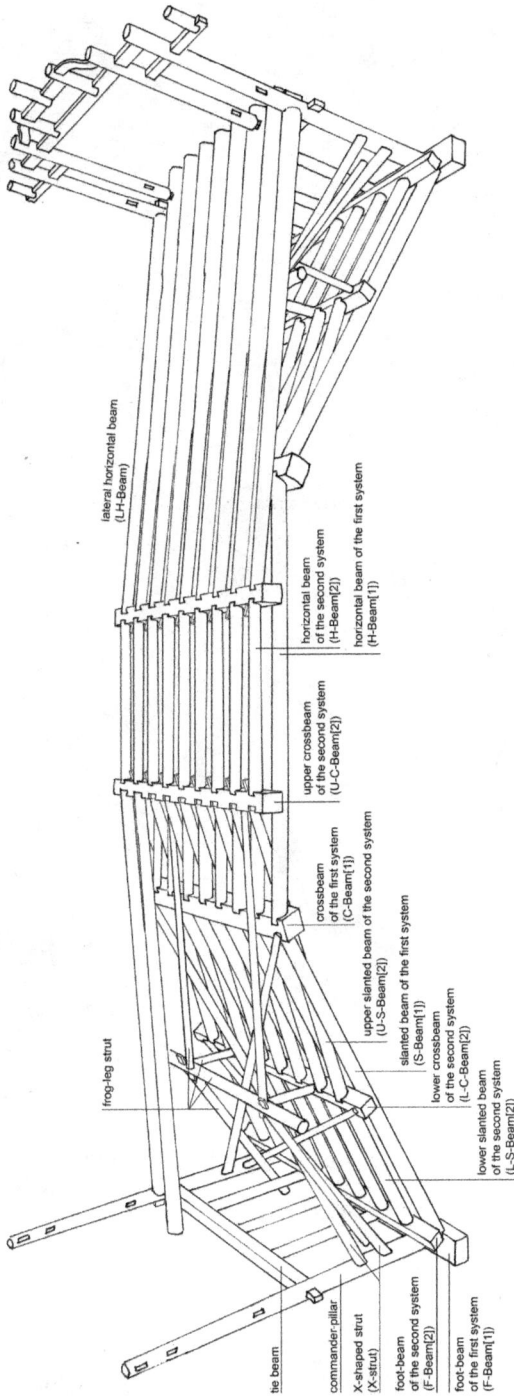

the beam

commander-pillar

X-shaped strut
(X-strut)

foot-beam
of the second system
(F-Beam[2])

foot-beam
of the first system
(F-Beam[1])

frog-leg strut

lower slanted beam
of the second system
(L-S-Beam[2])

lower crossbeam
of the second system
(L-C-Beam[2])

slanted beam of the first system
(S-Beam[1])

upper slanted beam of the second system
(U-S-Beam[2])

crossbeam
of the first system
(C-Beam[1])

upper crossbeam
of the second system
(U-C-Beam[2])

horizontal beam
of the second system
(H-Beam[2])

horizontal beam of the first system
(H-Beam[1])

lateral horizontal beam
(LH-Beam)

Figure Intro.II.4 Typical (under-deck) structure of the woven arch bridge with the terminology of structural members.

deck beam system

deck beams

frog-leg struts

top beam

upper leg

commander-pillar frame

lower leg

middle leg

X-shaped struts

tie beam

commander-pillar

middle struts

horizontal beam of the second system (H-Beam[2])

upper slanted beam of the second system (U-S-Beam[2])

second system of the woven arch

lower slanted beam of the second system (L-S-Beam[2])

upper crossbeam of the second system (U-C-Beam[2])

lower crossbeam of the second system (L-C-Beam[2])

foot beam of the second system (F-Beam[2])

first system of the woven arch

horizontal beam of the first system (H-Beam[1])

crossbeam of the first system (C-Beam[1])

slanted beam of the first system (S-Beam[1])

foot beam of the first system (F-Beam[1])

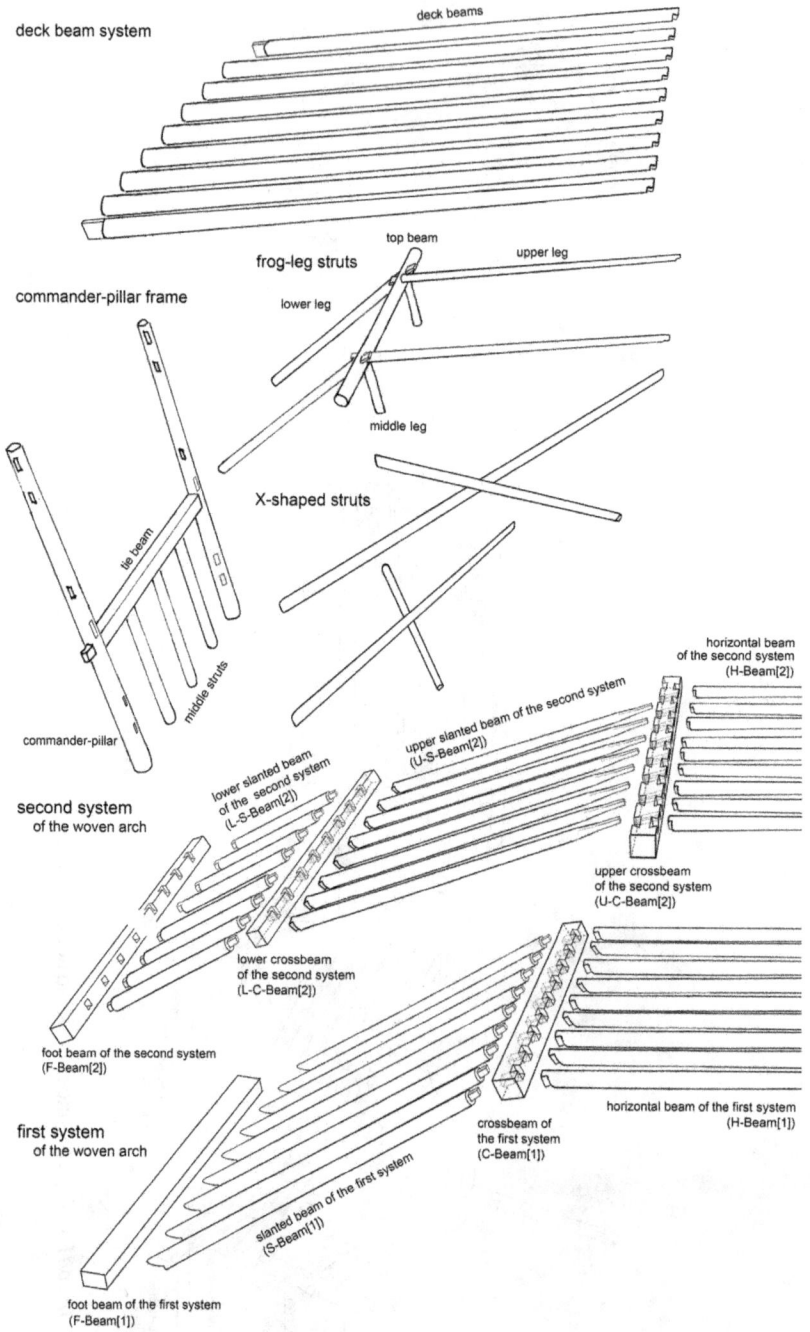

Figure Intro.II.5 Exploded view of the woven arch of typical MZ bridges.

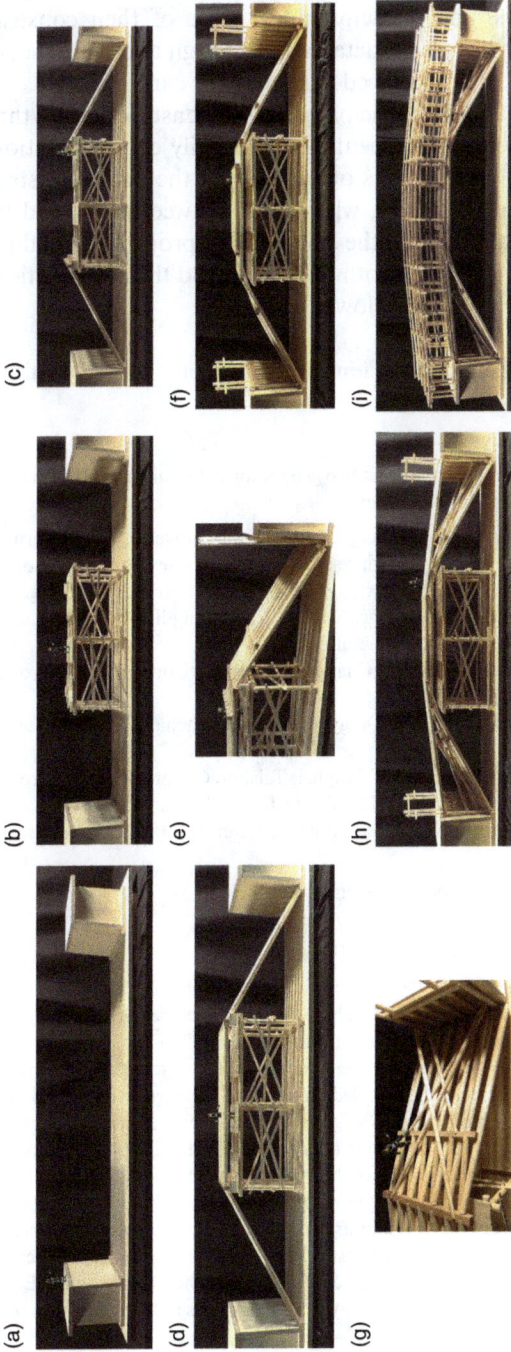

Figure Intro.II.6 Construction processes demonstrated using a model of the Jielong Bridge, 1:20 scale. (a) Building the bridge abutments on the riverbanks. The projecting part of the abutment serves as the bearing for the first system. The bearings sometimes take the form of a pair of foot beams (F-Beams), and at other times they are in the form of a square layer of stone. (b) Building the scaffolding in the middle of the river. (c) Setting up the slanted beams and crossbeams of the first system (S- and C-Beams[1]). (d) Completing the first system with horizontal beams (H-Beams[1]). (e) Setting up the slanted beams (and the crossbeams connecting to them) of the second system (F-, L-S-, L-C-, U-S-, U-C-Beams[2]) and the post-frame at the sides. The lateral posts of the frame reaching from the arch feet to the bridge corridor are called the "commander pillars." (f) Completing the second system with horizontal beams (H-Beams[2]). (g) Setting up the X-shaped struts (X-struts), seen from above. (h) Completing the under-deck structure with the lateral horizontal beams (LH-Beams) and the frog-leg struts. (i) Building up the bridge corridor.

bridges (71 of the ca. 105 historical bridges) and an in-depth archaeological study of the oldest example (case study of Rulong Bridge), he examined and revealed the technical features of the silent witnesses. By interviewing the traditional bridge carpenters[3] and participating in some of their construction projects, the author also learned in detail all the design and construction procedures, thanks to the great trust placed in him by the carpenters.[4]

During his research, the author participated in the construction of three bridges, all of them built by bridge carpenters with family career traditions. In each of these projects, the author was on-site during the entire construction period of the under-deck structure, which took between three and five weeks of intensive activity, and during the third bridge project, he qualified as a bridge master and took charge of both the design and the construction.

The three bridge projects are listed below:

Guanyin Bridge, Shengshuitang, China (hereinafter "Shengshuitang Bridge" 生水塘桥)

3 The family-trained bridge carpenters (with at least two generations of family members involved in the bridge building career) interviewed include:
   - Zhang Changzhi (张昌智) and Peng Fodang (彭佛党) from Xiukeng Village (formerly "Xiajian"), Zhouning County, Fujian Province (Xiajian Masters, seventh generation).
   - Zheng Duojin (郑多金) and Zheng Duoxiong (郑多雄) from Xiaodong Village, Wu Dagen (吴大根) from Choulinshan Village, Township of Kengdi, Shouning County, Fujian Province (Kengdi Masters, sixth generation).
   - Huang Chuncai (黄春财) from the town of Changqiao, Pingnan County, Fujian Province (Changqiao Master, third generation).
   - Wu Fuyong (吴复勇) from Daji Village, Qingyuan County, Zhejiang Province (second generation).
   - Dong Zhiji (董直机) from the Township of Lingbei, Taishun County, Zhejiang Province (partly self-taught and partly taught by a Kengdi Master).
   - Wei Shunling (韦顺岭) from the town of Daxi, Pingnan County, Fujian Province (second generation).

The list of interviewed non-family-trained bridge carpenters include:
   - Zeng Jiakuai (曾家快) from Taishun County, Zhejiang Province.
   - Liu Fanseng (刘繁森) from Zhenghe County, Fujian Province.
   - Zhang Yijin (张以进) from Qingyuan County, Zhejiang Province.

4 The author feels obliged to explain that even though he demonstrates the detailed design and construction methods in this book, or "reveals the technique secrets" of the carpenters, he is not betraying their trust. This is because on the one hand, as will be discussed and explained later on in this and subsequent chapters, the revealed secrets, as a set of rules of thumb of the carpenters, belong to a time before modern knowledge and technology. These secrets are losing their power nowadays: even without the intervention of academia, the family-trained carpenters are more willing to teach their colleagues now than they used to be before. Furthermore, all the carpenters interviewed are fully aware of the author's profession as a scholar and that therefore, the content of their conversations with him are very likely to be revealed by appearing in print. This was indirectly commented on by Master Wu Dagen, the carpenter master who made the greatest contribution to the knowledge of carpentry in this book – he said: "Even if you learned it all, you would never be able to take my job!" (This goes both for the author and the readers of this book.)

Master: Wu Fuyong (吴复勇) from Daji village, Qingyuan County, Zhejiang Province.

Timeline: 13 October – 26 November 2012.

Location: Shengshuitang Village, Township of Longgong, Qingyuan County, Zhejiang Province, China.

Span: ca.14 m.

Huilong Bridge, Dongtang, China (hereinafter "Dongtang Bridge" 东塘桥)

Masters: Zheng Duoxiong (郑多雄) and Wu Dagen (吴大根) from Xiaodong Village, Township of Kengdi, Shouning County, Fujian Province, hereinafter referred to as the "Kengdi Masters."

Timeline: 27 November – 23 December 2013.

Location: Dongtang Village, Township of Jingnan, Jingning County, Zhejiang Province, China.

Span: ca. 16 m.

Rainbow Bridge, Regensburg, Germany (hereinafter "Regensburg Bridge")

Master and architect: Liu Yan (刘妍, the author)

Master: Zhang Changzhi (张昌智) from Xiukeng Village, Zhouning County, Fujian Province. As per local tradition, this family is referred to hereinafter as the "Xiajian Masters," "Xiajian" being the old name of their home area.

Timeline:

Prefabrication: 18 October – 31 October 2014, Ningde, Fujian, China.

Construction: 14 June – 18 July 2015, Nepal-Himalaya-Pavillon, Wiesent (Regensburg), Bavaria, Germany.

Span: 7.5 m.

4   Building a woven arch bridge
Local knowledge

This chapter contains a complete description of the entire project process of the construction of a typical wooden arch bridge in the MZ Mountains. The description is divided into individual steps according to the project timeline: organization and preparation, material and working tools, design and calculation, processing the members, and erection of the structure. This is supplemented by a section discussing the understanding and considerations of the bridge masters regarding their structural designs. Readers who are unfamiliar with this type of bridges are advised to look through Section 4.5 (Construction) before Section 4.3 (Design), to get a general understanding of the building process before diving into the detailed rules which determine the forms and scales of the relative structural members.

This suggested step, which constitutes a reversal of the actual construction procedures, is indeed the natural process for a lay-person to approach acquiring the knowledge of construction technology of a construction site. In other words, an apprentice is first given tasks (construction) to become familiar with the entire job, and then is trained in processing material and structural elements, and finally becomes a competent (master) carpenter when he has mastered the knowledge of design. Accordingly, in the three bridge projects in which the author participated, he also followed these reversed steps as a focus of observation during his on-site investigation – he studied with a focus on material processing and construction during the first project, and only thereafter, during the second and third projects, was he able to delve into the secrets of design.

4.1 Organization and preparation

4.1.1 The role of the project director

Bridge construction is closely connected to traffic and transport, and therefore the economy and the wealth of a place; thus they are of profound significance to the local government and the local people. This situation is even more critical in the MZ Mountains, since the land is broken up by dense hills and valleys. Villages sparsely occupy the rather limited narrow plains – as do the bridges scattered about, which are their links to the outside world.

A bridge project involves at least one nearby village. The project director can be either an individual, a group of individuals or the whole community. In the first and second instances, he or they could be a squire or local official (but in his own name, rather in official capacity), or perhaps villagers of a specific religion (Buddhists or worshipping local deities); and in the last case, the board of directors could be an entire family or village, monastery or local government.

No matter the status and number of the advocates, a board consisting of a greater or lesser number of members will be formed to manage the project. The board of directors is in charge of financing, the tendering process, material preparation, craftsmen's accommodation and other services for the entire project.

Finance comes in the form of a collection of donations, from local squires, officials, notable families and common villagers. Donating towards the construction of a bridge is considered to confer great merit and great virtue, in both social and religious terms, and the names of sponsors are always recorded on/with the bridge – either written onto the bridge corridor beams or cut into the stone stele – and are expected to last forever.

Important sponsors, those who donate significantly large sums of money, have their names (and the name of their hometown) at the top of the list or taking up an entire beam in the middle of the corridor. The names of smaller donors are written in a smaller size and share less significant beams with the names of others. On some bridges at a crucial traffic node, names can be found from a huge variety of origins, spanning a large part of the region, indicating the power and influence of the bridge in the transport network of the time. Donations in other kinds were also possible, in the form of labour or timber.

This concept of the more or less autonomously formed bridge project committee is still common today. Quite a number of village bridges today are organized in this rather unofficial way. In more developed areas in the MZ area (such as Wenzhou, Taishun, and Fu'an), a bridge is sometimes constructed according to the wish and financed by a single entrepreneur. Nowadays, local governments could also play an important role. However, even in the case of a generous major sponsor, small donations from local residents or travellers are always welcome. The traditional belief in the merit conferred by donating to bridge-building is still deeply rooted in the local culture of the mountain people.

The bridge project in Shengshuitang, for example was organized by four families. The total cost was about 800,000 Yuan (approx. 100,000 Euro in 2012). Multiple approaches were utilized to collect money, from villagers and local companies, etc. The organizers called their collecting "begging" for money. Donations were not limited to money or material: labour was also a common form of donation. A day of work on the construction site was equal to 160 Yuan (around 20 Euro) for that bridge.

The Dongtang Bridge project was the project of an entire village. The board of directors comprised almost the entire village committee (the officially elected leading members of the village). The total cost was around 700,000 Yuan (approx. 87,500 Euro at that time). Most of the wood came from the woods collectively owned by the village. The trees had been planted 40 years previously, and were mature enough to become construction timber, and they formed part of the conditions for and reasons behind the village deciding to build a bridge. In addition, wood donations were also received from individual villagers. Naturally, timber purchase from outside was still necessary, because the largest building members required huge logs.

4.1.2 The role of the master carpenter

In a large part of southern China, the master builder who is in charge of a specific construction project is called a *"shengmo"* (绳墨, master in charge, literally: "string and ink") or a *"zhumo"* (主墨, literally: "the master in charge of the ink (lines)") of the project, who is also the project contractor. This name comes from their woodworking tool, the inkpot/ink-line, with which they make the linear drawings for design or the guiding lines for the woodworking (on this more later, Figure 4.11). *Shengmo* as a title for the master builder means "ruler" in its original and derivative meaning (i.e. "master" and "measuring tool").

In the past, it was less likely for MZ bridge projects to be put out to tender in our modern way. The construction of these bridges was (and is) so specialized, that only a rather limited number of carpenter families/groups were qualified to do the job. In the case of especially challenging projects (e.g. with a large span or at a dangerous site), a carpenter or carpenters from specific families of repute were invited in from other counties or even from other provinces.

Working in a family-based profession, the master carpenters normally worked in groups, with one chief master and one or more deputy masters. Together, they formed the core of the construction crew. When invited to take on a project, there were basically two approaches: they would come with their entire crew to undertake the entire carpentry work for a bridge, or they would come alone and act as project leaders, cooperating with the local group of carpenters, who worked under the instruction of the master(s). In the latter case, they could only take charge of the construction of the under-deck structure. The construction of the corridor (since this is principally the same as a common dwelling construction) was undertaken by the local carpenters.

After reaching an agreement on the scope of the project, a contract was signed between the board and the master, and a guarantor – similar to the way it is done nowadays – stated the responsibility of both sides and the amount of material and payment. In the contract for the Jielong Bridge in 1916 (Figure 4.1), the stated duty of the masters (brothers from the Zhang family, Xiajian Masters) included the scale of the bridge (relating to the scale

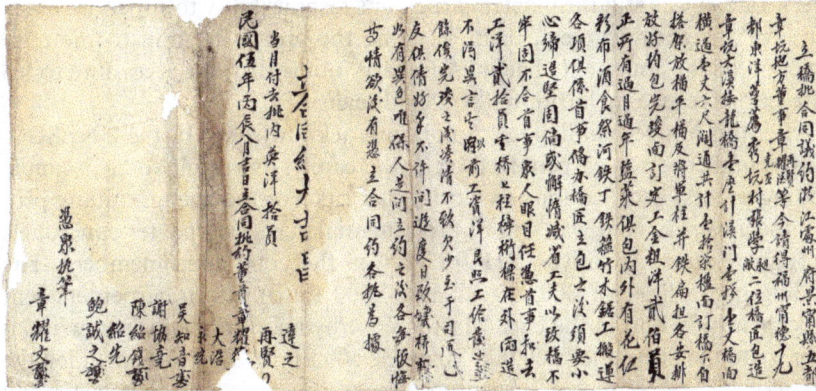

Figure 4.1 Project contract for the Jielong Bridge, between the master carpenters from the Zhang Family (Xiajian Masters) and the board of directors (from the Zhangkeng Village), 1916.

Source: Shouning Museum.

of the corridor,[1] which is 16 Chinese feet in width, 17 bays[2] in length) and the procedures for which they were responsible (from scaffolding to the under-deck structure, to the commander-pillars and tie beams).[3] A deposit (200 Yuan) was paid before the construction commenced. The board of directors provided the everyday food and organized the feasts during ritual times and festivals, and the payment ("red envelope") for the carpenters. It was also the board's duty to provide building material (bamboo and timber, iron nails and bands, etc.) and the workers for the unskilled tasks (common carpentry including sawing and other construction work). It was set out that the carpenters must work diligently, and that the bridge must be stable. Furthermore, the contract stated that if any damage was done to the bridge through negligence on the part of the carpenters, then 20 Yuan would be docked from their payment. The rest of the salary was to be paid on completion of the construction. If the carpenters proved to be incompetent, the guarantor would be held responsible.

Although the carpentry was the direct responsibility of the master carpenters and other craftsmen, including the stonemason for the abutment

1 The span/scale of the under-deck structure is not mentioned. The contract gives no direct information about the structural form of the bridge.
2 A "bay" is the basic space unit of Chinese architecture. A bay is the width between two pillars seen from the elevation or the space between four pillars seen from the plan. A bridge of seventeen bays has eighteen crosswise pillar frames along the direction of the corridor.
3 It is thus clear that the bridge corridor is not the responsibility for the bridge masters in this project. This is proved by the ink inscription on this bridge, that the corridor was built by local carpenters.

construction, and the plasterer for the corridor, were hired directly by the directorate, it was still the master carpenter's responsibility to supervise every phase of the entire project, from selecting the bridge location to the construction of the abutment. The masons carry out their work according to his instructions regarding location, form, and scale.

After this bridge-building technology was acknowledged by the UNESCO, a large number of new carpenters took up this career. As a result, today, a simple bidding process is called for new bridge projects. For the Shengshuitang project, the advertisement was sent out to potential master carpenter-candidates through relations (relatives and friends) of the committee members. The bidding had a limit of 300,000 Yuan (37,500 Euro). Seven carpenters sent back their bids, and in the end, Master Wu Fuyong won the project with an offer to do the project for 260,000 Yuan (32,500 Euro). This sum included the salary of the entire crew of carpenters, who were recruited by the master and were paid a daily wage. The master is the contractor in the modern sense.

For the Dongtang project, the committee sent out only a few "invitations to tender." The master carpenters who were awarded the contract for the project come from a neighbouring county: Zheng Duoxiong and Wu Dagen are brothers-in-law and are the descendants of a famous nineteenth-century bridge master. They have a traditional family reputation and are famous in their own right as well, as having the honourable title of bridge-building master from the provincial government at the time of the project. Their contractors' payment was 208,000 Yuan (the equivalent of 87,500 Euro at the time).

4.1.3 Selection of the bridge location

Wooden arch bridges can be sorted into two types, depending on their location and function: those which serve purely as transit routes for traffic and are normally built over rivers or valleys at a distance from the nearest settlements; and those which are closely connected to settlements, which, in addition to serving as traffic routes, also function as public spaces and relate to local *feng shui*.

In either case, the selection of the bridge location is crucial for the success of a bridge structure. Natural rock is the best choice as the foundation. Sometimes bridges are built to make the best possible use of the natural rock, sacrifice other aspects for the convenience (Figure 4.2). Even *feng shui* – the geomancy of construction in harmony with the natural environment, the art believed to influence the welfare of local residents, and the ruling factor in location selection for all other types of Chinese architecture, loses some of its power where a bridge project is involved. The bridge master has the highest authority to decide the location of a bridge; higher than a geomancer.

Generally speaking, the reasonable span scope for a woven arch bridge is from 20 m to 40 m. Bridges smaller than 20 m span could be achieved through other simpler forms of structure (struts structure or cantilever structure). The largest clear span of MZ bridges is ca. 43 m, both historically and in modern

Figure 4.2 Yangmeizhou Bridge (杨梅洲桥). Shouning, Fujian, 2013.
In order to make use of the natural rock as the foundation, the bridge was built above a river pool where the water is more than 20 m deep, whereas the position just dozens of metres upstream is very shallow.

practice. In locations where there is a wide river with convenient aspects (e.g. gentle river bed not surrounded by high gorges), multi-span bridges with one or more abutments inside the river would be built. The longest historical wooden arch bridge is the Wan'an Bridge (Figure 4.3) in Pingnan, Fujian. The bridge has six spans and a total length of almost 100 m. It was (re)built in 1932 by the Hung family (Changqiao Masters).

Both the Shengshuitang and Dongtang bridges are reconstruction projects. Up to the first half of the twentieth century, both villages had an old wooden bridge at the location where the new bridge was eventually built. Around the second half of the twentieth century, the old bridges were destroyed and were replaced by a new stone/concrete bridge, which is still in good condition and serving the traffic today. Both the new wooden bridges stand just several metres away from the stone/concrete bridge: they have no traffic function at all, but rather were built to promote the *feng shui* of the villages.

In the centre of the bridge corridor, there is always a shrine containing statues of local or Buddhist deities. Therefore each bridge acts as a small temple as well. Unlike the common village temples, where the deities are

Figure 4.3 Wan'an Bridge (万安桥), Changqiao, Pingnan, Fujian, 2009.

usually so placed that they face the village, the shrine of a bridge always faces the direction of the oncoming water, regardless of the location of the village.

4.2 Material and tools

After the location of the bridge has been decided and the scale of abutments confirmed, the basic scale of the bridge is then determined considering the wish of the board of directors and the site location. The bridge master will then calculate the required amount, size and length of timber. The board will prepare the material accordingly.

4.2.1 Selection of wood species

4.2.1.1 Different species of "Shanmu"

The wood species *Cunninghamia lanceolata*, called *Shanmu* (杉木) in Chinese, although commonly rendered as "China fir" in English, is not a species of fir (from the pine family), but belongs to the cypress family. *Shanmu* is native to Asia. Thanks to its strength-to-weight ratio and rapid growth, it is the predominant construction material in most parts of southern China.

The MZ area is one of the most famous *Shanmu* production areas in China. It is therefore no surprise that it is also the main kind of timber used in bridge construction. "The role of *Shanmu* in construction in southern China is the same as the role of rice in the kitchen."[4] Another attribute for bridge

4 Saying by a local carpenter from Shouning County, in the course of the Dongtang Bridge project, December 2013.

construction is *Shanmu*'s water-resistance. In the (exaggerated) words of the local carpenters, "*Shanmu* won't rot in a thousand years." Even coffins are made of *Shanmu*!

Another wood species used in bridge construction is *Cryptomeria fortunei*, which also belongs to the cypress family. In the MZ area, this species of wood is known as *Wenmu* (榅木) or *Liushan* (柳杉: "willow fir"). Less commonly, it is known as *Shuishan* (水杉: "water fir"). The wood of this species grows taller and larger than *Shanmu*, and is used as an alternative when large *Shanmu* logs are in short supply. However, the wood of the *Wenmu* is looser than that of *Shanmu*, and thus is considered of lower quality and a less ideal choice.

Another type of wood used, belongs to the genus *Keteleeria,* from the pine family is called *Youshan* (油杉: "oil fir"). This wood is harder than *Shanmu*, and higher in oil content, thus takes longer to dry, but it also has better water-resistance. It is considered to be better than *Shanmu*, but it is less plentiful and therefore more expensive.

In the Shengshuitang project, the common structural members are made of *Shanmu*, but the important "commander-pillars" are made of *Youshan*. In the Dongtang project, where large-scale *Shanmu* logs were in short supply, they were used only for the "commander-pillars," while the other beams are made of *Wenmu*.

4.2.1.2 Crossbeams/harder wood

The crossbeams of the woven arch are the most critical building members in the MZ bridges. These beams work as an integration of a group of joints connecting the slanted and the longitudinal beams, and they bear a number of cuts for the wood joints. They must be large in scale as well as hard and firm.

Pinewood is commonly used for crossbeams. The wood must be from old trees over a hundred years of age: it is then hard and with a high oil ratio, which provides better water- and termite-resistance. Because of its resistance to decay, pinewood is also used for the abutment foundation in bridge construction.

Hardwood is also commonly used for crossbeams. In the Shengshuitang Bridge project, the local people decided to use chestnut wood, even though several old pinewood logs had been ordered and were already on hand at the construction site. Two particular species belong to the chestnut family: *Castanea henryi*, known as *Zhuili* (锥栗), which is hard and good for underwater, and is the favoured species in that area (also commonly used for dwelling foundations), and *Castanea seguinii*, known as *Maoli* (毛栗), less favoured but also considered suitable for the crossbeams.

However, if it is available, *Shanmu* remains the preferred wood species for crossbeams. The large square crossbeam requires a huge wooden log which is often beyond the available scale of *Shanmu*. Carpenters have to turn to other wood species as a less-than-ideal alternative. The Changqiao Master

Huang Chuncai insists on using *Shanmu* for the crossbeams with no tolerance for expediency.[5] His persistence relies on the favourable timber supply in his hometown of Pingnan, Fujian, one of the best *Shanmu* production areas in China today.

4.2.2 Wood processing

4.2.2.1 Time of felling

As in most German-speaking areas, winter felled wood[6] is also considered to be the best quality in many parts of southern China. Carpenters prefer to work with wood felled in winter and spring, when the material contains the lowest level of water. This is not only to minimize the tendency to decay, but to make work easier, since the lumber is much lighter in weight.

However, the MZ bridge carpenters prefer wood felled in a different season. To their way of thinking, spring-felled wood contains too little oil and is therefore apt to split due to dryness. Thus, they consider August or September the best time to fell wood for bridge construction. Trees felled in this season have a membranaceous tissue between the bark and the sapwood: the cambium. Thanks to the cambium, the bark is easier to remove, and the wood is protected from splitting due to dryness. "This thin skin on the wood functions the same way as a skincare cream works for the face," says a carpenter on the Dongtang Bridge site.

4.2.2.2 Dry wood vs. wet wood

After the wood arrives, it must be dried. Theoretically, the drier it is, the better, especially for the timber of the beams in the wooden arch, since these beams are subjected to a great bending force. Dry wood is much stiffer than wet wood. With a lower moisture content level, the wood is less pliable and thus sags less under the load. Consider two beams of the same size, where one is dry and the other is wet: their material features are so different that in the eyes of the carpenters, they are almost two entirely different materials.

In the past, woods for bridge projects would be felled one year ahead of the scheduled start of construction. Nowadays, such a long period of drying is almost impossible. Wood is prepared and dried in a much shorter time, and is usually still wet when it is used. Sometimes, the wood for posts and beams is felled right at the beginning of a project, which leaves only a few weeks for drying.

Bearing this in mind, the bridge masters have to design the structural members in larger dimensions. If 20 cm is the proper diameter for dry wood

---

5 Master Huang Chuncai, in an interview with the author on 18 December 2013 in Pingnan, Fujian.

6 In the German-speaking area, the optimal felling timepoint for needle-wood is considered to be between September and December; for beechwood, it is between January and February (Mooslechner 1997).

for example the carpenter will select wet wood with a 24 cm diameter.[7] This is also one of the main reasons that new bridges today have visibly larger beams than old bridges.

Another reason for the preference for dry wood is to prevent the wood splitting and shrinkage. Wooden architecture in southern China can be seen as a form of *Fachwerkbau* (timber frame structure) but without using the *Strebe* (inclined struts). Unlike the European *Fachwerk*, which gains stability through triangular-shaped structural members, the Chinese pillar-beam frame gains stability through the firmness of the wood joints. If built when the wood is still wet, the wood will shrink and the joints become loose, and the structure will become deformed over time. It is thus better to process the wood joint using dry wood.

4.2.2.3 Debarking

When the tree trunks arrive at the construction site, they are first debarked and laid out in the open air for drying. The bark of the sunlit side of a tree trunk grows tighter than the shady side, requiring stronger tools to debark it. Two kinds of tools are applied for the debarking process: the spade knife with long staff (Figure 4.4) is effective for the shaded side of the wood. It is a labour-saving tool, but less powerful than the second tool, the curved knife with handles at both ends (Figure 4.5). This more powerful tool is used to debark the sunlit side, where the bark is harder to remove. However, the disadvantage of this curved knife is that working too close to the timber means that clothes easily get smeared with wood oil – this is not a problem with the spade knife.

The above-mentioned cambium, the thin membrane between the bark and the sapwood, also protects the wood against rot. During the processing of the timber of the under-deck beams, it is especially important to ensure that this layer of tissue remains undamaged. Experienced carpenters will accept and utilize the natural form of beams without cutting them as much as possible. Curved wood is used for the slanted beams of the woven arch. The convex side of the beam is placed upwards, for the benefit of the statics of the arch structure. In the woven arch, since the beams are interlocking, the natural shape will cause a problem with the design. Experienced bridge carpenters are able to ensure the tightness of every intersection position where beams meet each other, without cutting off the convex part of the timber, no matter how irregular the shape of the wood shape is.

4.2.2.4 Removing the gnarls

During and after the debarking process, the gnarls need to be removed. A tree grows more branches on its sunlit side, thus, the timber has more gnarls and

---

7 According to master Wu Fuyong, during the Shengshuitang project.

Figure 4.4 Debarking the trunk with a spade knife. Dongtang Bridge construction site.

Figure 4.5 Debarking the trunk with a curved knife. Dongtang Bridge construction site.

grows harder on this side: the gnarls are thus stronger, and need to be removed using an axe before the electric times. The gnarls on the shady side are much weaker, and can be easily taken off with the spade knife during the debarking process.

According to the carpenters, the gnarls on *Shanmu* are very hard and therefore quite difficult to remove, compared with many other wood species, even including some hardwoods. In the past, the hard gnarls would be removed with an axe. Chisels or planes could not be used for this, as they would be easily damaged. Nowadays, the gnarls are usually removed using an electric saw.

4.2.2.5 The direction of the wood

In many regions in southern China, including the MZ bridge area, there are rules in the woodworking professions concerning the direction for positioning the wood. For houses and temples, for example the pillars should stand like a natural tree, i.e, the root end of the tree trunk should be set as the bottom end of the pillar, the beams should have their root end pointing to the central bay of the building, where the figures of deities are usually located (and beams in the central bay should have their root end pointing to the right-hand side of the deity). The bridge corridor, with its shrine in its central bay, is also considered to be a temple in some sense, and therefore it is built according to the rules for temples.

For the bridge structure, the direction of wood is decided by the water and beams should be laid following the direction of water, that is the root end should be facing the headwater, and the tip end facing downstream.

Even after the trunk has been processed into a regular (circular or square) shape, carpenters can easily determine which are the root and treetop ends, according to the flow of the grain, the colour of the wood and the direction of the gnarls.

4.2.3 Woodworking tools

4.2.3.1 Tools for measuring and lofting

4.2.3.1.1 THE "LU BAN RULER" (CARPENTER'S SQUARE)

The carpenter's square is commonly known as a "Lu Ban ruler" (鲁班尺, or Lu Ban square) in most areas in China, named after the legendary ancestor of woodworking craftsmen, Gongshu Ban (or "Lu Ban") who lived around the fifth century BC. Its form varies from region to region. In the MZ area, it is a right-angled ruler (square) with a side length of 1 x 2 *chi* (the *chi*, or "Chinese foot," is approximately 30 cm: see below for details). On the shorter side, made of a square section wood, the length is divided into 10 *cun* (1 *cun* = 1/10 *chi*), with each *cun* being further divided into 10 equal sections. The longer side of the Lu Ban ruler is usually made of a bamboo chip, and is marked at 20 *cun* intervals.

Chi is the unit of length of the traditional Chinese measurement system. In the past, it changed over time and from region to region. In the last dynasty (Qing Dynasty, 1644–1912), the official length of the *chi* was approximately 32 cm. In 1919, the traditional measurement system was replaced by the metric system, which set the *chi* at 33.3 cm (1/3 of a metre), and this measurement has been kept until today. In the MZ mountains, however, the length of the *chi* varies from county to county, its length ranging from 26.0 cm to 33.4 cm[8]. Thus, when carpenters from different areas work on a project together, it is the *shengmo*'s measurements that are used in the construction.

---

8 Similar as the measurement system, the high mountains and less-developed transportation/ communication systems have resulted in the MZ mountains being an area with the highest

Figure 4.6 A Lu Ban square, an axe and an inkpot. Dongtang Bridge construction site.

Figure 4.7 A carpenter drawing a line on the beam end with a Lu Ban square. Dongtang Bridge construction site.

4.2.3.1.2 SCALE-STAFF

The scale-staff, literally "the staff of *zhang*" (*zhanggan*, 丈杆, 1 *zhang* = 10 *chi*) is a combination of ruler and guidance for construction. It is a dominating tool for building construction in southern China. The scale-staff in the MZ area is a long wooden stick, square in section. Its three sides bear ink marks

diversity of dialects: it is common for people from neighbouring counties not to be able to understand each other, and having to communicate by speaking in Mandarin.

Figure 4.8 The scale-staff at the Regensburg Bridge construction site.

as processing instruction for pillars. The length of the scale-staff is normally a bit longer than the longest pillar in the structure (Figure 4.8).

1:10 models of the scale-staff of the Dongtang Bridge could be used as an illustration of its principle (Figure 4.9). On the first side are the graduations of *chi,* numbered from bottom to top. Every mark is followed by an auspicious word (e.g. *chi* 1 – Wealth, *chi* 2 – Happiness, *chi* 3 – Luck, *chi* 4 – Prosperity, etc.). On the next two sides are instruction for the processing of the pillars. These two sides are perpendicular to each other, representing the two perpendicular sides of the pillars (in the crosswise and longitudinal directions of the corridor). Marks are made indicating the position, scale and form of the cuts (mortises) on the pillars through which they connect the (tie) beams in the two directions. Information for different pillars is integrated onto a single scale-staff, distinguished from each other by the different rows of marks on one side. Bearing all information relating to the design and processing procedures, the scale-staff integrates a ruler, a series of plan drawings and templates for building construction.

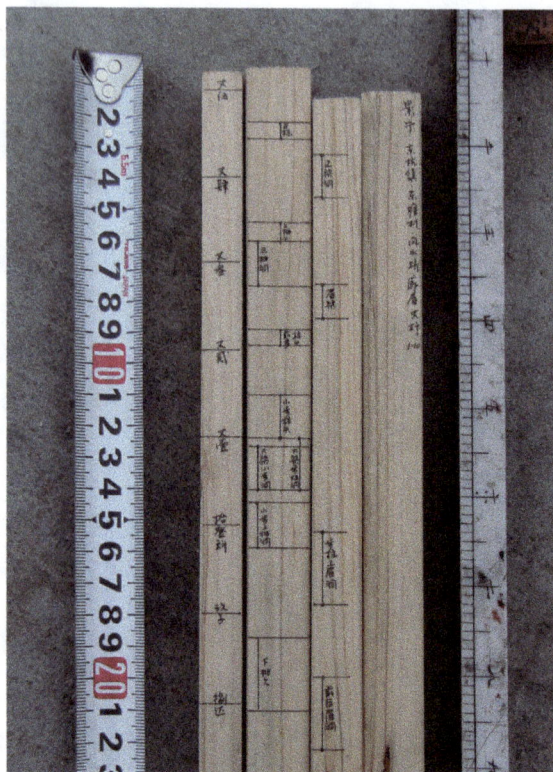

Figure 4.9 Model of the scale-staff of the Dongtang Bridge (scale 1:10).
The four staff models show the upper part of the four sides of the staff. Marks on three sides give the instruction for processing the different building members. (The fourth side is kept blank, but on this model, contains information of the project.)

Source: Made by the author during the bridge construction. 2013.

For the construction of simple buildings, carpenters can make the calculations for the respective positions and scale in their head and mark the scale-staff directly, without the help of sketch drawings or calculation aides (e.g. an abacus or paper). The ability to mark a scale-staff independently is the final criterium for a carpenter apprentice to complete his apprenticeship. In projects involving buildings with a complex form, carpenters may also draw sketchy linear plans to assist their planning and calculation.

On completion of the construction, the scale-staff is usually put into the roof structure of the building, where, according to a superstitious tradition, it will act as the protector of the building and bring good luck to the owner (Figure 4.10).

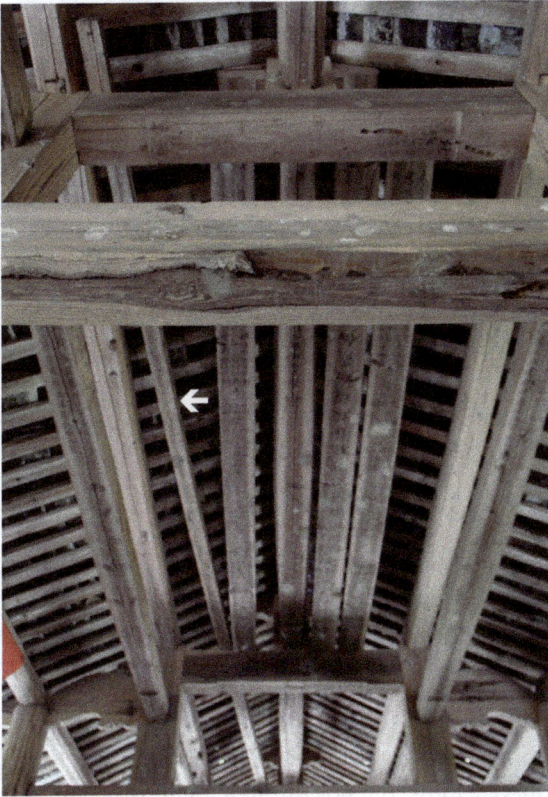

Figure 4.10 The scale-staff is kept inside the roof structure of the house (here the bridge corridor). Dan Bridge (单桥), Shouning, Fujian, 2013.

4.2.3.1.3 TEMPLATES

As the scale-staff serves as the processing templates for the corridor members, the beam members for the under-deck structure have their own templates as processing assistance. Members of the woven arch are inclined by themselves and by their joints. The countless mortises and tenons/tails and sockets all have different angles and scales. To guide the processing, the bridge master makes templates for each type of joint, giving the outline of the wood joints, along with the correct angle and scale. Using the templates, he is able to layout the guiding lines for processing the building members on the timber uniformly and effectively.

Some of the templates are prepared during the design phase (i.e. in advance), while others are made during the construction process, in-situ, as needed. The templates are usually made of material easily available on-site, such as wooden boards (Figure 4.42), cardboard or bamboo.

Figure 4.11 An inkpot with a bamboo brush. A wooden arch bridge construction site in Ganzhushan, Qingyuan County, Zhejiang, 2013.

4.2.3.2 Tools for marking and drawing

The tools for drawing lines and making marks on wood is an inkpot/ink-line and a bamboo brush.

The inkpot (Figure 4.6, Figure 4.11) enables the carpenter to draw a straight line onto the wood. Every carpenter makes his own inkpot – it can almost be considered a symbol of his profession (as the word *shengmo* – "string and ink" – indicates; the Lu Ban square is another symbol). The inkpot consists of a wooden pot containing cotton which has been soaked in ink, and a roll of string. The string is inserted into the pot through a pair of small holes on either side of the circular wall, and thus absorbs ink from the cotton. Onto the end of the string is tied a piece of wood, bone or horn with a pointed iron shaft at one end, and this can be temporarily fixed onto the wood. By pulling the string taut and then releasing it, a straight line can then be drawn in ink. (Figure 4.12).

The bamboo brush is dipped into the inkpot to make ink marks. The brush is simply a piece of bamboo, with a wide, flat end as the ink "brush." It is used to draw straight lines along the Lu Ban square or the templates.

If errors are made in drawing the lines, then carpenters will mark the final, correct line with an "X." It is interesting to note that despite that fact that an "X" commonly indicates selection in some western cultures (including the German-speaking area), it means "wrong" in the Chinese culture and as such, would usually mean the opposite of chosen ("not selected" or "to be removed"). It is only in woodworking that the "X" mark carries the meaning of "selected."

The bamboo brush is too stiff to write with or to draw curved lines. Therefore, any actual writing on the wood, for example the names and positions of building members, is done with a traditional ink brush.

(a)

(b)

Figure 4.12 A carpenter is using an inkpot to make ink lines on a beam. Processing site of the building members of the Regensburg Bridge. Ningde, Fujian, China, 2014. (a) The end of the string is temporarily fixed onto the piece of wood using the pointed iron shaft. (b) Ink lines are drawn onto the beam.

4.2.3.3 Tools for levelling

Levelling is an important measure during construction, used to check that the bridge members are at the correct level and in the correct position. Nowadays, carpenters use a transparent rubber tube (Figure 4.13) which is the cheapest reliable way to do this today: when the tube is filled with water, and when both ends of the tube are kept open (under the same atmospheric pressure), the surface of the water on both sides of the tube must be at the same level. In the past, bamboo was used for the same purpose: a long bamboo pole was split along its longitudinal axis, and the walls of the bamboo joints were removed

Figure 4.13 Checking the horizontal level of the beam ends with a transparent rubber tube. Dongtang Bridge construction site.

except the pair at the ends. It was then filled in with water and become a levelling pipe.

In the case of a large-scale bridge over water, another method for levelling is simply to measure and check the distance from the members in question to the water surface.

Vertical levelling is done using the inkpot-string, which in this case functions as a plumb (Figure 4.14).

4.2.3.4 Tools and devices for wood processing

The axe is the most commonly-used woodworking tool for carpenters in China. It is used for felling trees, splitting, mortise digging and surface planing. Even for the steps requiring finer tools, pre-processing with an axe was often done before electricity was available. For example people would praise a carpenter for his craftsmanship by saying that he planed the surface of the wood so well with the axe, that there was no need to further plane it with a woodworking plane.

Figure 4.14 The vertical levelling is checked using the inkpot string. Dongtang Bridge construction site.

The ability to use an axe is a fundamental skill for any carpenter. In the past, when a carpenter was looking for a job, the project overseer would test his skill through axe-work. In Taishun County, Zhejiang, there is a young bridge carpenter (Zeng Jiakuai) who is famous for being able to peel eggs with an axe.

Chisels are used for working on wood joints. Four chisels of different sizes are sufficient for the basic toolkit of a carpenter in southern China. The narrower, thicker ones are used for digging mortises and sockets, and the bigger thinner ones for planing the surface of the joints (Figure 4.15).

The common woodworking plane and frame saw in China are quite similar to those in Europe (Figure 4.16).

Nowadays, electric tools are usually used even in traditional construction, because of their great efficiency. Wooden boards are cut with electric band saws. In Qingyuan County, the author met a sawyer and his adapted "tractor bench saw" three times, on three different construction sites (Figure 4.17). He travels the county with this powerful machine, going wherever he is needed. He earns 200 Yuan (25 Euro) per hour, the equivalent of a day's wage for a common carpenter in the same county.

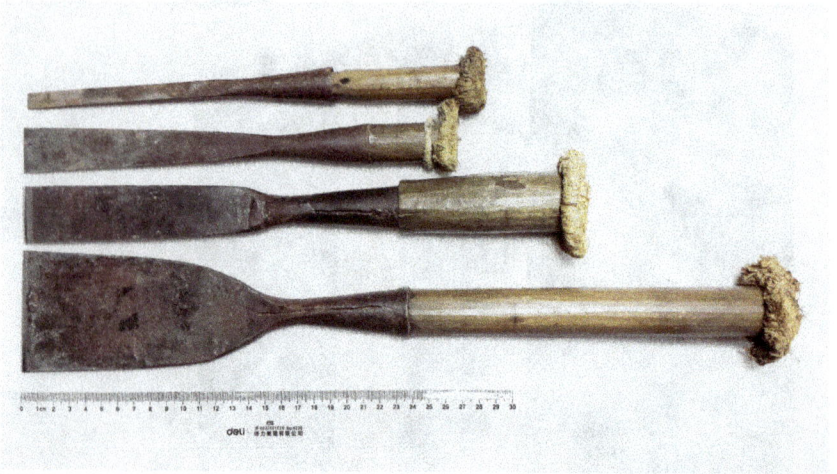

Figure 4.15 Typical Chinese carpenter's chisels in the MZ area.

Source: From the author's collection (as his "apprenticeship gifts" from Master Wu Fuyong).

Figure 4.16 Typical working tools of a country carpenter: drill, plane, square and saw. Hexi Village, Zhouning, Fujian, 2014.

Figure 4.17 The adapted "tractor bench saw." Shengshuitang Bridge construction site, 2013.

4.3 Design of the woven arch and the production of its building members

4.3.1 Basic principles

The peculiarity of the MZ bridges is embodied by the "woven" arch structure. It is composed of two arch systems: the three-sided arch is the first (primary) system, and the five-sided arch is the second (secondary) system. The greatest design challenge lies in the correct calculation of the length of the arch-beams, so as to ensure the tight weaving of the systems. Errors in design could result in an immediate deformation, or even in a failure of the structure during the construction.

The looseness also happens after years of shrinkage. The two systems come apart from each other on one side of the bridge, and compress each other on the other side, whereby the complete structure deforms and leans to one side (Figure 4.18). The more or less departed status of the two systems can be seen by the naked eye on many old bridges. This causes an uneven distribution of force. Luckily, most of these bridges are still structurally stable. The Wenxing Bridge in Taishun, Zhejiang was famous for its deformation (Figure 4.36, Figure 6.38) in the past century before it was destroyed by flood in 2016.

Figure 4.18 Lanxi Bridge (兰溪桥), Qingyuan, Zhejiang, 2012.
After the reconstruction and repair of the bridge, the leak between the U-S-Beam[2] and the C-Beam[1] was filled with wooden blocks.

Accurate calculations before construction commence and a proper means of control during construction is the key to keeping the woven structure tight, and crucial to the stability of the arch structure. Every bridge carpenter family has his own (secret) technical tradition in the form of a series of proportions, construction measures, and means which have proven to be reliable.

Today, only the two oldest bridge carpenter families stick to these traditional design methods, the descendants of the Kengdi and the Xiajian Masters. Both families have proven bridge construction histories longer than two centuries. They are the only families in the MZ area today which took up this career longer than three generations.

In this traditional design method, the structure is displayed with a linear sketch in ink lines, in which beams are expressed by their axis lines, on the floor or on a wooden board such as a door leaf. The length and angle of certain specific members are then measured using the drawing (Figure 4.20).

Since most woven arch bridges have a common range of spans (between 20 m and 35 m), family-trained bridge carpenter families have a shared knowledge of a series of fixed proportions or length of each section of the structure. They apply them to all the building conditions with only minor adjustments being made when necessary.

The design method and the corresponding numbers make up one of the core technical secrets of the bridge carpenter families. The information is stored in the minds of the core members (with or without the help of some sketchy drawing) and handed down within the families. When Master Wu Dagen (descendant of the Kengdi Masters) showed the author the design steps, he did so away from the other carpenters.

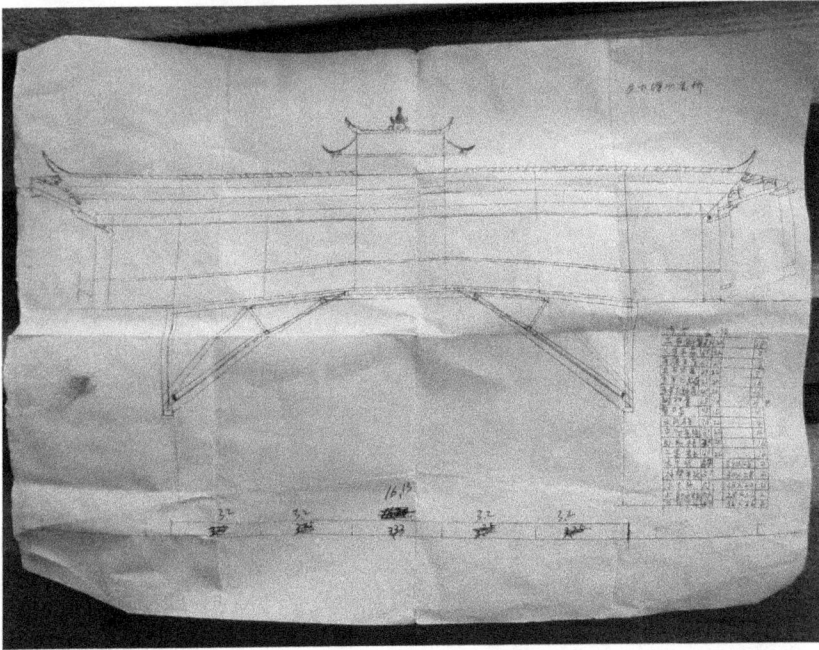

Figure 4.19 "Modern" plan of a bridge design by Master Wu Fuyong.
It is a combination of cross-sections and elevations. 2012.

The dimensions of members derived from the traditional design are mostly rough scales. The precise measurements of many elements must be decided during construction, since they need to be measured directly from the developing structure.

In recent years, with the advent of new standardized requirements for official projects, modern drawing methods are being used. Bridge designs are accompanied by modern drawing skills (Figure 4.19). The dimensions of building elements are therefore more precisely calculated right from the design stage. With the encouragement provided by the new wave of the cultural heritage revival movement, more and more carpenters are taking up the career of woven-arch-bridge-builders. These "modern" bridge carpenters may learn from the old bridges some commonly-used scales and proportions, but they do not stick to such fixed numbers as were used for the former.

4.3.2 Traditional design method: longitudinal-section-based design

4.3.2.1 The traditional design plan

The design for the arch structure includes determining the dimensions of all building members as well as the scale and angle of the wooden joints. The key factors include the length of the longitudinal beams and the angle of the mortises in the crossbeams.

Figure 4.20 Design plan of the wooden arch bridge, drawn on the floor.
Master Wu Dagen was making the template for the S-Beam[1] according to the drawing. Dongtang Bridge construction site.

In the traditional design method, the carpenters draw the linear (axis-line) plan, mostly in a scale of 1:10, with the inkpot and bamboo brush on a hard floor or a wooden board to determine the length of beams (Figure 4.20). The length and slope of the first system are decided beforehand while that of the second system is determined during the drawing stage. The entire drawing is made following a series of rules based on family-inherited knowledge (Figure 4.21). The two families with the longest tradition still plan the design according to the same principle. The only difference is in the detailed means and figures of the specific rules. The rules or experiences for determining the factor of scales and slopes, as family-inherited knowledge, remain more or less constant within a carpenter family down the generations. They are thus seen as the factors in our research on the bridges' technical pedigree. The steps for making the plan are explained below.

4.3.2.2 First arch system: the basic length and the slope

The first arch system of a wooden arch bridge is composed of a pair of crossbeams (C-Beams[1]), a pair of foot beams (F-Beams[1], which in some cases are replaced by stone bearings), horizontal beams (H-Beams[1]) in the middle and slanted beams (S-Beams[1]) at sides. There is always an odd number of longitudinal beams of the first system (S-Beams[1] and H-Beams[1]): occasionally eleven, more often nine and in smaller bridges, seven. Odd numbers are loved in Chinese culture.

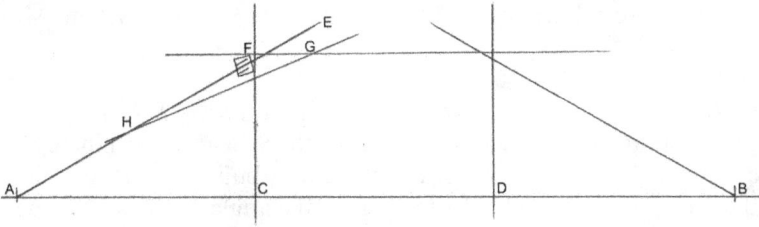

Figure 4.21 Linear (axis-line) plan of the wooden arch following the traditional design method.

- Draw a horizontal line representing the horizontal level of the arch feet. The clear span is calculated according to the scale (mostly 1:10) and marked on the line with point A and B.
- According to [Rule I], divide AB into three equal parts and mark the division points C and D. Draw vertical lines through C and D.
- According to [Rule II], determine the slope of the first arch system (e.g. 6/10), and draw the axis line of the S-Beam[1] AE accordingly.
- Draw the rectangular cross-section of the crossbeam (C-Beam[1]) in the correct scale on AE, at the outer side of the vertical line through point C, according to [rule V].
- Draw a horizontal line through the top point (F) of the crossbeam, representing the top line of the H-Beams[1], and the height of the bottom of the U-C-Beams[2].
- Determine the position of the L-C-Beam[2] according to [Rule III], mark it with H and draw the line HG as the axis line of the U-S-Beam[2]. It should be a radius-distance away from the crossbeam drawn.
- The point G is the position of the U-C-Beams[2]. Check if it is in the correct position [Rule IV].
- After the position of the U- and L-C-Beams[2] have been determined, the master carpenter is able to make the template for the beams and begin the processing.

[Rules] mentioned are interpreted below.

For the design of the first arch system, the key factors are the length proportion of the longitudinal beams (Rule I), the slope of the slanted beams (Rule II) and the scale and inclination of the crossbeams (Rule V).

[Rule I] The proportion of longitudinal beams of the first system (S-Beams[1] and H-Beams[1])   In bridges with a large span, the S-Beams[1] and the H-Beams[1] are approximate of the same length, since they both are close to the length limit provided by natural trees.

Carpenters from both longest-tradition families determine the length of the beams in the same way. They divide the clear span into three equal parts, and position the intersection (the C-Beam[1]) slightly outside the division points, to get approximately equal-length slanted and horizontal beams.

Master Huang Chuncai from Pingnan (the Changqiao Master) does things differently. In his design and practice, the H-Beams[1] are apparently longer than S-Beams[1]. In a bridge with a 20 m span, for example as he says, the middle horizontal beams are about 8 m long. The difference in proportions of the first system will not only change the appearance of the bridge, but also relates to the design rules of the second system. More on this later.

[Rule II] The slope of the slanted beams of the first system (S-Beams[1])   The height of the arch is decided by the slope of the S-Beams[1]. Chinese building tradition has a specific term for the slope of a building part (e.g. the slope of the roof): it is defined by the tangent of the angle of the related member/part, and is "N (number)-tenths-water (level)" (N *fen-shui* 分水). As in the case of the slope of the S-Beams[1], if it is "6/10" (six-tenths-water, 6 *fen-shui*), the tangent of the angle between the beam and the water level is 6/10. The most commonly used slope is from 5.5/10 to 6/10, although a range between 5/10 and 6.5/10 is also found. The larger the span, the gentler the slope should be.

There are also deviations from this among carpenter groups and from region to region. Wooden arch bridges in Taishun County are much steeper due to meet the particular local aesthetics (Figure 6.37–40). Carpenters yield to regional customs within the range of what is technically permissible when they are engaged in projects in these foreign regions. For example when the Kengdi Masters built the Xuezhai Bridge (Figure 6.17) in Taishun in 1857, they built it much more steeply than their usual projects in other counties.

The Xiajian Masters mostly use a 5/10 slope and with variations between 4.5 and 6. The Kengdi Masters use 6/10 for bridges spanning 15 to 30 m. Smaller bridges must be steeper, so their middle H-Beams[1] are therefore longer to avoid the conflict between the beams of the upper parts. Larger bridges are flatter, and take a 5/10 slope at the most (otherwise the required length of wood would be too huge).

Now that wooden arch bridges have become so popular (after being inscribed on the UNESCO Intangible Heritage List), they are also being built at locations where the required span is much smaller than the common bridge of this type. In these cases, the handed-down rules are sometimes invalid. An extreme example of this is the project at Regensburg. The bridge span is only 7.5 m. Initially, the design was made by the traditional carpenters. Master Zhang Changzhi first tried using a 6/10 slope, but this was not successful, and the beam members still conflicted even when he gradually increased the slope bit by bit. The design problem was finally solved by the author with his computer, using drafting software (AutoCAD) and with an unusual slope 7.5/10, resulting in an extraordinary proportion of the horizontal beams, the 3.8 m long H-Beams[1] in the middle cover more than half of the entire span (Figure 4.32).

Unlike the limited family-inherited bridge carpenters, carpenters who take up bridge-building as a profession in modern times describe the slope in modern geometry terms, using the degree system. Wu Fuyong uses "no less than 35 degrees," Dong Zhiji says "from 33 to 45 degrees, the best is 36 or 37 degrees."

4.3.2.3 The proportion of the second arch system

The second arch system is composed of two pairs of crossbeams (U- and L-C-Beams[2]), a pair of foot beams (F-Beams[2]), a group of horizontal beams (H-Beams[2]) in the middle and two groups of slanted beams (U- and L-S-Beams[2]). The S-Beams[2] lie between the S-Beams[1], and are thus one less in number than the first system.

The key factor in designing the second system is to decide on the position of the crossbeams ([Rule III] and [Rule IV]), which in turn determines the proportion of the longitudinal beams and – like for the first system – the scale and inclination of the uppermost crossbeams (Rule V). The key factor among them is the position of the L-C-Beams[2] ([Rule III]), it is defined by the length of the L-S-Beams[2] to the length of the S-Beams[1] (roughly AB/AC in Figure 4.22). Hereinafter, this ratio will be written as "position-ratio (of the L-C-Beams[2])."

[Rule III] The location of the lower crossbeam of the second system (L-C-Beam[2]) The position-ratio of the L-C-Beams[2] among the Xiajian Masters is 2/3[9] or 6/10.[10] Relatively speaking, this is a higher position compared with the Kengdi Masters. The reason the carpenters gave for choosing a higher location is that the upper part of the S-Beam[1] is thinner and more pliable, making it easier to carry out the crucial "crossbeam-rammings" step (Sec.4.5.3.4.2 below) – in which the S-Beams[1] are bent as much as possible to weave the structure tight.

On bridges built by the Kengdi Masters, the position of the L-C-Beams[2] is noticeably lower (in other words, the L-S-Beams[2] are much shorter). The position-ratio of the historical bridges constructed by this carpenter group is 1/2 to 2/3. The youngest carpenters from this family use an even lower ratio: only 1/3 in the case of Master Wu Dagen and his Dongtang Bridge (Figure 4.93). In a model located in their home and made by Master Zheng Duoxiong (Master Wu Dagen's brother-in-law), this proportion is 3.6/10 (Figure 4.22).

The Kengdi Masters choose the lower position for the L-C-Beams[2] intentionally, as they say, to ensure a closer distance between the uppermost crossbeams of the structure (the U-C-Beam[2] and C-Beam[1]). There is a huge bending moment in the H-Beams[1] on which rest the U-C-Beams[2]. Positioning the U-C-Beams[2] close to the ends of these H-Beams[1] will effectively reduce the burden (bending moment) inside the H-Beams[1].

The positioning of the L-C-Beams[2] is exaggerated by the latest generation of the Kengdi Masters (Wu Dagen and Zheng Duoxiong) compared to the way of their ancestors. Such a low position is seldom seen in the earlier works of this carpenter group. However, the difference in the position-ratio between the Xiajian and Kengdi Masters is still obvious when observing their works down the centuries (Figure 4.23), namely, in the bridges of the Xiajian Masters, the

9 According to Zhang Changzhi.
10 According to Peng Fodang.

(a)

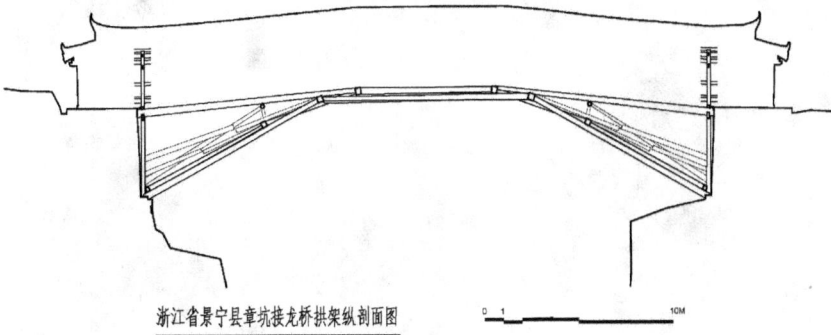

浙江省景宁县章坑接龙桥拱架纵剖面图

0 1   10M

(b)

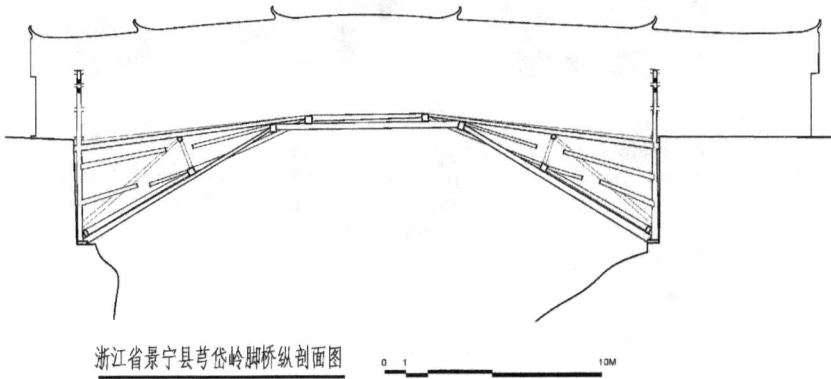

浙江省景宁县芎岱岭脚桥纵剖面图

0 1   10M

Figure 4.23 Comparison of typical historical works of Xiajian- and Kengdi Masters. The work of Xiajian Masters has higher L-C-Beams[2]. (a) A typical example of a bridge built by the Xiajian Masters (by Zhang Xuechang, 1917). Longitudinal cross-section. Jielong Bridge, Jingning, Zhejiang. (b) A typical example of a bridge built by the Kengdi Masters (by Xu Bin'gui, 1883). Longitudinal cross-section. Lingjiao Bridge (岭脚桥), Jingning, Zhejiang.

the distance between the top two pairs of crossbeams (U-C-Beam[2] and C-Beam[1]) should be no larger than 1.5 m. In the case of the Dongtang Bridge, since it is a small-scale bridge, this distance is exactly 1.5 m.

With the traditional design method, the exact length of the "weaving-beams" (U-S-Beams[2]) can only be roughly calculated during the design period: their exact length can only be measured at the required points on the structure during the actual construction process (Figure 4.24). It is the same for the last longitudinal beams of the second system, the H-Beams[2], whose

Figure 4.24 Determining the exact length of the U-S-Beams[2] using a long wooden stick (scale-staff). Dongtang Bridge construction site.

The L-C-Beam[2] is temporarily kept in position by wooden dowels or iron clamps. The carpenter uses the scale-staff to check the position of the U-S-Beams[2], mark their required length on the staff, followed by the expected position of the U-C-Beam[2] on the horizontal beams. This position is then checked using [Rule IV].

length has to be measured and determined in-situ after the other parts of the second system has been erected.

4.3.2.4 Sizes and angles relating to crossbeams

Of the squared cross members, two groups are particularly significant for design: the C-Beams[1] and the U-C-Beams[2]. They are the largest and uppermost crossbeams in the structure and thus of greater importance; and unlike the foot beams and the lower crossbeams (which are simply laid on the surface of their supporters), the uppermost crossbeams are also problematic with regards to angles, while their own incline has also a close relationship with the angles of the joint cuts (mortises and sockets) inside.

Let us start by looking at the C-Beam[1], whose problems are the most obvious. The crossbeam is inclined by itself roughly following the slope of the weaving-beams (U-S-Beams[2]) which lay right behind it. From a layperson's view, the inclination of the crossbeam is in itself already rather problematic. Theoretically, for a better force transfer, the crossbeam should press onto the surface of the weaving-beams as closely as possible, which means keeping neat parallel angles between the two. However, the inclinations of these two kinds of members from the two systems are mutually affected. If the aim is to keep them highly coordinated, even a slight adjustment would result in a change in the entire system, and lead to endless adjustments and re-adjustments.

To make it even more complicated, the crossbeam has on both sides many joint cuts (mortises for tenons and sockets for dovetails) for jointing the longitudinal members of the first system. The difference in their slope causes different angles of the joints inside the crossbeam.

Equipped with only a sketchy draft, the carpenters simplified the design problem greatly and redefined the geometrical requirements of the beam relationship. The crossbeam does not need to press onto the weaving-beam with its entire lower surface, but only a corner. To ensure a good force transfer, the pressing corner must be the inner corner (towards the mid-span of the bridge) which the carpenters call the "chin" of the crossbeam (Figure 4.25). Therefore, the crossbeam does not need to follow the exact slope of the slanted beams, but only to make sure that its inclination is flatter than the slanted beam. Thus, despite the change in the scale and slope of the arch structure, the crossbeams can be made according to a series of fixed scales and angles.

Figure 4.25 The "chin" of the C-Beam[1] presses on the S-Beams[2] underneath. Dongtang Bridge construction site.

[Rule V] The angles of the crossbeam in the first system (C-Beams[1])   Both the Kengdi Masters and the Xiajian Masters use fixed scales, proportions or angles for designing and processing the crossbeams and their joints.

In the Kengdi Masters' design, the slope of the first arch system changes depending on the scale of the bridge (see [Rule II]). Despite the change in the slanted beams, they set the crossbeams in at fixed angles. The C-Beam[1] is placed with a slope of 3/10.[12] The mortise cuts are drawn and calculated accordingly (Figure 4.26, Figure 4.27).

The Xiajian Masters solved the design challenge even by making the design more rigid. Thanks to their great reputation for bridge-building, most of their projects are large-spanned structures, making a uniform structure design feasible. The most recent generation tend to apply uniform scales and cuts for C-Beams[1] in almost every project regardless of the scale and conditions of the bridges[13] (Figure 4.28).

[Rule VI] The angles of the crossbeam in the second system (U-C-Beams[2])   Similarly to the C-Beam[1], the U-C-Beam[2] also poses design challenges in terms of slopes. It has angled mortises for the U-S-Beams[2] and its underside is in contact with the H-Beams[1]. As with the design for the C-Beam[1], the Kengdi Masters also have rules on angles and the Xiajian Masters on scales.

The Kengdi Master (Wu Dagen) sets the U-C-Beams[2] horizontally and makes the mortise with a slope of 4/10, which is roughly the same slope as the S-Beams[2] (Figure 4.29). Since the real slope of the (tenons of the) slanted beams may vary slightly from this slope, the U-C-Beams[2] are generally horizontal with (only) a slight inclination (Figure 4.30).

The Xiajian Master makes the U-C-Beam[2] exactly the same as the C-Beams[1]. As a result, the U-C-Beams[2] of their bridges are always noticeably inclined. The advantage of these measurements is not only the convenience in terms of design, but also the economy in the dimensions of the timber used. By reducing the angle between the penetrating tenon and crossbeam itself, the inclined form allows the U-C-Beams to be smaller in scale (lower in height) (Figure 4.31; cf. Figure 4.32, a and b).

The construction of the Regensburg Bridge is a rather interesting case in point (Figure 4.32). As mentioned above, after Xiajian Master Zhang Changzhi's design of this rather small-scale structure failed, the author took over the task of designing the bridge. This project was carried out a year after the Dongtang Bridge in which the author had made a careful and in-depth study on the craftsmanship of the Kendi Masters (Wu Dagen and Zheng Duoxiong). With the deep impression left on him by the former project, the

---

12 According to Master Wu Dagen.

13 However, this rigidity of their 'one-size-fits-all' design approach is what caused the masters to fail in the design of the Regensburg Bridge, whose main feature is its rather small-scale structure.

(a)

(b)

Figure 4.26 Kengdi Master Wu Dagen is determining the direction of the mortise in the C-Beam[1]. Dongtang Bridge construction site. (a) The C-Beam[1] will be set on a slope of 3/10, which is designed to be its final position in the structure. Subsequently, vertical and horizontal lines are marked on the beam ends as basic lines, and the position and direction of the mortises will be defined from these basic lines. (b) From the basic lines, the axis line of the mortises will be drawn on the beam according to the slope of the S-Beams[1]. The edge lines of the mortises will be marked parallel to the axis line.

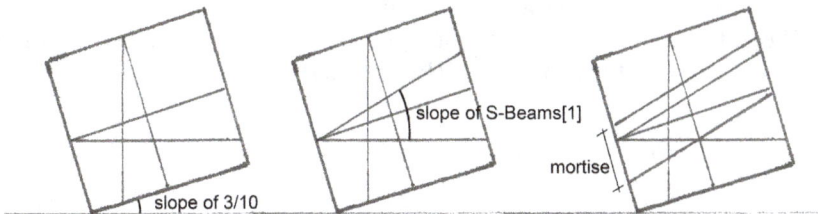

slope of S-Beams[1]

mortise

slope of 3/10

Figure 4.27 Method of Master Wu Dagen (Kengdi Master) for determining the slope of the mortises for the S-Beams[1] in the C-Beams[1].

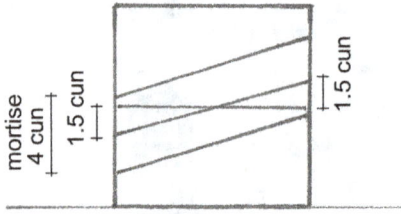

Figure 4.28 Method of the Xiajian Master for working out the slope of the mortise of the S-Beam[1] in the C-Beam[1].

Figure 4.29 U-C-Beam[2] with ink lines on the beam end.
The lines mark the mortise (highlighted) in 4/10 slope. Dongtang Bridge construction site.

author unconsciously designed the U-C-Beams[2] according to the Kendi Masters' method, with horizontally-laid U-C-Beams[2]. As explained above, the horizontally-laid crossbeams require a higher section, the U-C-Beams[2] of this bridge thus produce an oblong-shaped in section.

The difference in the angle of the U-C-Beams[2] is also distinctive in the historical works of the two families (Figure 4.33).

4.3.3 Bridge width

The width of the bridge is defined by that of the corridor, which is most commonly 5–6 metres in MZ area. The arch is of the same width at its top, while wider at the bottom, and thereafter appears a trapezoidal form on the cross-section (Figure 4.34).

Figure 4.30 The Kengdi Masters' U-C-Beams[2] are laid almost horizontally. Dongtang Bridge construction site.

Figure 4.31 Another impressively small-scale bridge under construction showing the rather inclined U-C-Beams[2]. Ganzhushan Village, Qingyuan County, Zhejiang Province.
The bridge is built by a non-family carpenter who had worked with Master Wu Fuyong.

The difference in width between the top and bottom, from the one hand, is a result of the construction of commander-pillars: The commander-pillars play the role of common corridor pillars at their upper part and thus define the width of the corridor by the distance between them. They stand almost vertically and reach downwards to the arch feet at the lower end. Since the

(a)

(b)

Figure 4.32 The slope of the U-C-Beams[2] of the Regensburg Bridge.
This bridge is actually a hybrid of the methods used by the two traditional bridge carpenter families and was built using modern drafting tools. (a) Although the bridge was
built by Xiajian Master Zhang Changzhi and bears his name, the woven arch structure
was designed by the author, partly according to the design method used by the Kengdi
Masters – so the U-C-Beams[2] are laid horizontally, according to their rules, and are
oblong in shape. (b) If the U-C-Beams[2] were designed and made according to the
traditional rules of the Xiajian Masters, they should be placed on an incline, keeping
the same angle as the axis of the mortise (for the tenon of the U-S-Beams[2]) with the
C-Beams[1] and be square in shape.

(a)

(b)

Figure 4.33 Comparison on the form of the C-Beams(2) between typical historical works of Xiajian- and Kengdi Masters. (a) A typical example of historical bridges built by the Xiajian Masters (built in 1917 by Zhang Xuechang). The U-C-Beams[2] are visibly inclined. Jielong Bridge, Jingning, Zhejiang. (b) A typical example of historical bridges built by the Kengdi Masters (built in 1883 by Xu Bin'gui). The U-C-Beams[2] are horizontal. Lingjiao Bridge, Jingning, Zhejiang.

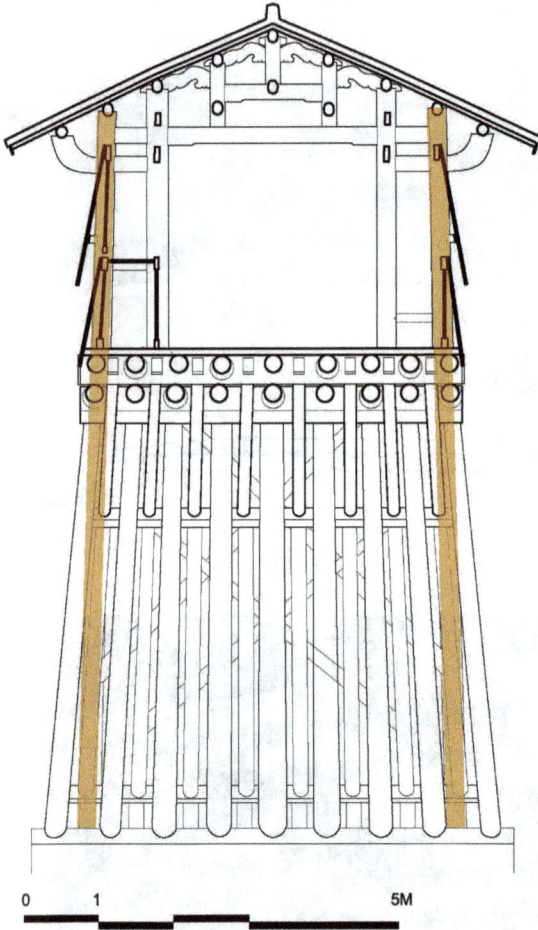

Figure 4.34 The position of the commander-pillars in the cross-section of a typical MZ bridge: hidden right behind common corridor pillars at the upper part and the arch at the lower part in the cross-section. Yangmeizhou Bridge, Shouning, Fujian.

woven arch is of the same width as the corridor at its top, if it were to keep the same width from top to bottom, the outermost S-Beams[1] would be in conflict with the commander-pillars at their feet. By making room for the commander-pillars, S-Beams[1] move outwards at their feet.

More importantly, the trapezoidal form of the arch has a significant role in structural stability, especially to increase the later stability against wind pressure and the threaten of rushing flood. It's a measurement resulting from long practical experiences. This feature is absent in some earlier bridges (e.g.

Rulong Bridge, Figure 5.9 c) or bridges built by inexperienced carpenters in more recent times.

4.3.4 A closer look at the LH-Beams

In the design of the second arch system, there is a problem worthy of attention. In the most favourable conditions, the lateral horizontal beams ("LH-Beams") are connected to the woven arch on the topmost crossbeam (U-C-Beams[2]) with dovetails. In bridges with a steeper arch, when the arch is much higher than the abutment, the C-Beams[1] might stand in the way of the LH-Beams. Sometimes carpenters have to hack off the bottom part of the LH-Beams to make them fit (Figure 4.35), but this is unfavourable, as it damages the structure.

In an area where a steep arch is an aesthetic preference, in Taishun County, for example, the wooden arch bridges get around this problem by using other structural solutions. Here, the LH-Beams simply rest their top ends on the C-Beam[1] (instead of the C-Beams[2]), without dovetails or other forms of constructional joints. This solution, however, brings with it disadvantages. The structure tends to deform when the woven arch becomes looser after ageing. This tendency is restricted in a typical bridge, in which the LH-Beams can block the top of the C-Beams[1], preventing them from turning around.But if the LH-Beams simply rest on the C-Beams[1], they are not able to restrict this movement. The Wenxing Bridge is an extreme example of this problem

Figure 4.35 Part of the LH-Beam has been hacked off, since the C-Beam[1] stands in its way. Shengshuitang Bridge construction site.

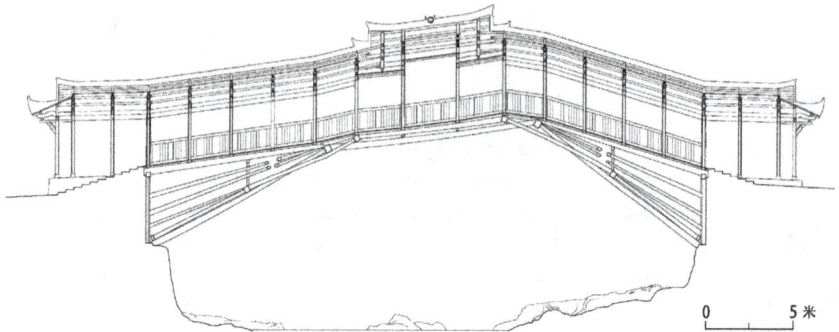

Figure 4.36 Longitudinal section of the Wenxing Bridge, Taishun, Zhejiang. The LH-Beams are laid onto the C-Beams[1].

Source: Zhou 2016.

Figure 4.37 The utilization of naturally-grown wood. Shunde Bridge (顺德桥), Longquan, Zhejiang, 2012.

(Figure 4.36, Figure 6.38): before it was destroyed by flood in 2016, it was famous for its deformation.

4.3.5 *The art of utilizing naturally-grown wood*

Naturally-grown tree trunks have a non-uniform shape and scale, and are often curved, to a greater or lesser degree. To utilize them in the most suitable and appropriate way calls for the important experience of the carpenter. The best carpenter masters are masters of material, able to utilize the irregularly-shaped members to get the best mechanical properties, or even approach the construction as a work of art (Figure 4.37).

4.4 Processing the wooden arch elements

The processing of the beam members (of the woven arch, is roughly equivalent to the processing of wood joints. The longitudinal beams bear tenons or tails at the ends, while the crossbeams bear mortises and sockets to hold the former (Figure 4.38).

As a basic rule, the horizontal beams (of both woven arch systems) are connected by dovetails, as determined by the construction process. The slanted beams have a (straight) tenon at the top, but more options (tenon or dovetail) at the lower end (depending on their origin in terms of their technical pedigree). The reasons for the choice of joint type among the different groups of carpenters are discussed at the end of this chapter. In this section, the focus is on the processing principle and method.

4.4.1 Longitudinal beams

4.4.1.1 Slanted beams of the first arch system (S-Beams[1])

As a rule, the S-Beams[1] have a "duck-beak" at their lower end (Figure 4.40, b), by which they sit on the edge of the stone bearing or the foot beam. Their upper ends bear a tenon which is connected to the C-Beam[1] (Figure 4.40, a).

When the carpenters talk about the scale of the beam members, the length of an S-Beam is defined by the length of its lateral axis of the main beam body (Figure 4.39, AB), namely the distance from the inner corner of the duck-beak to the middle point of the "shoulder" of the tenon measured from the lateral surface of the beam (Figure 4.21 along with line AE, between point A and the crossbeam, Figure 4.41, b-L).

The shape, especially the angle of the tenon and the duck-beak, is given by templates (Figure 4.42) which are prepared in advance according to the design drawing made by the master carpenter (Figure 4.20). The Steps for processing an S-Beam[1] is given in Figure 4.41.

4.4.1.2 Horizontal beams of the first arch system (H-Beams[1])

The H-Beams[1] have a dovetail at both ends. Since the C-Beams[1] are inclined, the dovetails are also at non-orthogonal angles with the crossbeam. As a rule, the sockets of the C-Beams[1] are kept right-angled (perpendicular to their surfaces), and the tails are sawn inclined. This is not only to ease the processing work, because sawing an inclined tail is easier than chiselling an inclined socket, but also, more importantly, it is done so as to form a trapezoidal shape of the entire beam (seen from a side, Figure 4.43). Thanks to this shape, the horizontal beams can keep pushing the crossbeams during the process whereby they are knocked into their sockets, and so make the joints even tighter.

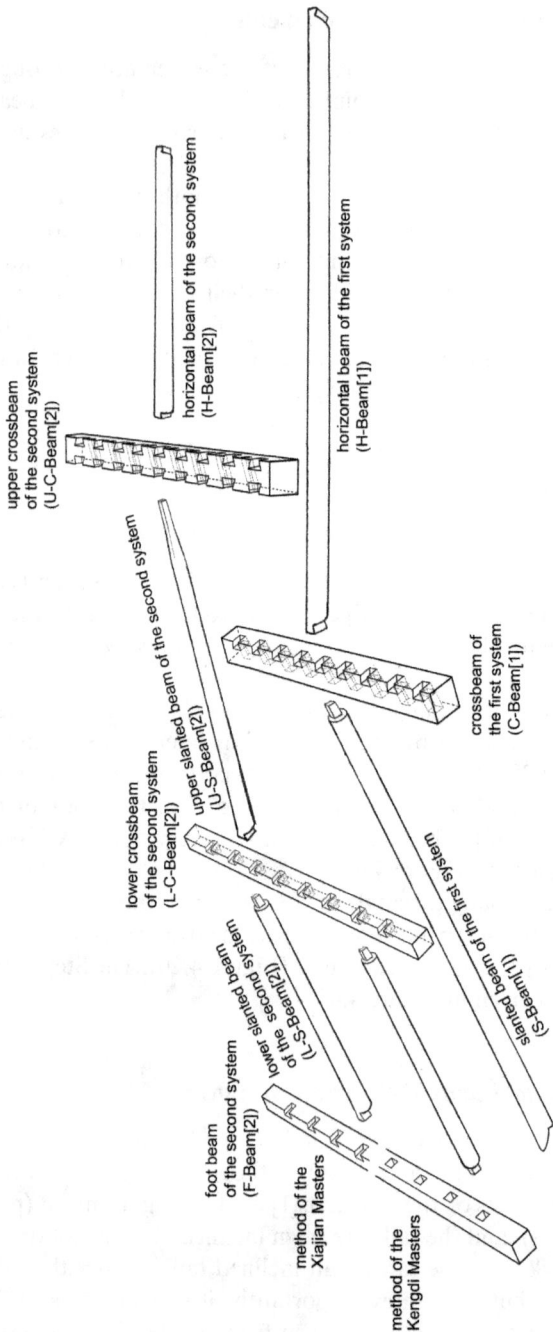

Figure 4.38 Structural Members of the woven arch in the MZ bridges.

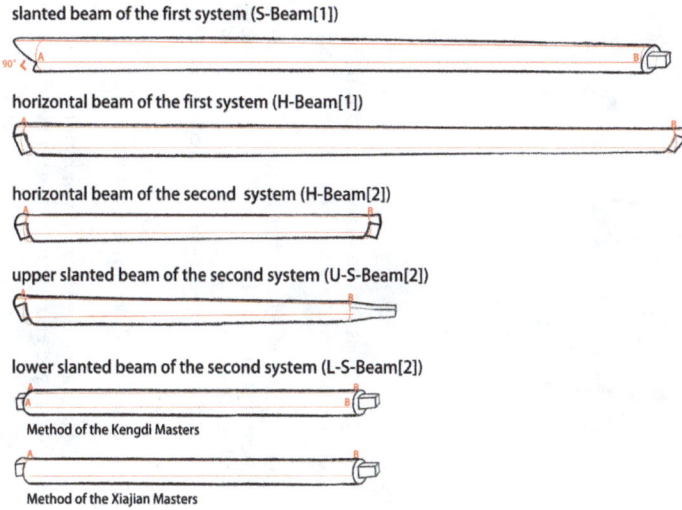

Figure 4.39 Working axes of the longitudinal beams of the woven arch.

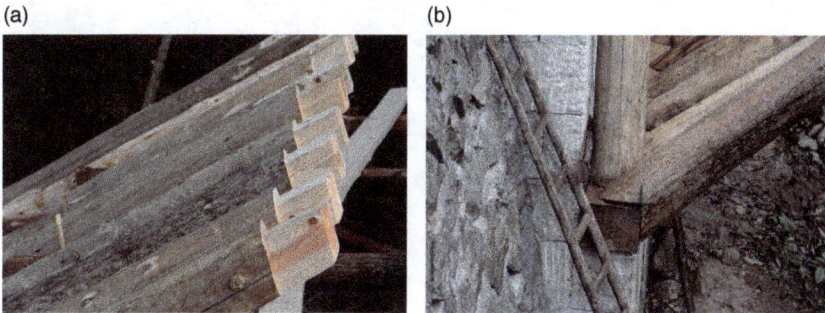

Figure 4.40 Joints at the ends of the S-Beams[1]. Dongtang Bridge construction site. (a) Tenons at the upper ends of the S-Beams[1]. (b) The duck-beak at the lower end of the S-Beams[1].

4.4.1.3 Slanted beams of the second arch system (L- and U-S-Beam[2])

The L-S-Beams[2] have a tenon at the top end. Their lower ends have either a dovetail or a tenon, depending on the carpenter group: the Xiajian Masters only use dovetail, while the Kengdi Masters only use tenon. This technical feature remains rather stable within a particular group of carpenters (on this more see Sec.4.6.2.1). The processing is similar to that for the S-Beams[1].

As a rule, the U-S-Beams[2] have a dovetail at the bottom end and a tenon at the top end. In the bridges built by family-tradition carpenters, the top tenons pierce through the crossbeam and extend even further out. Placing

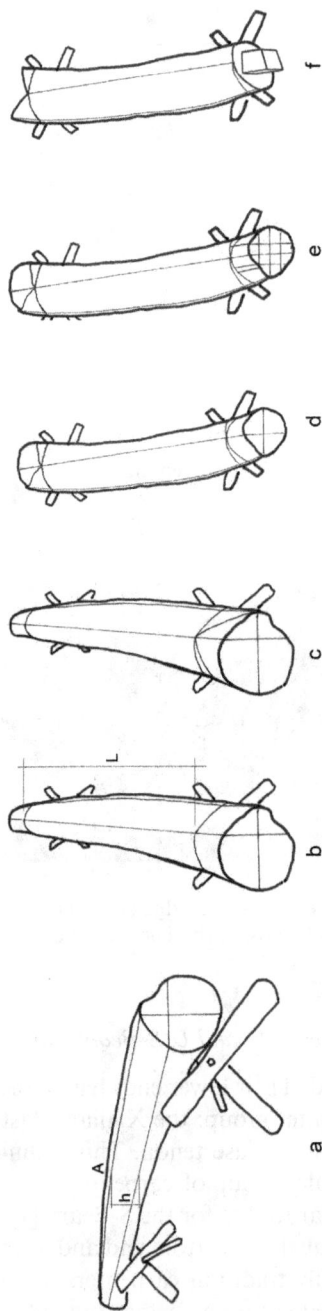

Figure 4.41 Steps for processing a S-Beam[1] from a naturally-grown and formed curved tree trunk:

a. First, the trunk is set on a pair of work-stands, known as the "wooden horses." A right-angled cross is then drawn on both trunk ends using the inkpot and the carpenter's square. The ends of the two end sections are then connected by ink lines. They represent the four axes of the beam: top, bottom and two lateral axes.

The cross does not need to be at the geometrical centre of the trunk section. The position is controlled by another factor: where a S-Beam[1] will bear the L-C-Beam[2] (Position A), also see Figure 4.22, position B), the vertical distance (h) between the vertex of the beam section and the lateral axis line should be equal on every S-Beam[1]. Only thus will the L-C-Beam[2] be supported evenly by every S-Beam[1] and the force be distributed evenly among the longitudinal beams. Given this situation, naturally-curved wood requires special attention.

b. The length of the beam body (L) is defined by two circles drawn around the trunk at two trunk ends. The circular lines are perpendicular to the four axes. At the ends of the trunk, from the circular lines outwards, there must be enough length left for the tenon and duck-beak.

c. Then at the thicker (root) end of the trunk, the profile of the duck-beak is drawn following the shape or the tracks given by a template.

d. At the thinner (tip) end, the inclination of the tenon shoulder (the ring-shaped cutting surface at the end of the tenon) is given by a template.

e. The shape of the tenon is drawn along the axis according to the scale calculated using the carpenter's square.

f. Last, the duck-beak and the tenon are cut to shape using a saw and a chisel.

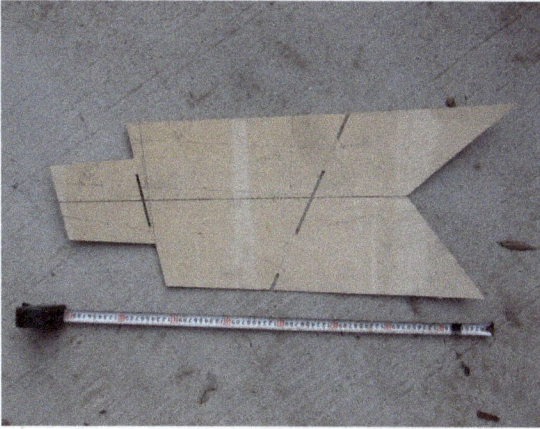

Figure 4.42 A template for S-Beams[1], made by Master Wu Fuyong for the Shengshuitang project.

Information regarding the duck-beak, the inclination of the S-Beam[1] tenons and of the H-Beam[1] tails (indicated by the linear cuts in the middle) are combined on this single template. The ink axis line represents the lateral axis line of the beams.

Figure 4.43 The horizontal beams are trapezoidal in shape thanks to the inclined tails. Shengshuitang Bridge Construction site.

Figure 4.44 The tenons on top of the U-S-Beams[2] go through the U-C-Beam[2]. Dongtang Construction site.

the U-C-Beams[2] onto the tenons is the most crucial step during the construction (Sec.4.5.3.4.2) as it is the step that enables the structure to be woven tight. The upper tenons are always adjusted in-situ to ensure the tight contact between the members.

It is for this reason, the family-tradition bridge masters make the tenons longer and cut them to fit after the woven arch is completed. The Kengdi Masters make the tenons with a smooth root part (without tenon shoulder) to push the crossbeam as deep as possible (Figure 4.44).

Similar to the S-Beams[1], the position where the U-S-Beams[2] bear the C-Beams[1] must be dealt with carefully. At this position, the wood needs to be flat and straight. If there is a snarl at this position, for example it should not be removed by an axe, so as to avoid breaking the snarl fibre – instead, it should be removed (or thinned) using a plane.

4.4.1.4 Horizontal beams of the second arch system (H-Beams[2]) and lateral horizontal beams (LH-Beams)

H-Beams[2] are similar in form to the H-Beams[1], but they are shorter. The LH-Beams have a dovetail at their mid-river ends (the tip ends of logs). The abutment-ends are only roughly cut to a uniform thickness and then simply laid on the tie beam of the commander-pillar frame.

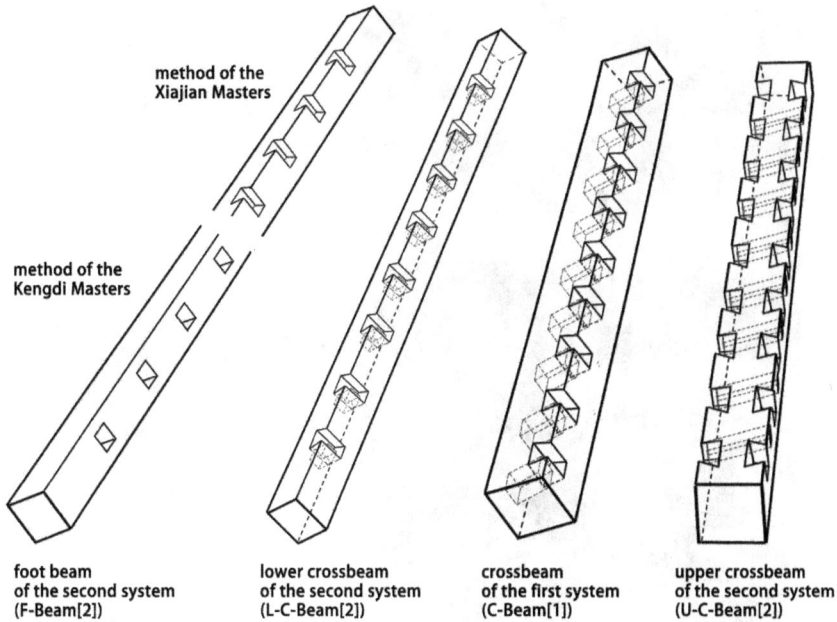

method of the
Xiajian Masters

method of the
Kengdi Masters

foot beam
of the second system
(F-Beam[2])

lower crossbeam
of the second system
(L-C-Beam[2])

crossbeam
of the first system
(C-Beam[1])

upper crossbeam
of the second system
(U-C-Beam[2])

Figure 4.45 Form and shape of the cross members in the woven arch of MZ bridges.

Because of the many uncertainties throughout the construction process, the H- and LH-Beams are processed only after the other parts of the woven arch have been set up. Their required lengths are measured in-situ and cut accordingly.

4.4.2 Crossbeams

In the MZ bridges, crosswise members include crossbeams and foot beams (Figure 4.45). They connect and fix the longitudinal beams, jointed to the latter through mortise&tenon joints or dovetails, and many of them are inclined.

The L-C-Beam[2] bear mortises on one side (for L-S-Beams[2]) and dovetails on the other (for U-S-Beams[2]). These two types of cuts connect inside with each other.

The C-Beams[1] are similar in joint cuttings to the L-C-Beams[2], bearing mortises (for S-Beams[1]) at the one side and dovetails (for H-Beams[1]) at the other side. These two types of cuts connect inside with each other.

The U-C-Beams[2] bear the most cuttings: sockets for dovetails at the two sides on the top (for the H-Beams and LH-Beams), and mortises for the

Figure 4.46 Built by Xiajian Master Zhang Changzhi, the irregular F-Beam[2] fits
snugly into the angle between the arch foot and the lateral pillar.
The Xiajian Masters call this form of F-Beam "halfmoon." Regensburg Bridge
construction site.

tenons from the side (for U-S-Beams[2]), which penetrate the crossbeam and
in between the dovetails.

The F-Beams[2], can be made out of smaller wood. Mostly they are square
in section, but when smaller wood pieces are used, the lower part of the foot
beams can be irregular (Figure 4.46). They bear either mortises (Kengdi
Masters) or dovetails (Xiajian Masters).

At the feet of the three-sided-arch, the bearings can be either wood or
stone. When the bearings are made of wood, these foot beams (F-Beams[1])
bear mortises on the top surface for the posts stand. The species of wood is
the same as the crossbeams.

The top two pairs of crossbeams (C-Beams[1] and U-C-Beams[2]) are the
largest of the beams. The C-Beams[1], as the crossbeam of the main system,
are crucial to the load-bearing of the entire system. The U-C-Beams[2] bears
the most joint cuttings. The larger the number of cuts, the greater the danger
that a crossbeam will split, therefore it should be on a larger scale. Iron bands
or wires are applied to stop them from splitting (Figure 4.47). Because of

(a)

(b)

Figure 4.47 The crossbeams are wrapped with iron to avoid splitting. (a) Dongtang Bridge construction site. (b) Lingjiao Bridge, Jingning, Zhejiang.

the many cuttings in the U-C-Beams[2] and the danger of damage during construction, the Changqiao Masters makes only the mortise in this pair of beams at the beginning. The sockets for the dovetails are cut in-situ once these beams are set in their correct location (on this more see "crossbeam-ramming" in Sec.4.5.3.4.2).

Traditionally, the cuts were made by hand, using an axe and a chisel. Nowadays, for larger mortises and sockets, they are first cut by an electric saw or drilled using an electric drill for ease of processing. The various angles of the cuts mean that carpenters have some means to control and check if the shape is right (Figure 4.48).

Figure 4.48 Different means for control and check the shape and inclination of the cuttings. (a) To deal with the inclined mortises in the C-Beam[1], Master Wu Dagen lays the crossbeam on a slope according to the inclination of the mortises, thus the mortises are worked vertically. Dongtang Bridge construction site. (b) To check the direction of the mortises during the cutting and chiselling steps, Master Wu Fuyong uses wooden boards to highlight the direction of the inclined holes. Shengshuitang Bridge construction site. (c) The scale and inclination of the mortises are checked against the template. Shengshuitang Bridge construction site.

4.4.3 Other struts

The rest of the structural members in the under-deck structure include the X-shaped struts and the frog-leg. Their exact length is measured in-situ once the woven arch has been erected and they are made afterwards.

4.5 Erection of the wooden arch

4.5.1 Rituals

There are many rituals related to building projects, and they range from simple to elaborate, depending on local customs and the generosity of the owner or the person who commissioned the project. Rituals are held to honour the local deities, and to ask for luck and success from the gods. Festivals are also an opportunity for craftsmen to get special payments from the project director/owner. The name of this special payment literally means "red envelope" (*hongbao*红包), since the money is packed inside such an envelope. This special payment is always in lucky number (e.g. 6 and 8) according to the Chinese culture.

In this section, the main rituals will be introduced in chronological order. Some of the rituals are indispensable, others are now much simpler or even

Figure 4.49 Making a wooden horse. Regensburg Bridge construction site, Ningde, Fujian, 2014.

missed out altogether, depending on local conditions. The most important and in any case indispensable ritual is the *dong*-festival, the festival for topping out, and on this occasion, carpenters also get their biggest extra payment.

4.5.1.1 Erection of the "wooden horse"

The "wooden horse" is made of out of two logs in an X-shape and an additional strut. Two wooden horses serve as a workbench (Figure 4.49). They are always made in-situ, symbolizing the beginning of the project (Figure 4.50).

4.5.1.2 Marking the scale-staff

Marking the scale-staff is a sign that the woodworking is about to begin (Figure 4.51). As the scale-staff has four sides, only one side needs to be marked at this stage: the side showing the marks of *chi* (Chinese foot) and the corresponding lucky words. Other sides of the scale-staff are only marked later, as necessary, during the construction.

Figure 4.50 The wooden horses serve as the workbench.
The colourful ribbons are attached to bring luck. Dongtang Bridge construction site.

4.5.1.3 Positioning the first building element

The day for the positioning of the first building element of the wooden structure is of great significance. The Chinese word for the beginning of the construction process is *xiatu* (下土), literally "down to the earth." Before setting
the first building element, the master will first worship the local earth god
(village god /god of that particular area), the god of water and the perceived
ancestor of woodworking craftsmen (Lu Ban). The date and hour for this
ritual are selected (as for most of the ceremonies) according to the Chinese
lunar calendar. In the Shengshuitang and Dongtang projects, work on this
special day began in the *maoshi* (卯时) time period, which goes from 5 to 7
a.m., somewhat earlier than on usual workdays.

For a wooden arch bridge, the first building element is normally a middle
slanted beam of the first system (Figure 4.52). This could be dealt with expedi-
ently. In the Shengshuitang Bridge project, the scaffolding was not erected in
time, making the setting of the slanted beam impossible, so the commander-
pillars frames were erected for the ceremony.

In the Dongtang Bridge project, it was already mid-winter, and in the early
morning, there was always slippery hoar frost on the surface of the scaffolding
timbers, making it dangerous to walk on them. Therefore, the first beam was
put into its correct position in advance in the evening before, paying special
attention not to let the beam touch the ground. For this purpose, a small piece
of wood was put under the beam foot (Figure 4.52). During the ritual in the
next morning, this piece of wood was simply removed and the beam was set
onto the ground.

Figure 4.51 The top of the scale-staff is wrapped in a piece of red cloth for luck. Dongtang Bridge construction site.

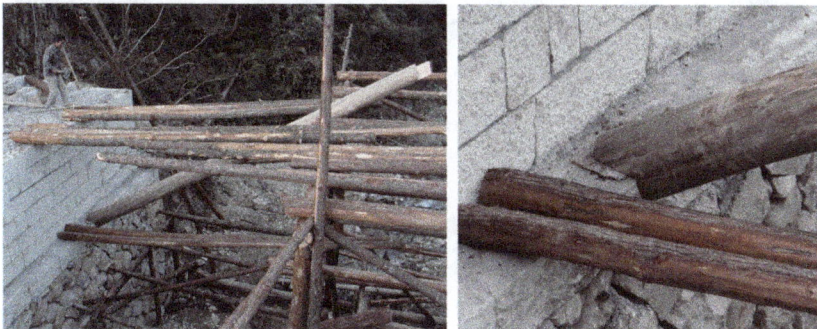

Figure 4.52 The slanted beam prepared for the *Xiatu* ceremony the next morning. Dongtang Bridge construction site.

4.5.1.4 Closing the first system

Helong (合龙), literally "closing the dragon," refers to the moment when the first arch system is closed by adding in the horizontal beams on the top, which connect the two sides of the bridge. *Helong* is a symbol of the end of the most dangerous stage of the work (Figure 4.53).

4.5.1.5 Felling the tree for the building element "dong"

The term *dong* (栋) refers to the middle top beam of a building. Depending on the custom of different areas in China, it could be either the top purlin or the beam which is parallel to the top purlin but underneath it (Liu 2016). This

Figure 4.53 "Closing the dragon." Bridge Master Zheng Duojin. Demonstration bridge construction site, Shouning, Fujian, 2001 – part of a project for the documentary film "Searching for the Rainbow Bridge."

Source: Photo: Gong Jian.

building element is highly respected and almost worshipped. The *dong* is the symbol of the house.

The tree for the *dong* is very carefully chosen, felled, processed, and kept (Figure 4.54). The felling of the tree is also marked by a special ritual.

4.5.1.6 Topping out, or the dong-festival

Positioning the *dong* is marked by the most important festival for a building project in China. People offer food and liquor to the *dong*, and hold a special religious ceremony for it (Figure 4.55). They play music and set off fireworks. The *dong*-festival marks the end of the woodworking phase of the construction project, and is also a festival praying for luck for the house and its owners/occupants (Figure 4.56). In the past, the carpenters were paid a daily wage during the construction, but at the end of the *dong*-festival, they received their special payment in a red envelope.

4.5.2 Abutment

Before the woodwork starts, the masonry work for the abutment is crucial. Bridge abutments are traditionally made of stone. Natural rock is preferred

Figure 4.54 During the entire construction period, the tree trunk to be used for the *dong* stands on the riverbank, decked with a red cloth, 2013.

Figure 4.55 The *dong*-festival for the bridge. The *dong* (in this bridge, the ridge purlin) was kept covered by a red cloth up until the festival. The ritual took place beside the bridge. Shengshuitang Bridge construction site.

Figure 4.56 The *dong* of the Regensburg Bridge.

The *dong* always displays information regarding the construction project (especially the date of the *dong*-festival, names of the board members, and sometimes names of the carpenters). In the case of the Regensburg Bridge, the name of the "architect" (the author) and the carpenter masters are written here (in this particular bridge, the *dong* is the beam under the ridge purlin).

(Figure 4.2, Figure 4.57), not only to save labour, but also to have the most stable foundation. When natural rocks are not available, manmade abutments are built (Figure 4.58–Figure 4.60). Made of either natural-shaped pebble stone or of ashlar blocks, the abutments are built in the form of a block of stone walls and filled with rubble (Figure 4.61).

If the riverbank is too soft, pinewood posts are knocked into the earth and used as the pile foundation. After the Yuqing Bridge, a large wooden arch bridge with three spans in Wuyishan, Fujian, was destroyed by fire, its abutments were also badly damaged. However, as the bridge master in charge of the reconstruction project, Master Zheng Duoxiong told the author, under the abutments there were numerous vertical pine posts buried in the earth that were exposed during the reconstruction.[14]

The construction of the abutment is done under the instruction of the bridge master. At the very beginning, the master will ask local people to tell

14 Master Zheng Duoxjong, in an interview with the author on 5 November 2014 at the Yuqing Bridge reconstruction site, Wuyishan, Fujian.

Figure 4.57 The wooden arch bridge is built on natural rock. Shunde Bridge, Longquan, Zhejiang.

Figure 4.58 Abutment made of unprocessed stone. Rulong Bridge, Qingyuan, Zhejiang.

Figure 4.59 Rubble-stone abutment consisting of rubble stones. Yangxitou Bridge (杨溪头桥), Shouning, Fujian.

Figure 4.60 Abutment made of ashlars. Dachikeng Bridge (大赤坑桥), Jingning, Fujian.

Figure 4.61 Broken abutment after a fire. The space between the stone walls is filled with rubble. Yuqing Bridge (余庆桥), Wuyishan, Fujian, 2014.

him about the highest water levels in the past years. This will decide the height of the abutment, since all the timber members of the wooden arch must stand above this water level.

Normally the abutments are completed before the woodwork begins. However, the Kengdi Masters use a different method: to facilitate the construction of the wooden arch, they first erect the abutment to only half of its height, the rest being finished after the woven arch has been completed (Figure 4.62).

4.5.3 Construction of the under-deck structure

4.5.3.1 Scaffolding

4.5.3.1.1 PLATFORM-SCAFFOLDING

In constructions across narrow and shallow rivers, scaffolding can be built out of dense posts and beams put together to form a platform to serve as a safe construction aid (Figure 4.63). Carpenters can work on the platform while they are constructing the woven arch.

The height and position of the platform (especially the position of the top edge beam of the scaffolding) must be carefully decided. During the construction process, it should support the S-Beams[1] while leaving enough space for the U-S-Beams[2]. The ideal position of this top edge beam is under the upper part of the S-Beams[1], and a little lower than the intersection of the S-Beams[1] and [2]. Since it is difficult to calculate exactly, it is better to have

Figure 4.62 During the construction of this bridge, the abutments were first built to only half their height. Dongtang Bridge construction site.

it lower than too high. If the scaffolding is not high enough, the arch-beams can still be raised by packing in edge blocks. If it is too high, nothing can be done but to adjust the height of the scaffolding beams, which will cause great danger during construction. Lives have been lost in accidents in such situations, according to some of the older masters.[15]

As for other design rules, carpenters also have their own ways of deciding the correct scale and height of the scaffolding. In the Dongtang project, the top edge of the scaffolding was 1.5 m away from the C-Beam[1], its height was calculated according to the slope of the S-Beams[1] (Figure 4.64).

4.5.3.1.2 SWING-FRAME-SCAFFOLDING

In the past, when bridges were being built above high gorges or deep water, erecting dense posts-and-beams scaffolding was almost impossible. Therefore, to save resources, the bridge carpenters developed a clever but dangerous scaffolding system known as "swing-frame-scaffolding" (the name given to it by the carpenters literally means "water-post-frame" 水柱架). Its main structure consists of four huge posts, two on each side, combined with a pair of

---

15 In a project in his youth, Master Zhang Changzhi saw a carpenter fall to his death from the scaffolding when he was trying to adjust the height of the scaffolding during construction. Interview with Zhang Changzhi in 2014.

Figure 4.63 Iron tube scaffolding. Shengshuitang Bridge construction site.
The string connecting the riverbanks is used to check the height of the scaffolding and to keep it level.

Figure 4.64 The top edge of the scaffolding is 1.5 m away from the C-Beam[1]. This beam lies right under the intersection of the S-Beams[1] and the U-S-Beams[2]. Dongtang Bridge construction site.

Figure 4.65 The swing-frame was dragged by rope using traditional cranes from the river bank. Construction site of a demonstration bridge. Shouning, Fujian, 2001.

Source: Photo: Gong Jian.

beams to form a frame (Figure 4.65). The posts go down into the river bed, and reach up to some 2 m higher than the arch structure. If the bridge is going to be high or the water is deep, two or three tree trunks are connected head to head to make up the required height for the post.

The frame contains two levels of crosswise beams. The lower-level one, which is at the level of the woven arch, also serves as the top edge beam of the platform-scaffolding: it supports the S-Beams[1] and leaves space for the U-S-Beams[2] in between the S-Beams[1]. The upper-level one, at the top of the frame, serves as a "rope roller." This is because, unlike the platform scaffolding, the post-fame-scaffolding provides no work-area in the middle of the river, and therefore the building members in the middle of the woven arch have to be dragged up by rope from the river.

For the swing-frame-scaffolding, a (crosswise) two-post frame is installed on the ground. It is then transported by the river on bamboo rafters if there is broad enough water surface under the bridge. Stones are tied in advance to the bottom of the posts to sink them into the desired position, and they are then pulled up from the bank by rope. The post-frame can also be pulled up from the bankside at places where the water is shallow and narrow.

Building a bridge with the swing-frame-scaffolding requires great skill and can be life-threatening. It is not easy, even for experienced bridge carpenters. In the construction of the Yangmeizhou Bridge (Figure 4.2) in Shouning, Fujian in 1937. At first the project was appointed to a member of the Kengdi Family Masters, Master Wu Daqing from Choulinshan Village in Kengdi area; Wu Daqing was capable of building wooden arch bridges, but he dared not carry out the construction above the deep water. The project board had to invite another bridge master from Xiajian to construct the woven arch. – It must be emphasized here that the project site was also in the Township of Kengdi and the leader of the board were also Kengdi locals. It was a "home project" for the Kengdi Masters. By admitting their lack of capability and inviting another master to come from another county, they were actually admitting the dominant status of the Xiajian Masters in this profession at that time.

An old Xiajian Master, Master Zhang Xuechang, over 60 years of age at the time, brought with him three groups of bridge carpenters. One group were good swimmers and worked in the water to build the swing-frame-scaffolding. The second group were the bravest craftsmen and worked on the scaffolding (they had no fear of height and they also knew how to swim, which would save them if they fell off the scaffolding in the worst case), and the rest worked onshore. After the completion of the scaffolding, the first group of carpenters went back home. The Xiajian Master built up the woven arch and left the construction site with his people. The Kengdi Masters then finished the rest of the bridge with his local carpenters.

However, even for the well-trained and experienced Xiajian Masters, working on the swing-frame-scaffolding could be extremely dangerous. In fact, Master Zhang Changzhi witnessed a fatal accident in his youth, when another carpenter fell off the scaffolding to his death.[15]

4.5.3.2 First arch system – the three-sided arch

4.5.3.2.1 CONSTRUCTION USING THE PLATFORM-SCAFFOLDING

Now let's first go back to our modern constructions with the platform-scaffolding.

The erection of an MZ bridge requires the C-Beams[1] to be put onto the platform first. To this aim, people lay thin logs side by side to connect the riverbank and the platform. Together, they serve as a log bridge for carpenters to reach the platform in mid-river and as a rail track for transporting the building members (Figure 4.66, Figure 4.67). After the heavy crossbeams are moved onto the platform, the S-Beams will be positioned (Figure 4.68, Figure 4.69). In order to keep the scaffolding balanced, the S-Beams must come one by one from both sides of the bridge.

Figure 4.66 The C-Beam[1] is moved (pushed/pulled by sliding) to the platform in the middle of the river rope and sticks, on the temporary wooden rails. Shengshuitang Bridge construction site.

Figure 4.67 The C-Beam[1] is rolled to be moved onto the platform on the temporary rails. Dongtang Bridge construction site.

(a)

(b)

Figure 4.68 Positioning the S-Beams[1]. Shengshuitang Bridge construction site. (a) The wood at the top edge of the stone abutment functions as the rotating axis. A group of people working together with ropes to rotate the S-Beams[1] down to the foot of the arch. (b) Then the beam is moved slowly along the wooden rail to the middle platform using ropes. It will be positioned towards the corresponding mortise in the crossbeam.

Once the slanted beams have all been settled into position, the crossbeam is put onto their tenons (Figure 4.70).

The tips of the tenons reach inside the dovetail sockets. They will be chiselled into shape and adjusted in-situ (Figure 4.71).

(a)

(b)

Figure 4.69 Positioning the S-Beams[1]. Dongtang Bridge construction site. (a) The
S-Beam[1] is carried to the scaffolding. (b) Then its end is positioned
downward to the arch foot by rope.

Once the S-Beams[1] and the C-Beams[1] have been connected, the H-Beams[1] are put into position. They have a dovetail at both ends and are pressed from above into the sockets of the C-Beams[1]. The exterior pairs of H-Beams are installed first (Figure 4.72), then the middle one, with the rest last (Figure 4.73).

Figure 4.70 Positioning the C-Beams[1]. Dongtang Bridge construction site. (a) Before positioning the crossbeam, carpenters check the height of the tenons using a string. (b) The crossbeams are capped onto the tenons of the S-Beams and knocked into position using a tree trunk. (c) For better structural stability, the joint of the crossbeam and the S-Beams are adjusted in-situ.

Figure 4.71 The tip of the tenon is chiselled in-situ inside the C-Beam[1] to fit the form of the socket. Dongtang Bridge construction site.

Since the S-Beams[1] are usually first laid lower than their final position, and thus the distance between the C-Beams[1] is not wide enough, wooden wedges are added to adjust the beam height (Figure 4.72).

Before the first arch system is complete, the bridge master must check the height and the alignment of the arch. A rubber tube (Figure 4.74) and strings can be used for this aim. The author observed a clever method used by Master Wu Fuyong on the Shengshuitang Bridge site:

As mentioned before, the frame of the S-Beams is of the same width as the corridor at its top. The top axis of the S-Beams[1] is thus vertically aligned with the handrail of the corridor. Master Wu Fuyong uses this rule to check the position of the S-Beams[1] through the joint holes of the handrail which have been cut through the commander-pillars. (Figure 4.75)

4.5.3.2.2 CONSTRUCTION USING THE SWING-FRAME-SCAFFOLDING

Bridge construction on dangerous sites using the swing-frame-scaffolding is a special technique belonging almost exclusively to the Xiajian Masters, at least in the recent century before the modern era. The above-described Yangmeizhou Bridge is a good example of how, in 1937, the Xiajian Masters were working as a professional and systematically organized group of craftsmen to master the swing-frame-scaffolding construction. Among the bridges surveyed by the author, most of the bridges built over deep waters have been built by the Zhang family of the Xiajian Masters. Of these bridges, the most fabulous include the Jielong Bridge (Figure Intro.3), the Yangmeizhou Bridge (Figure 4.2) and the Houlong Bridge (Figure 4.76).

Figure 4.72 Positioning the exterior pair of beams H-Beams[1].
The height of the S-Beams[1] is adjusted by adding wooden wedges if necessary.
Shengshuitang Bridge construction site.

Figure 4.73 If the joints are too tight, as in this case, they will be adjusted in-situ.
Dongtang Bridge construction site.

Figure 4.74 Master Wu Dagen (left) checking the horizontal alignment of the arch with a rubber tube. Dongtang Bridge construction site.

Following the description given by Master Zhang Changzhi, who took part in the construction of the Houlong Bridge in the 1960s, and the descriptions of Master Dong Zhiji and Master Zheng Duojin, who witnessed the construction of the Yangmeizhou Bridge[16], the author was able to reconstruct and demonstrate the building process of the wooden arch bridge using swing-frame-scaffolding through the model below (Figure 4.77):

During construction, the carpenters climb up the scaffolding via the S-Beams. To facilitate the carpenters' path on the beams of the first system, bamboo pegs are nailed onto the S-Beams[1] as steps. These pegs can still be seen in some historical bridges (Figure 4.78).

4.5.3.3 Commander-pillars

Before starting the second arch system, the frames of the commander-pillars must be put into position, since the F-Beams[2] will lean on them.

The commander-pillars are the only building elements in a wooden arch bridge that connect the under-deck structure and the corridor. Therefore they

---

16 Both these masters were only children at the time. Dong Zhiji was 12 years old, while Zheng Duojin was only nine. They were given simple assistant tasks, and observed with children's curiosity. More on the story of the Yangmeizhou Bridge in Sec.6.2.4.

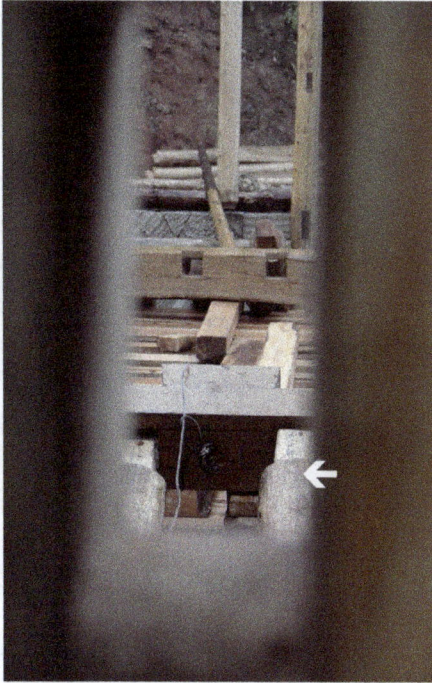

Figure 4.75 Master Wu Fuyong's method for checking the alignment of the three-sided arch:

He looks through the mortise which goes through the commander-pillar, these holes are cut in advance for the handrail beams of the corridor. The mortises in the pillars of both sides of the bridge should be in line with the middle axis ink line of the exterior S-Beam[1]. Shengshuitang Bridge construction site.

define the direction of the wooden structure and are of great significance for the stability (Figure 4.79). There are also bridges whose pillars at this position do not reach the corridor. In such cases, these "half-commander-pillars" stop at the level of the deck beams (Figure 4.80)

There are three main reasons for bridges to use half-commander-pillars. First, it might be due to a division of labour between the bridge carpenters who are responsible for building only the under-deck structure and who usually come from outside, and the common carpenters who build the corridor and who, as a rule, are local villagers. (Although this kind of division of labour does not necessarily result in the use of half-commander-pillars. The afore-mentioned Jielong Bridge has complete commander-pillars, despite the fact that the under-deck structure and the corridor were built by different groups of carpenters.) The second reason for half-commander-pillars is the lack of logs large enough to build full commander-pillars. The Dongtang Bridge is a case in point. The third reason is that the original commander-pillars were

Figure 4.76 Houlong Bridge (后垄桥), Zhouning, Fujian, 2014.

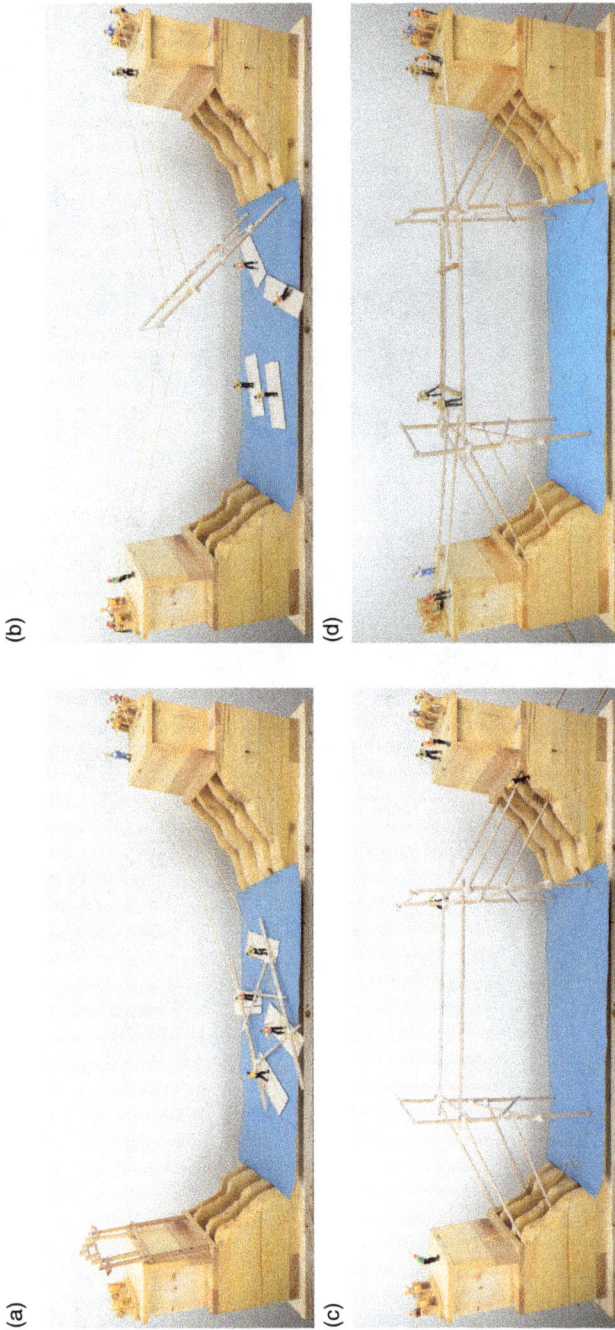

Figure 4.77 Building processes of the first system of the Yangmeizhou Bridge (construction using swing-frame-scaffolding) demonstrated by a wooden model (1:20 scale). (a) The scaffolding post-frame is brought onto the water on bamboo rafts. (b) The posts have stones tied to their feet. The rafts are then overturned to sink the post feet into the river bed and then they are pulled into position by rope. (c) The entire frame is fastened by ropes or struts which are fixed onto the shore. (d) The C-Beams[1] are transported to the middle of the river, where they are laid on the scaffolding beams and await the S-Beams of the woven arch.

Source: Made by the author.

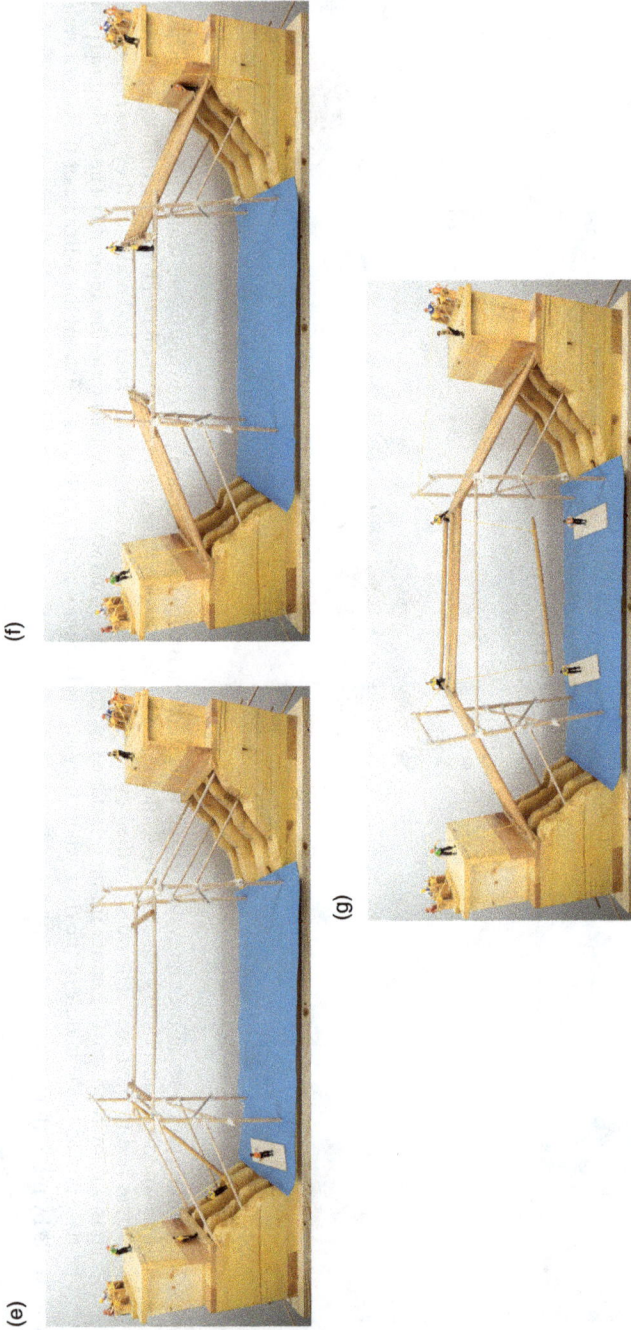

(e)

(f)

(g)

Figure 4.77 (e) The S-Beams[1] are in place, the C-Beams are raised and cap onto the tenons of the S-Beams. Once they are connected into position, they form stable frames with the S-Beams, and the scaffolding is less shaky. The C-Beams are wide enough for the carpenters to walk on. Before installing the horizontal beams, carpenters have to check several crucial parameters. They measure the height of the crossbeam down to the water level, and check if it is uniform at both ends of both crossbeams. Only after the crossbeams are kept and fixed at the same level is the required length of the horizontal beams calculated on- site and then made. (g) The H-Beams[1] are transported to the middle of the river by rafts, lifted by rope and placed into their sockets. Once the first system is complete, a rough bridge form frame spans the river.

Figure 4.77 (e) The S-Beams[1] are transported by raft on the water and lifted from there into position by rope. (f) When the

Figure 4.78 The bamboo pegs on the S-Beams[1]. Yangmeizhou Bridge, Shouning, Fujian.

Figure 4.79 The F-Beams[2] at the feet of the arch lean on the commander-pillars. Shengshuitang Bridge construction site.

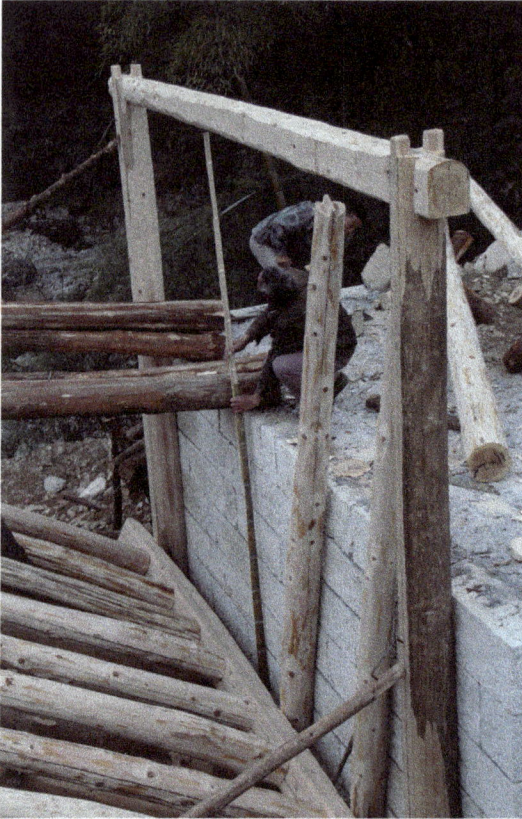

Figure 4.80 Tenons are made in the top ends of the half-commander-pillars to form connections for the LH-Beams which rest on top of them. Dongtang Bridge construction site.

removed from later reconstructions, for example in the Rulong Bridge, which will be discussed in detail in Chapter 5.

After the erection of the frames of the commander-pillars (the pillars and the tie beams), small posts are set in under the tie beams.

4.5.3.4 Second arch system – the five-sided arch

4.5.3.4.1 GENERAL PROCESS OF BUILDING THE FIVE-SIDED ARCH

Almost all bridge masters[17] erect the five-sided arch from bottom to top: that is first the F-Beams, then the L-C- and L-S-Beams, the U-C- and U-S-Beams, and finally, the H-Beams.

17 Except for Master Wu Fuyong, see Sec.4.6.1.5 below.

In the construction process of the second arch system of the woven arch bridge, there is a crucial working step: the *choudu* (抽度) as the Xiajian Masters call it, or "crossbeam-ramming" (*niutou zhuang* 牛头撞) as the Changqiao Masters call it. There follows a description of the typical construction sequence for the five-sided arch as demonstrated by the Kengdi Masters (Figure 4.81).

4.5.3.4.2 THE *CHOUDU* / CROSSBEAM-RAMMING STEP

Choudu or crossbeam-ramming is done to push together the topmost crossbeams (U-C-Beams[2]) and the weaving-beams (U-S-Beams[2]) as closely as possible, so as to weave the structure as tightly as possible, and it is the most crucial step in building the second arch system – perhaps even of the entire bridge construction. All kinds of methods are used to achieve this aim, and different carpenter families/groups have their own way of doing this.

The Xiajian Masters mainly use ropes to pull the crossbeams tight. They twist the rope with a wooden bar to tighten it (Figure 4.82). Nowadays, the carpenters use an iron roller and iron chain to achieve this aim (Figure 4.95, a). The Kengdi and Changqiao Master use a tree-trunk rammer to knock the crossbeam. Additionally, both groups use crowbars made of iron or wood as auxiliary tools in this process (Figure 4.81, j).

Master Huang Chuncai from the Changqiao Family has a special way of verifying the effect of the crossbeam-ramming. As mentioned before, wooden wedges are often inserted between the S-Beams[1] and the supporting scaffolding beam to adjust the height of the S-Beams[1]. During the crossbeam-ramming, when the woven arch is pressed tighter, the beams become bent and the structure rises up a little. The arch then stands by itself, departing away from the supporting scaffolding. In the process, the wooden wedges fall into the water. For Master Huang Chuncai, the falling of the wedges is the sign that the U-C-Beams[2] have reached their correct position.

In the Changqiao Masters' constructions, the crossbeam-ramming is so forceful that the sockets on these beams should not be cut before this step. Otherwise, there is the danger that the forceful ramming might damage the crossbeams with the many cuttings. Before the crossbeam-ramming, only the mortises are cut into the crossbeams. The sockets on both sides of the U-C-Beams[2] are only processed in-situ after this pair of beams have reached their correct position.

4.5.3.5 *The exterior LH-Beams or the edge-beams*

The Xiajian Masters call the four exterior LH-Beams "edge-beams." They can either be positioned at the same time as the H-Beams[2] or shortly thereafter. In any case, they must be positioned before the middle struts and other LH-Beams. In bridges with commander-pillars, the end tenons of the edge-beams go through the commander-pillars (Figure 4.83). In bridges with half-commander-pillars, they are put on top of the pillars (Figure 4.84).

Figure 4.81 Construction steps of the five-sided arch. Dongtang Bridge construction site. (a) If the F-Beams[2] are square, their edges are adjusted to hold the commander-pillars. (b) The L-C-Beam[2] is temporarily fixed onto the S-Beams[1] with bamboo pegs, in a position somewhat higher than its final position. (c) After the L-C-Beams[2] have been put into their temporary positions, the L-S-Beams[2] are put in. In Kengdi Masters bridges, the L-S-Beams[2] have a tenon at both ends which are stuck into the mortises in the crossbeams and foot beams. Before they are connected to the crossbeams, they are supported by thin pieces of wood at their upper ends to ensure they do not drop off. (d) When all the L-S-Beams[2] are in position, the L-C-Beam[2] is moved onto the top of the tenons of the L-S-Beams[2] and swallows them. (e) The connections between the L-S-Beams[2] and the L-C-Beam[2] are tightened using a crowbar. (f) ... and a hammer.

Figure 4.81 (g) Once the lower part of the five-sided arch is finished, the exact length for the U-S-Beams[2] is decided, calculated in-situ from the structure. (h) The planned position of the U-C-Beam[2] is marked on the H-Beam[1]. The ink mark is a checkmark indicating the final position of the crossbeam. (i) Positioning the U-S-Beams[2]. The dovetails at the lower ends are inserted into the sockets of the L-C-Beams[2]. The upper ends are temporarily supported by small pieces of wood to ensure they do not drop off. (j) Once the U-S-Beams[2] are in position, the U-C-Beam[2] is moved towards the beams and cap the tenons. Then comes the important step called "crossbeamrammings" or *choudu*, to push the crossbeam as much as possible to tighten the joints. Iron crowbars are used and a tree trunk is used as a hammer. (k) Only after the crossbeams are set in position, is the required length of the H-Beams[2] and the LH-Beams measured and decided in-situ. (l) The length of each H-Beams[2] is measured in-situ, and the beams are constructed and adjusted accordingly.

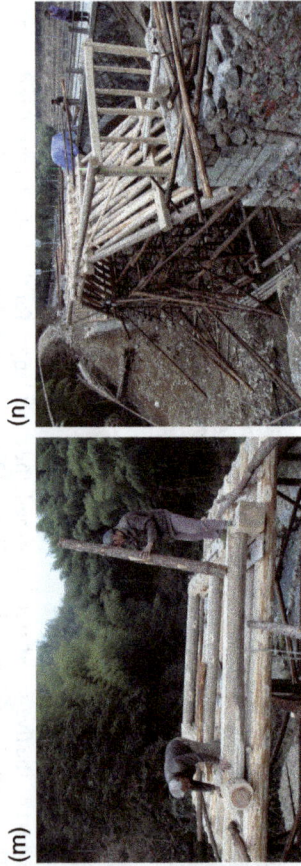

Figure 4.81 (m) Positioning the H-Beams[2]. Their dovetails are knocked into the sockets with a tree trunk. With the H-Beams[2] in place, the woven arch is now complete. (n) After the woven arch has been completed, the scaffolding needs to be partly disassembled. It must not have any further contact with the woven arch, to prevent any possible damage to the bridge structure in case of flooding in bad weather.

Figure 4.82 Xiajian Maters pulling the structure tight using rope and a wooden bar. Regensburg Bridge construction site.

Figure 4.83 The edge-beams are connected to the commander-pillars by mortise& tenon joints. Shengshuitang Bridge construction site.

Figure 4.84 An edge-beam lies on the half-commander-pillars frame. Dongtang
         Bridge construction site.

4.5.3.6 X-shaped struts

The X-shaped struts (X-struts) are literally called "scissor-struts" in Chinese.
They connect the commander-pillars and the crossbeams of the woven arch.
They help to increase the stability of the commander-pillar-frame and therefore that of the corridor as a whole. In larger bridges, two pairs of X-shaped
struts are commonly used, connecting the C-Beams[1] and L-C-Beams[2],
respectively, with the commander-pillars (Figure 4.85). In smaller bridges,
a single pair of X-shaped struts might be enough, as is the case with the
Shengshuitang and the Dongtang bridges (Figure 4.86). Large bridges could
even have three pairs of struts (Figure 6.46).

4.5.3.7 Lateral horizontal beams (LH-Beams)

The rest of the LH-Beams are put into position after the X-shaped struts
(Figure 4.87), thus closing the under-deck structure (not yet completed).
(Figure 4.88, Figure 4.89)

4.5.3.8 "Frog-leg structure"

After all the LH-Beams are in place, come the middle struts of the LH-
Beams (Figure 4.90). They are called "frog-leg struts" or "horse-leg struts"

Figure 4.85 Two pairs of X-shaped struts. Yangmeizhou Bridge, Shouning, Fujian.

(a)   (b)

Figure 4.86 The joints for the X-shaped struts are processed in-situ. Dongtang Bridge construction site. (a) The bottom end of the strut is connected to the half-commander-pillar through mortise&tenon. (b) The top end of the strut is connected to the crossbeam via a simple shallow cut, and will be fixed in place later with an iron nail.

due to their folded form. A set of frog-leg struts is commonly composed of a crosswise beam at the top (top beam), a group of vertical posts standing on the L-C-Beam[2] ("mid legs"), a group of slanted struts ("lower legs") pressing on the abutment, and a group of flatter slanted struts ("upper legs") connecting to a topmost crossbeam of the woven arch (Figure 4.91). In smaller bridges, where the HL-beams are relatively short, there is no need for

Figure 4.87 The tails of the LH-Beams are knocked into their sockets with the help of tree-trunk hammers. Dongtang Bridge construction site.

Figure 4.88 Closing the under-deck structure of the Shengshuitang Bridge.

Figure 4.89 Closing the under-deck structure of the Dongtang Bridge.

Figure 4.90 Erecting the frog-leg structure. Dongtang Bridge construction site.
After all the LH-Beams are in position, the top beam of the frog-leg structure is temporarily fixed onto the LH-Beams. The exact length of the struts will be measured and determined in-situ.

Figure 4.91 The composition of frog-leg structure

Figure 4.92 Completed under-deck structure, Shengshuitang Bridge construction site. This bridge has no frog-leg structure.

a complete frog-leg structure, and one or another group of struts could be omitted. For example in the Rulong Bridge (see Chapter 5) there is no vertical struts. In the Shengshuitang Bridge, the complete frog-leg structure is missing (Figure 4.92).

After the installation of the frog-leg-struts, the under-deck structure is now complete (Figure 4.93).

Figure 4.93 Completed under-deck structure, Dongtang Bridge construction site.

4.5.4 Erection of the corridor

The last part to be erected is the bridge corridor. The structure of the corridor is the same as the common house structure in southern China. It is a type of timber frame (*Fachwerkbau* in German) but without diagonal struts (*Strebe*). Beams (or their long tenons) go through pillars, they function as ties and are also called that. The crosswise frames are assembled on the ground, erected one after another, and bound together using longitudinal tie beams (Figure 4.94).

4.6 Constructional perspective of the bridge carpenters

4.6.1 Structural understanding of the bridge carpenters

4.6.1.1 Does the structure of the MZ bridge constitute an arch?

In this book, we have so far accepted the term "woven arch" without question, because of its arch-shaped structure, but the term "arch" needs to be checked from other aspects. The classification of the woven arch structure between a beam and an arch structure from the mechanical engineering point of view is discussed in the concluding chapter of the book, whereas here, it is the carpenters' perception that will be explored. The answer to the question "Is the so-called woven arch an arch?" is "Yes."

In the terminology used by contemporary bridge carpenters, the under-deck structure of the woven arch bridge is called the "lower arch" (*xiagong*, 下拱). In the case of a wooden arch bridge built in 1909 – the Tingxia Bridge (亭下桥) in Gutian, Fujian – the carpenters building the corridor were literally referred to as "masters of the pavilion of the arch" (拱亭师) in the ink inscription on the bridge (Ningde 2006, 62).

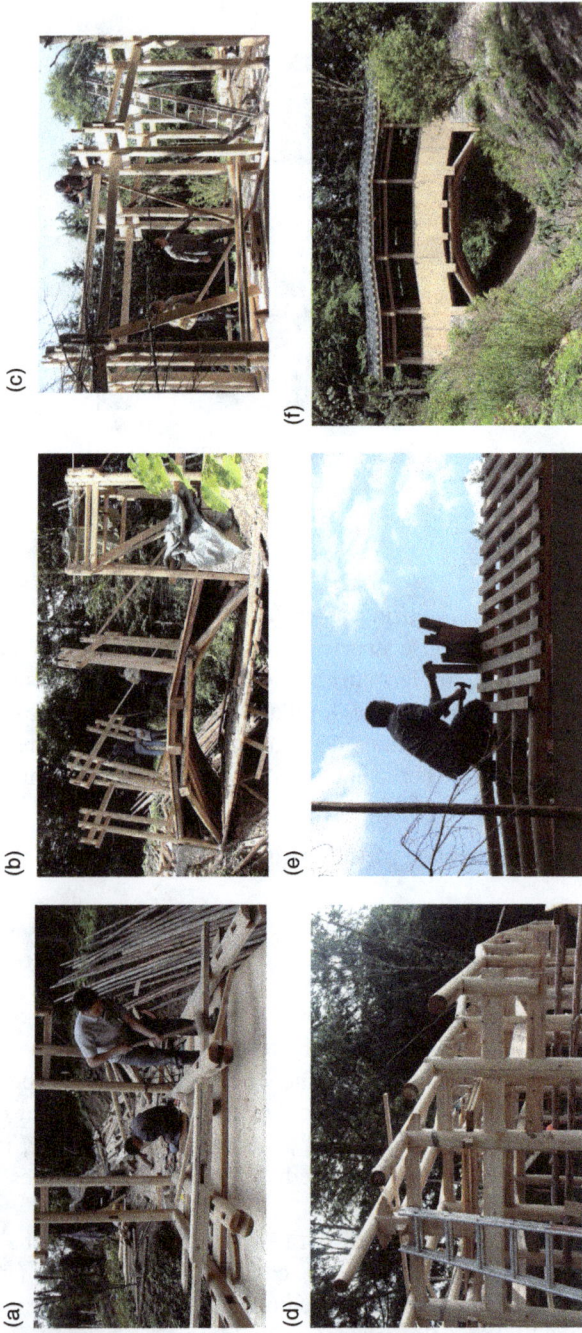

Figure 4.94 Building the bridge corridor. Regensburg Bridge construction site. (a) The cross-section frames are assembled on the ground. (b) Erecting the frames. (c) The standing crosswise frames are connected with longitudinal tie beams. (d) Putting up the purlins. (e) Nailing the rafters. (f) Once the final tile work has been completed, the corridor and therefore the bridge construction as a whole is also complete.

4.6.1.2 Relationship between the first and the second arch

In our narrative, the first system is called the main system of the woven arch, while the second the secondary system, and this is consistent with the carpenter's understanding.

In the terminology used by the Xiajian Masters, the main system is called the "the main/chief arch" (*zhenggong,* 正拱), and the secondary system is called "the secondary/deputy arch" (*fugong,* 副拱).

According to the description of the bridge carpenters, the three-sided arch is the main part to bear load; it is the main load-bearing system. The function of the five-sided arch is different. "In the five-sided arch there is no force," says Master Dong Zhiji, "there is no pressing force, but only bending force (*ao,*拗, see below)."[18] Master Zhang Changzhi says: "The force is only in the main arch. The function of the secondary arch is to keep the main arch stable, to avoid shaking and deformation."[19] Master Zheng Duoxiong tells that: "The five-sided arch can help the three-sided arch. For example if the force is 100 tonnes, the five-sided arch can share 30 tonnes of it."[20]

4.6.1.3 The choudu or "crossbeam-ramming"

The Xiajian Masters describe the function of the second arch system using a special verb: *chou* (抽), and hence call the step "crossbeam-ramming" *choudu* (抽度). When Master Zhang Changzhi talked about this, he made a movement with his arm as if he was pulling something out of the ground while bending it.

To clarify the verb *chou,* let us conjugate it according to English grammar.

Chou is sometimes translated as "pull," but it is more complex. *Chou* means to take/pull/extract/drag a slender object out of the pressure of the material surrounding it. In Chinese, for example we would say "The sword is *chou-ed* from the scabbard (unsheathed); The belt is *chou-ed* tighter around the waist (made tighter); The sprout of a plant *chou-s* out (pull itself out). And in basket weaving, a longitudinal piece of grass is *chou-ed* (pulled), while the transverse grasses are pressed towards each other, to make the woven textile tight.

So it is thus understandable that when they use the term *choudu,* the Xiajian Masters are describing the process of push the (long tenon of the) U-S-Beams[2] out of the U-C-Beams[2] while pressing/ramming the latter. The U-S-Beams[2] are therefore called *chou*-beams in their speech, and correspondingly rendered as "weaving-beams" in this chapter.

---

18 Master Dong Zhiji in an interview with the author, December 2013.
19 Master Zhang Changzhi in an interview with the author, December 2013.
20 Master Zheng Duoxiong in an interview with the author at the Dongtang Bridge construction site, December 2013.

(a)   (b)

Figure 4.95 Building method used by Master Wu Fuyong. Shengshuitang Bridge con-
struction site. (a) The first step is to weave the U-S-Beams[2] into the arch.
Here, steel cables and rollers are used to make the structure tight. (b) The
L-S-Beams[2] are assembled at the end. They have a dovetail at both ends.

This motion is sometimes referred to by other bridge carpenters as "*ao*"
(拗). This verb can be translated as "bend," because indeed the beams are bent
on their upper ends during this step of construction.

4.6.1.4 Relationship between the U- and L-S-Beams[2]

As mentioned before, the U-S-Beams[2], that is the weaving-beams are considered
the most crucial building elements in the woven arch structure. However, the
lower beams of the same arch system (L-S-Beams[2]), play a much less significant role. As Master Peng Fodang (a Xiajian Master) says regarding the
function of these beams: "On the L-S-Beams[2] and the L-C-Beams[2], there is
hardly any or very little force. If some of the L-S-Beams[2] get damaged or rot,
they can simply be removed and exchanged, without touching the other part of
the structure. The removal of these beams will not harm the entire structure."[21]

4.6.1.5 Wu Fuyong's method

Unlike all the other bridge carpenters, and disliking the process described
above, Master Wu Fuyong from Qingyuan, Zhejiang uses a rather different
construction method. When erecting the second arch system, he assembles
the structure from the top down: the weaving-beams and the corresponding
crossbeams are installed first, followed by the H-Beams[2]. The lower part of
the second system are put in only at the end (Figure 4.95).

Despite the uniqueness of his construction procedure, Wu Fuyong has very
reasonable grounds for his choice. When the weaving-beams are brought into
their correct position, they are not only self-standing thanks to the interlocking
mechanics, but also stiffen the entire structure: with the weaving-beams in

---

21 Master Peng Fodang in an interview with the author, December 2013.

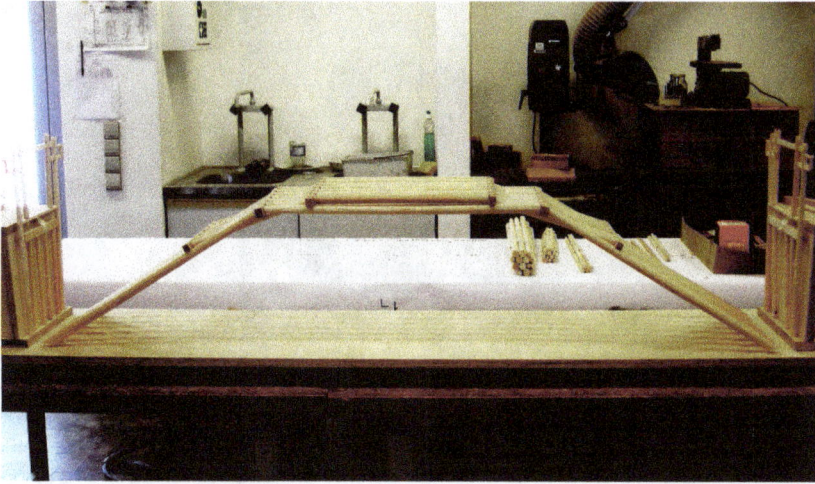

Figure 4.96 Model of the woven arch. When the U-S-Beams[2] are "woven" into the structure, the arch model becomes so stable that it rises and gets rid of the support of the bearings.

place, the woven arch can stand alone on its own feet without the help of scaffolding (Figure 4.96).

For this building method, Master Wu Fuyong makes dovetails at both ends of the L-S-Beams[2] (Figure 4.95, b). This visible technical feature is not found in any historical bridges, so it is safe to conclude that this method is an innovation developed by Master Wu Fuyong himself.

4.6.1.6 Beam shoulder and structural stability

A longitudinal beam has two "shoulders" (Figure 4.97). They are the "rest" (ring-formed) part of the wood at the root of a tenon or tail. In Chinese, the tenon is referred to literally as a "joint-head" (*suntou*, 榫头), while the rest of the cutting at the end of the head is referred to as the "shoulder." As the shoulder of a human body, which is one of the most powerful parts for carrying or lifting a load, the beam shoulders are a robust part of the structure of a woven arch bridge: by tightly holding the crossbeams, they serve to transmit the force, and to prevent the structure from rotating/deforming. However, most previous researchers considered these shoulders simply a "leftover" part of the wood joint, and therefore generally overlooked their structural function.

When discussing the lateral stability, the common understanding of most scholars who have received academic training in the field of architectural engineering attributes the stability of the woven arch to the X-shaped struts, according to the basic principle that the triangular form provides stability of

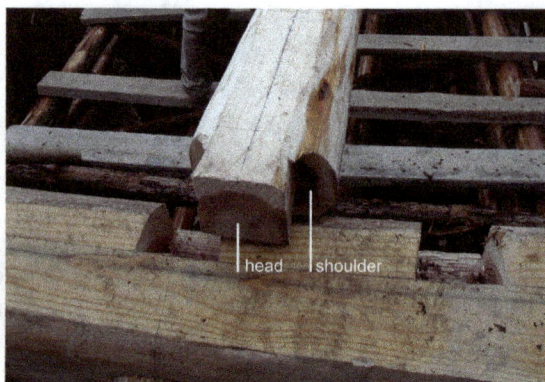

Figure 4.97 Beam shoulder of an L-Beam[1]. Dongtang Bridge construction site.

the structure (see Sec.3.2.2.1). This is not quite the same in the case of the MZ bridges, where the lateral stability of the woven arch relies mainly on the many beam shoulders, especially those of the huge beams of the first system. The X-struts, in this case, since they are connected to the commander-pillars, serve only as a source of lateral stability for these pillar frames and the corresponding part of the corridor. "The only purpose of the X-struts is to prevent the bridge corridor from shaking when people run across the bridge,"[22] says Master Huang Chuncai. Going into more detail, he adds: "A bridge has so many beam shoulders, and they work together to keep the bridge stiff. Not so with the X-struts, of which there are only a few: as soon as their joints get damaged or rot, they no longer function."[23] To achieve lateral stability, Master Wu Dagen uses the largest timber for the H-Beams[1], because shoulders in this position have the most crucial function of preventing the bridge from deforming.

Thanks to the shoulders, the lateral stability of the woven arch is so great, that the wooden structure of the MZ bridges

> does not fear flooding. The real threat of flooding is damage to the piers. When the (two spans of the entire six spans of the) Wan'an Bridge (Figure 4.3) collapsed in 1952 due to flooding, it was only when the piers were destroyed that the wooden structure collapsed.

says Huang Chuncai. The same thing happened to the Xuezhai Bridge (Figure 6.17), which was destroyed by flooding in the summer of 2016[24]: the stone abutment collapsed before the wooden bridge was torn to pieces.

---

22 Master Huang Chuncai in an interview with the author in December 2013 in Pingnan, Fujian Province.
23 Ibid.
24 The collapse of the Xuezhai Bridge was filmed by local people, and the video was spread through social media.

4.6.1.7 The function of the C-Beams[1]

Another difference in understanding between the scientifically-trained mind and that of the carpenter is regarding the structural function of the crossbeams. Whereas modern researchers emphasize their function for load-bearing and force transmission, to the carpenters, they are simply the jointing elements of the structure.

"The crossbeams function as the glue, they hold the H- and S-Beams together,"[25] says Master Wu Dagen, "the only thing to worry about is that the crossbeams may split, so we bind them with iron bands." The same understanding was also expressed to the author by a Xiajian Master, Peng Fodang.[26]

4.6.2 Selection of joint type

When engineers discuss the different wooden joints, their focus tends to be on the mechanical features. For example a dovetail can bear some tension, while a straight tenon cannot.

The traditional carpenters think in another way. When craftsmen are deciding what type of wooden joint to use, their first consideration is related to the ease of material processing and the execution of the construction. Before the age of electrical tools, all the holes for wood joints were cut with chisels. Carving out a (vertical) mortise is easier than the (inclined) dovetail sockets and therefore preferred by carpenters. And for the woven arch construction, a more deciding consideration is how the joints are then installed. Again taking a mortise/tenon and a dovetail (as they are at the ends of the bridge beams) as an example, the main difference is in the way they work with the joints: when a tenon is connected to a mortise, it moves along the axis of the beam; when a dovetail is put into its socket, it is pressed from above down to the cut. Therefore, the choice of joint type is indeed the choice of the construction method.

Different terms to distinguish between "mortise/tenon" and "dovetail" exist in the common language of Chinese craftsmen as well as in historical technical literature.[27] And although the terms vary from place to place, a basic pair of separate expressions for these two joint forms exist in many wood-working areas in China, and is used by many MZ carpenters. However, the Xiajian Masters do not have two separate words for these two types of joints. Instead, they refer the mortise/tenon and dovetail as "the (inwards) inserted" and "the downwards pressed," respectively.

25 Master Wu Dagen in an interview with the author at the Dongtang Bridge construction site. December 2013.
26 Master Peng Fodang in an interview with the author in Zhouning, December 2013.
27 The different terms for "mortise/tenon" and "dovetail" exist in the *Yingzao Fashi*, the official building book of the Song Dynasty (early twelfth century).

From a purely mechanical point of view, for each of the wooden joints of the MZ bridges, there is at least an alternative joint form; they could either be dovetail or mortise/tenon. Since the entire structure is subject to the bending and compressive forces, the change in joint form does not result in any noticeable structural difference.

However, when it comes to the actual construction of the MZ bridges, in most positions, the form of a wood joint of a specific position suits the conventional construction methods so well, that it would be unreasonable to change to the other form. If we take the H-Beams of the two systems for example: when these horizontal beams are put into position, their dovetails are pressed from above downwards into the sockets. If mortise&tenon were to be used instead, the crossbeams would have to be moved outwards to make space for the tenons, and then moved back inwards and the joints tightened – this is naturally not accepted by carpenters.

Similarly unchanged are the joints at the top ends of the slanted beams, which always have tenons, so that the crossbeams stay stable after being simply capped onto them. If dovetails were to be used instead, the entire building process would need to be changed.

Aside from these unchangeable joint forms in the established construction process, there are a couple of positions, where the joint form might have a theoretical or reasonable alternative. Positions of this kind (Figure 4.98) are discussed further on.

However, even in these places, different joint forms are seen only between carpenter groups/families. Within a specific group/family, a chosen joint form is seldom changed. Even the smallest change of joint forms has some effect on the building methods, and since the building methods are taught and handed down through a cooperative practice from generation to generation among the entire group/family, unless there is a convincing enough reason to change,[28] the joint forms became rather set over time. In fact, they could be seen as family features and used in the technical pedigree study of the MZ bridges (more on this in Chapter 6).

4.6.2.1 Joints on the F-Beams[2]

At the lower ends of the L-S-Beams[2], where they connect the F-Beams[2], both mortise/tenon and dovetail is possible. The Xiajian Masters use dovetail, while the Kengdi (and Changqiao) Masters use mortise/tenon. This technical feature is the same in the bridges erected by both sides and has not changed in over two centuries (cf. Figure 4.99 and Figure 4.100), from their earliest known works to their most recent projects, and can be seen as reliable factors in a technical pedigree study.

---

28 Such changes did happen in the history, as important technique progress. See Sec.6.3.2.

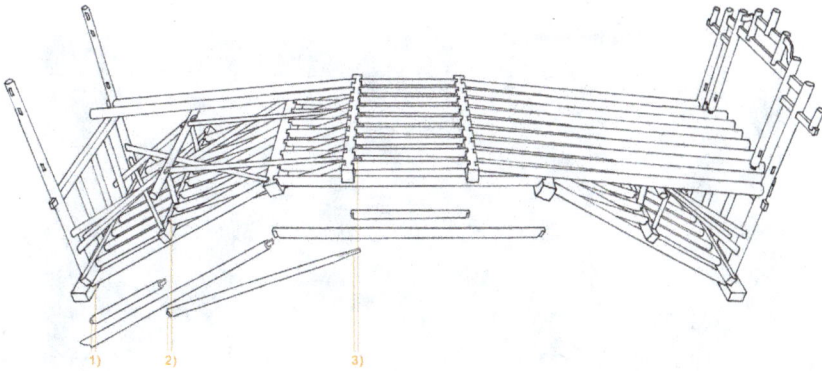

Figure 4.98 Relevant joints in this section.

Figure 4.99 Dovetail at the arch feet. From the earliest known work to the most recent bridge erected by the Xiajian Masters. (a) Xian'gong Bridge, Shouning, Fujian. Built in 1767, it is the first known bridgework of the Xiajian Master family. (b) Regensburg Bridge. Built in 2015 by Xiajian Master Zhang Changzhi.

According to Master Wu Dagen (a Kengdi Master), they chose mortise/tenon for this position because of the ease of processing.[29] When asked why other carpenters use dovetails here, Wu Dagen pondered and replied that

29 Master Wu Dagen in an interview with the author at the Dongtang Bridge construction site in December 2013.

(a)

(b)

Figure 4.100 Mortise/tenon at the arch feet. From the earliest known work to the most recent bridgework of the Kengdi Masters. (a) Xiaodong Bridge, Shouning, Fujian. Built in 1801, the first known bridgework of the Kengdi Masters. The bridge was repaired in 1939. Mortises from the original structure are left in the reconstructed L-S-Beams[2]. (b) The F-Beam[2] with mortises. Dongtang Bridge construction site.

in the case of larger bridges, where the L-S-Beams[2] are much longer and heavier, it is easier to put the dovetail into the socket from above than to install the tenon from the side.[30] This analysis is quite convincing in that, as mentioned in the design section (Sec.4.3.2.3 [Rule III]), because of the different design properties used by these two families, the L-S-Beams[2] in the bridgeworks of the Xiajian Master are visibly longer than those of the Kengdi Masters.[31]

The Changqiao Masters who also use mortise/tenon here, also have their specific reasons. During the crossbeam-ramming step, as mentioned above, the Changqiao Master will ram the topmost crossbeams (U-C-Beams[2]) very forcefully with big tree trunks to tighten the structure. As Master Huang Chuncai explains, if dovetails were applied at the arch feet, there is the danger that the tails would jump out of the sockets during this step.[32]

The Xiajian Masters have no fears for these dovetails to jump out, because they use ropes to pull the crossbeams into position (Figure 4.82). Since they don't use the forceful rammer trunks, they don't have the same worries as the Changqiao Masters have in this case.

4.6.2.2 *Lower end (foot) of the weaving-beams (U-S-Beams[2])*

At first glance, one would think that there could be alternatives of joint form at the foot of the weaving-beams (U-S-Beams[2]), as in the case of the L-S-Beams[2] discussed above. However, dovetails are used here almost exclusively.[33]

Master Wu Dagen explains that the reason for not using tenons in this position has to do with the ease of processing and the safety during construction.[34] As mentioned before, with a dovetail, the end of the beam can easily be inserted into the socket from above; whereas, with a tenon, the beam must be inserted from the side of the crossbeam. Before it is inserted, it has to hang in the air. The hanging of the weaving-beam could be dangerous, as the beam could fall, taking the people at its side with it and/or damage the entire structure. Besides, the tenon at the top end of this beam needs to be processed and adjusted in-situ (during the crossbeam-ramming process), and during this on-site processing, it would be all too easy for a tenon at the bottom to be pulled out of its mortise and again, fall and cause damage. Using a dovetail avoids the problems described above.

---

30 Ibid.
31 Note: When Master Wu Dagen made this analysis, he was not aware of this fact (that – as a rule – the length of this beam is different in works by these two families). This difference in design between the two families was only noticed by the author a year later, in 2014, when he had the opportunity to interview members of the Xiajian Masters.
32 Master Huang Chuncai in an interview with the author, in December 2013.
33 Except the bridges of Master Wu Fuyong, as discussed above.
34 Master Wu Dagen in an interview with the author at the Dongtang Bridge construction site, in December 2013.

4.6.2.3 Beam head (upper end) of the U-S-Beams[2]

All the family-trained bridge carpenters (esp. the Xianjian, Kengdi, and Changqiao Masters) emphasize the long tenons at the top of the weaving-beams (U-S-Beams[2]), that they must pierce through the crossbeam capping them (U-C-Beams[2]). During the crossbeam-ramming/*choudu* step, when the crossbeam is pushed as far as possible onto the tenons, these tenons must project out of the crossbeam as far as possible. For this purpose, the Kengdi Masters chop off the shoulders of these tenons and make a flat surface at the end of the tenons (Figure 4.44). When carpenters from other groups or families (e.g. the Xiajian Masters) keep the shoulders of the tenons, they adjust the tenon in-situ for the pushing.

These long (piercing) tenons are a vital symbol of the family-tradition bridge carpenters in terms of technique pedigree, which is a topic of Chapter 6.

4.7 A short summary of the mechanical features of the MZ bridges

(1) The woven arch of the MZ bridges has essentially two structural systems: a primary main arch and a secondary arch system. The secondary arch is used to reinforce the main arch.

(2) The longitudinal beams are connected to the crossbeams by woodworking joints, namely mortise&tenon and dovetail. The joints themselves are somewhat rigid. The shoulders of the joints, esp. of the H-Beams[1] contribute greatly to the rigidity and consequently to the stability of the structure.

(3) The stability of the woven arch is ensured mainly by the longitudinal beams (arranged in a trapezoidal form) and the shoulders of the beams. The X-shaped struts play only a small role in the stability of the woven arch, but play a more significant role in the stability of the commander-pillars-frame that connect with the corridor.

(4) The entire woven arch transfers the load mainly in form of pressure stress, while great bending stress occurs in the longitudinal beams.

(5) In the *choudu*/crossbeam-ramming step, the longitudinal beams (esp. U-S-Beams[1] and S-Beams[1]) are knocked so hard that they bend and tighten the two systems. This reduces the thrust of the arch, so it works as a form of prestressing force. The prestressing force will gradually be released over time after construction completed, due to the shrinkage of the timber and the gradual deformation of the structure.

(6) When the two systems become loose, this leads to them drawing apart on one side, and compressing each other on the other side, which causes an uneven distribution of force. All historical bridges have this deformation to a greater or lesser extent.

References

Liu, Yan

刘妍. "栋梁之材"与人类学视角的中国建筑结构史. 建筑学报, 2016(01): 48–53.

Mooslechner, Walter. 1997. *Winterholz*. Verlag Anton Pustet.

Ningde

宁德市文化与出版局. 宁德市虹梁式木构廊桥屋桥考古调查与研究. 科学出版社. 2006.

Zhou, Miao and Hu Shi, Wang Jixin

周淼; 胡石; 王际昕. 泰顺文兴桥木作技术研究. 文物. 2016 (05). 70–84.

5   The Rulong Bridge

A detective story

Now that we have explored the bridge-building technique and made it visible, we will briefly examine a rather old example.

5.1 The dragon-like bridge

Rulong Bridge (如龙桥), literally "dragon-like bridge," is the oldest and the only Ming Dynasty woven arch bridge still in existence today. It is located at the downstream side in Yueshan village (月山村 literally, moon-hill village), a small village at the southern boundary of Qingyuan County, Zhejiang Province. The village is named after a crescent-shaped hill. Along the Ju River, which runs past the village, there used to be six bridges within a radius of about two kilometres and four of them are still standing today, and in a quite good condition. Rulong Bridge stands at the tip of the crescent moon, a location regarded as the "dragon head" according to the precepts of *feng shui* (Figure 5.1, Figure 5.2).

With a drum tower at the northern end, two pavilions decorated with rich *dougong*[1] in the middle and at the southern end, as its name indicates, the design of bridge expresses the metaphor of a dragon (Figure 5.3). The tower plays the role of the raised head; the pavilions represent the crooked back and an upstanding tail.

On the middle top roof beam (*dong*) of the central pavilion, an inscription gives the construction's completion date: the bridge was finished in the fifth year (1625) of the Tianqi reign of the Ming dynasty (Figure 5.4). The local genealogy book has the same information regarding the time of construction.

The clear span of the Rulong Bridge is ca. 20 m, which is practically the smallest span necessary for a typical wooden arch bridge (Figure 5.5). The

---

1 *Dougong* is the name given to the bracket-block cluster on the top of columns in Chinese architecture supporting the roof. In the early dynasties (esp. the Tang and Song Dynasties), it was only allowed to be applied in high-ranking official and religious buildings. Its richness indicates the importance of a building. In the late (Ming and Qing) dynasties, *dougong* was used more commonly in vernacular architecture.

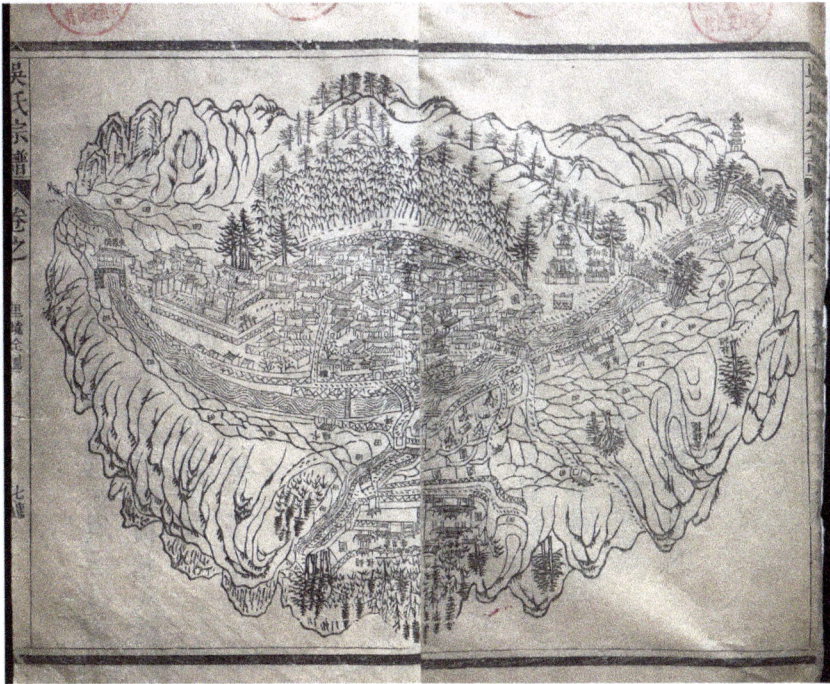

Figure 5.1 Drawing of the landscape around the Ju River.

Source: the genealogy book of the local Wu family, 1911. Preserved by a family member, Wu Desheng (吴德生).

woven arch structure is especially suitable for larger spans. Bridges smaller than 20 m can also be built as cantilever bridges (Figur 7.21) or in an easier strut bridge form, which can be considered a single three-sided arch, that is including only the first system of a woven arch structure (Figure 5.6)

Because of its small scale and early construction date, some construction characteristics of the Rulong Bridge are different from the typical form of the wooden arch bridges as described in the Introduction and the previous chapter.

The most distinct difference is the relationship between the deck beams and the woven arch. In a typical MZ Bridge (Figure Intro.4), the deck beams comprise three sections: the lateral horizontal beams of the two sides (LH-Beams), and the top horizontal beams of the woven arch (H-Beams[2]). The LH-Beams reach from the abutments to the topmost crossbeams of the woven arch (U-C-Beams[2]), and are connected to them by means of dovetails. In other words, in the middle section of the span, the H-Beams[2] play the role of the deck beams, thus, seen from the middle span, there are two layers of horizontal beams in total, namely the H-Beams[1] and [2].

Figure 5.2 Landscape around Rulong Bridge, Yueshan Village, Qingyuan County, Zhejiang Province.

Figure 5.3 Rulong Bridge. From the east.

Figure 5.4 The construction date for the bridge is inscribed into the top beam of the corridor. Rulong Bridge.

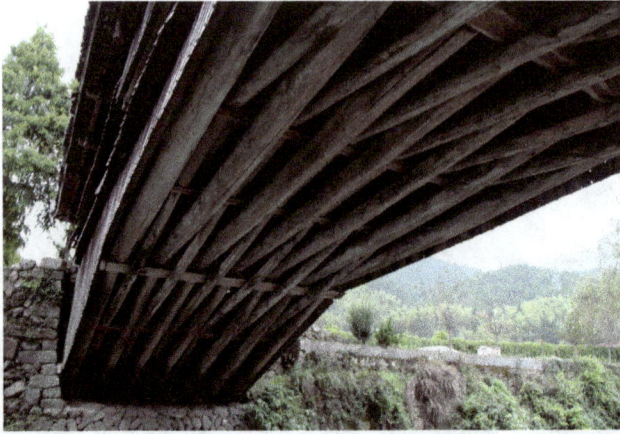

Figure 5.5 The under-deck structure of the Rulong Bridge. Seen from the northern side, looking towards the south.

Figure 5.6 A typical strut bridge with one three-sided arch. Shengxian Bridge (升仙桥), Shouning, Fujian.

In the case of the Rulong Bridge, since the span is narrow enough, the deck beams are not cut in the middle span but are composed of continuous 20 m-long logs, going from one side of the river to the other. As seen in Figure 5.7, the entire layer of deck beams rests on top of the woven arch. Seen from the middle span, there are three layers of horizontal beams: two of the woven arch and an additional top layer of the deck beam.

Other notable construction features include:

Figure 5.7 Model of the under-deck structure of the Rulong Bridge.
Source: Model and photo by Yu Yannan.

- The parallel beams of the arch. The arch is not shaped in a trapezoidal form, but keeps the same width at the top and the bottom of the arch (cf. Sec.4.3.3). This is a less stable form in terms of lateral stability, and indicates its less matured bridge-building technique.

- The proportions of the woven arch: The slope of slanted beams of the arch is rather gentle (only ca. 4/10 *fen-shui,* cf. Sec.4.3.2.2 [Rule II]), and the horizontal beams are distinctly longer than the slanted beams.

- The commander-pillars. Instead of using the usual type of commander-pillars, which reach from the arch feet to the corridor purlin, the Rulong Bridge applies half-commander-pillars that are cut at the height of the deck beams.

- The frog-leg struts. In a typical structure, the frog-leg struts are composed of the upper, middle, lower struts ("legs"), and the top beam. In the Rulong Bridge, only upper and lower legs are applied – the middle struts are omitted.

- Finally, there are nine groups of longitudinal beams in the first system (nine H-Beams[1] and nine S-Beams[1] on each side), and eight groups of those in the second system (in every section of the L-S-Beams[2], U-S-Beams[2] and H-Beams[2], there are eight beams). The H-Beams[2] are aligned with the S-Beams[2], while in between of the H-Beams[1] (Figure 5.10, c). This form is called "matching form," is a structural feature of earlier bridges and will be explored in more detail in the next chapter.

In the 1970s, during the construction of a local irrigation system, a water pipe was led through the under space of the bridge structure (through the triangular space between the deck beam and the woven arch of the southern side). The leaking water corroded the beams at the foot of the southern arch. To stop the structure from being further eroded, the villagers took away the water pipe, removed the rotten foot beams, cut off the damaged lower ends of the slanted beams, and simply laid the shortened S-Beams[1] on the stone base. As a result, the woven arch sunk considerably, and became separated

Figure 5.8 Gap between the woven arch and the deck beams, Rulong Bridge.

from the deck beams it supported. So that the deck beams would again have support, some wooden blocks were placed in the gap between the crossbeams and the deck beams (Figure 5.8).

The gap between the woven arch and the deck beams is so large, that there is enough space for an adult to crawl through it. Thanks to this unfortunate situation, the Rulong Bridge is the sole example of the MZ bridge that permits us to observe the top part of the woven arch without removing the deck boards. In 2012, this feature, together with the bridge's historical value, led the author to select the Rulong Bridge to carry out an archaeological case study (*Bauforschung*), based on a group of detailed hand-measured drawings.

5.2 The measured drawings

The measured drawing was made in July and August 2012, by the author and a colleague, Yu Yannan, then a MSc student at the Neubrandenburg University of Applied Sciences. The object of the study is the under-deck structure of the Rulong Bridge. The measured drawings, done in-situ, took us 26 days, as each step of the entire process was done by hand. Each of us took charge of half of the span, producing a set of drawings originally in 1:20 scale. The author took the southern part (left-hand side of the combined drawings) while his colleague took the northern part (right-hand side). After the fieldwork was completed, the two halves of the drawings were combined (Figure 5.9) and further processed (Figure 5.10) with a visualization program (Photoshop).

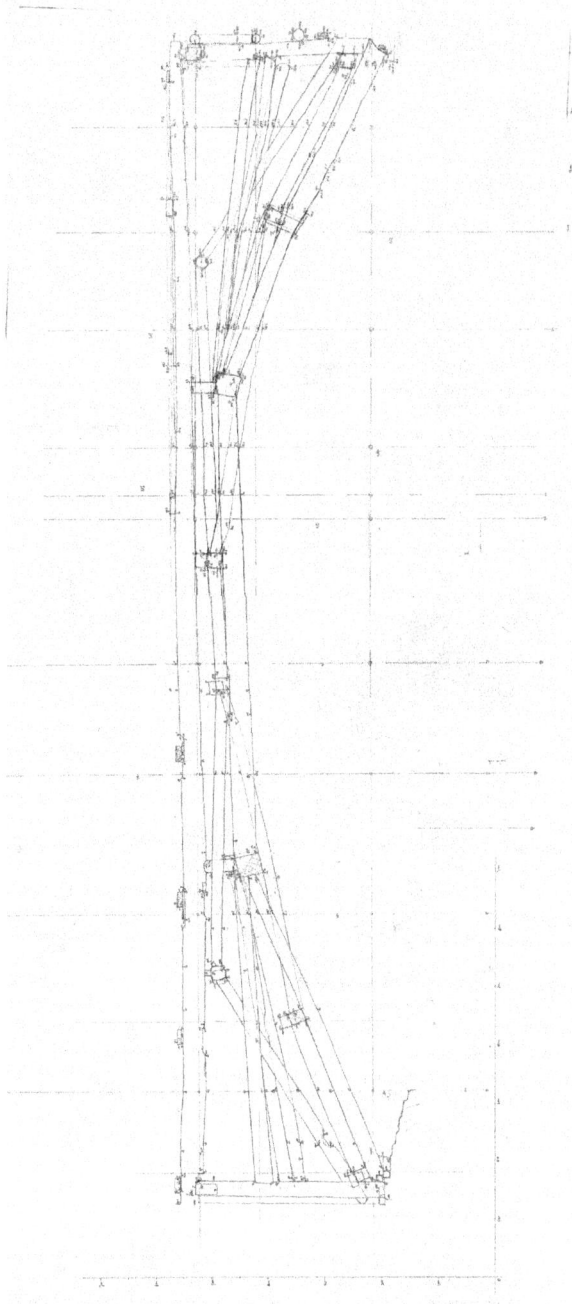

Figure 5.9 Plans of the Rulong Bridge, original drawing. (a) Elevation seen from the east (protective boards removed).
Source: Combined drawings of the author and Yu Yannan.

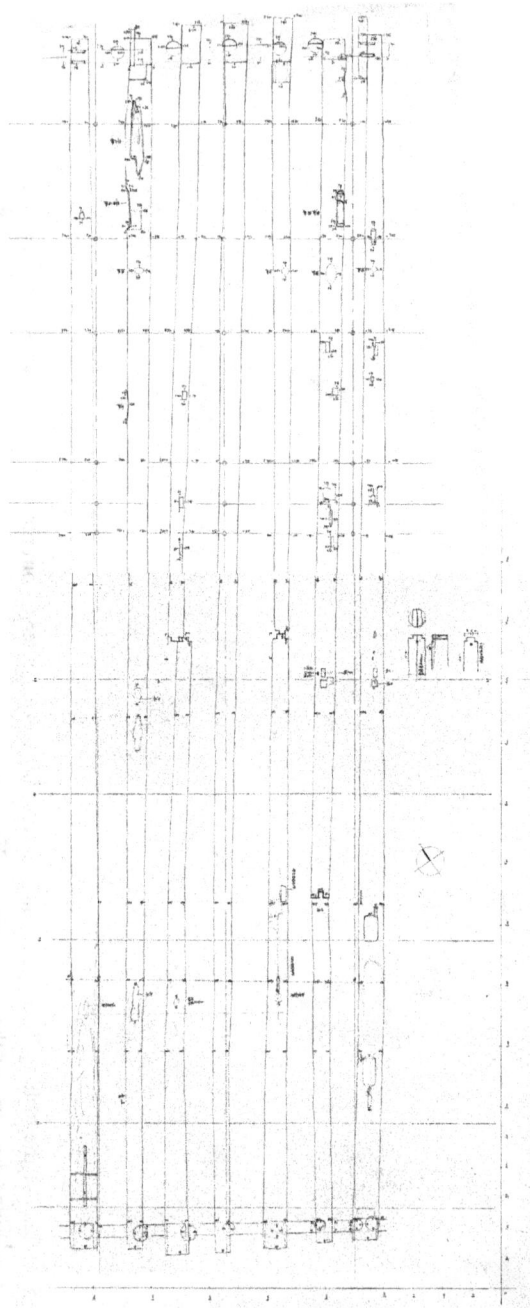

Figure 5.9 (b) Deck beams viewed from underneath.

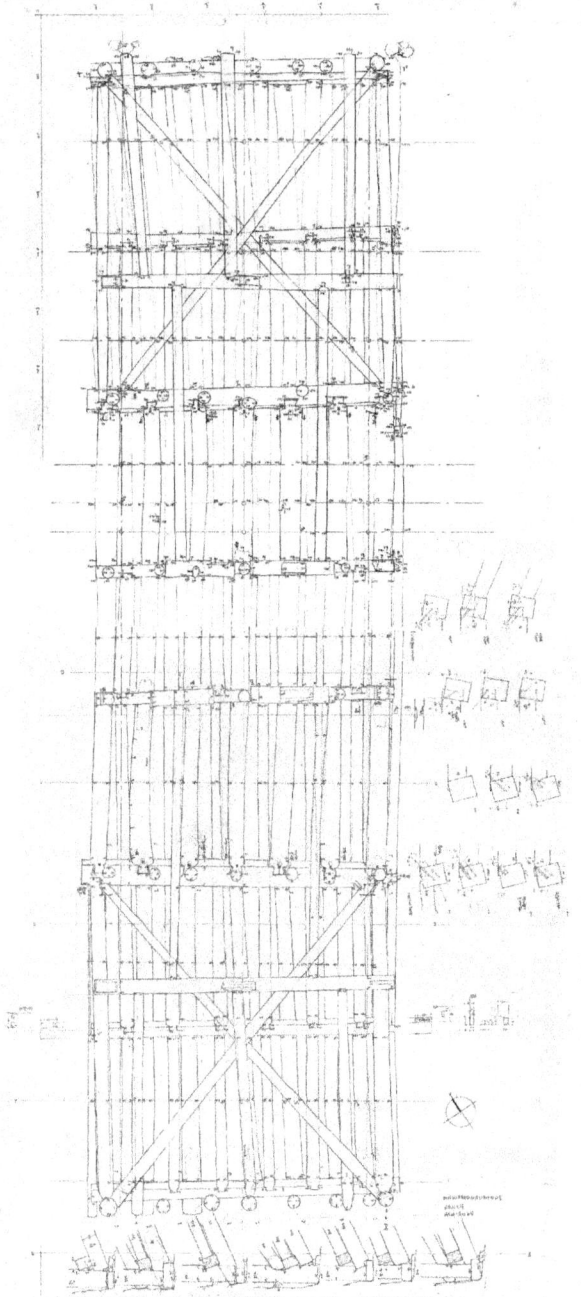

Figure 5.9 (c) Plan of the structure of the under-deck beams.

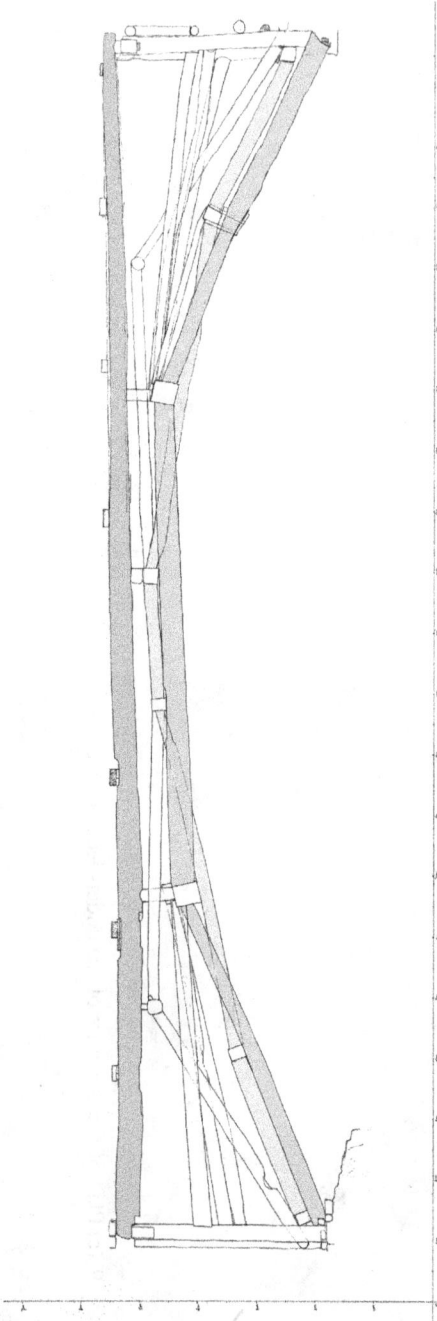

Figure 5.10 Reworked plans of the Rulong Bridge. Longitudinal beams (deck beams, horizontal beams and slanted beams of the woven arch) are shown in grey. (a) Elevation seen from the east (protective boards removed).

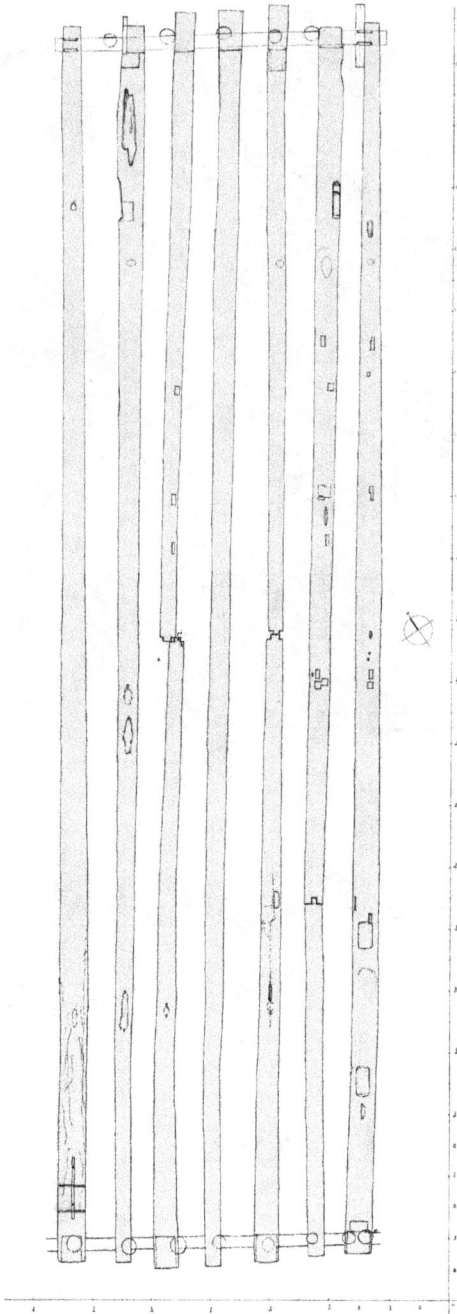

Figure 5.10 (b) Deck beams viewed from underneath.

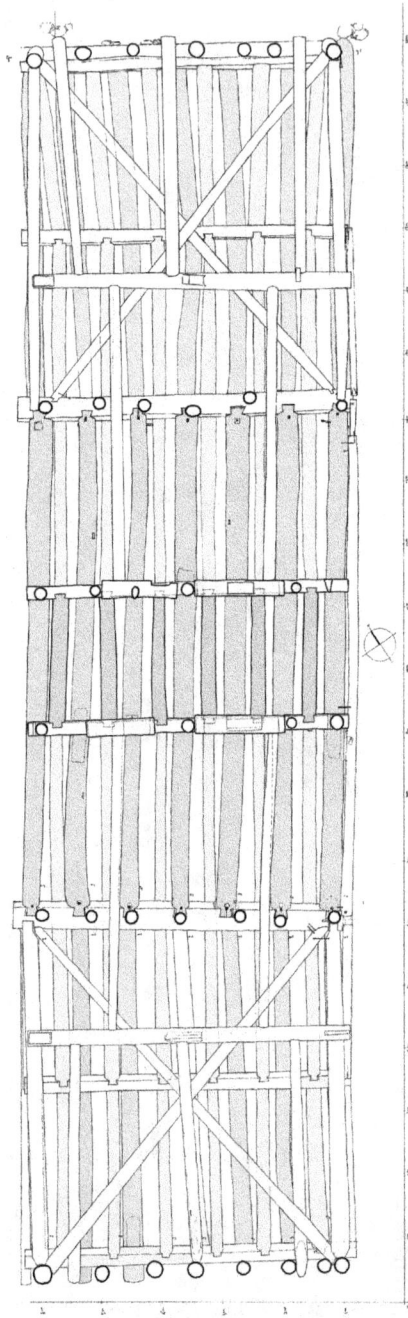

Figure 5.10 (c) Plan of the structure under-deck beams.

5.3 Solving the jigsaw puzzle

5.3.1 Initial clues

As soon as we crawled into the under-deck structure on the first day of field-work, we immediately noticed some chopping marks on the surface of the deck beams (Figure 5.11). Since the deck beams are built of thinner or thicker logs, the chopping marks are only found on the thicker ones, and are located at the same positions along the beams, not just on one side, but on both sides of the bridge (Figure 5.12). Some marks are at the bottom of the beams, while others are rotated at an angle.

These chopping marks are evidence that the structure has undergone some changes in the past. At some point, in a previous structure, there must have been a kind of cross-member perpendicular to these deck beams to support them. Since the upper surfaces of the deck beams are on the same level to support the deck boards, their undersides are higher or lower because of the difference in their diameter. As a result, part of the diameters of some of the thicker beams had to be chopped off, so as to enable them to sit on the same cross-member. At some point in a later reconstruction, when these cross-members were removed, and after the deck beams were removed and put back into position, the marks were unintentionally rotated since they no longer served any function.

The position of these marks in relation to the missing beam is just above the L-C-Beams[2]. This is where, in a typical wooden arch bridge, the top beam of the frog-leg struts lies. To be more specific: in a typical wooden arch bridge, a group of vertical posts – the middle struts of the so-called "frog-leg struts"– stand on the L-C-Beams[2] and support a crosswise (to the bridge) beam laid on them, which serves as a support for the deck beams at their mid-span. This top beam would usually be exactly where the beam is missing here. (cf. Figure Intro.4 and Figure 4.91). Connecting to the top beam, there are as a rule also the upper legs, which are stuck onto the topmost crossbeams of the woven arch (U-C-Beams[2]), and the lower legs, which are stuck onto the abutment at the arch feet.

In the current version of the Rulong Bridge structure, these vertical posts are missing, and only the upper and lower struts in the frog-leg strut group support the top beam (Figure 5.13). The position of the top beam is also ca. 1 metre away from the chopping marks.

With this basic information, it is easy to speculate that the missing pair of crossbeams were originally the top beams of the frog-leg struts, and that they were moved from their original position (chopping marks) to their current position during a repair, while the middle struts were removed. On the drawing, when we adjusted these top beams to the positions of the chopping marks, we got a Rulong Bridge with the most common frog-leg struts (Figure 5.14).

However, this version of the restoration was soon refuted by a surprising discovery we made on a deck beam (Figure 5.15, in the plan shown

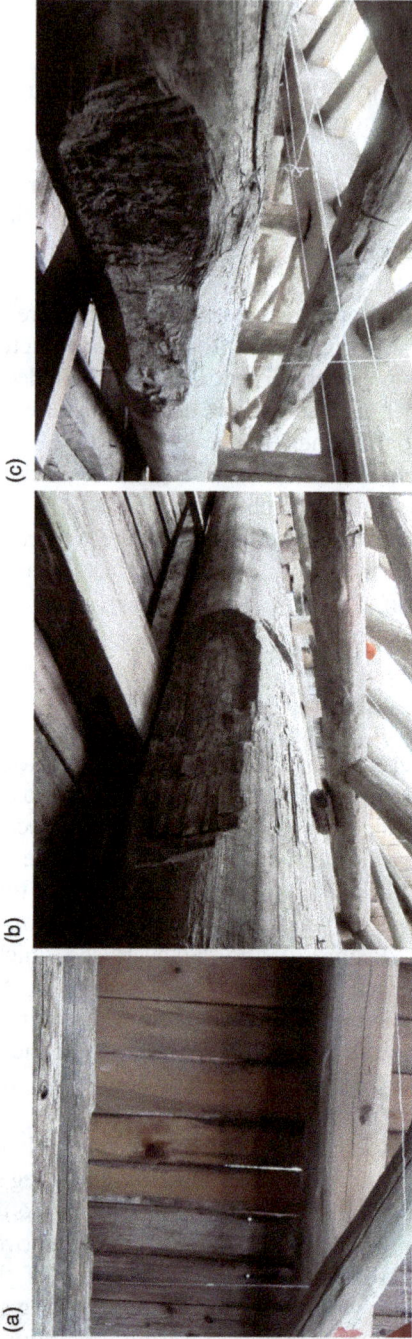

Figure 5.11 Chopping marks on the deck beams, Rulong Bridge.

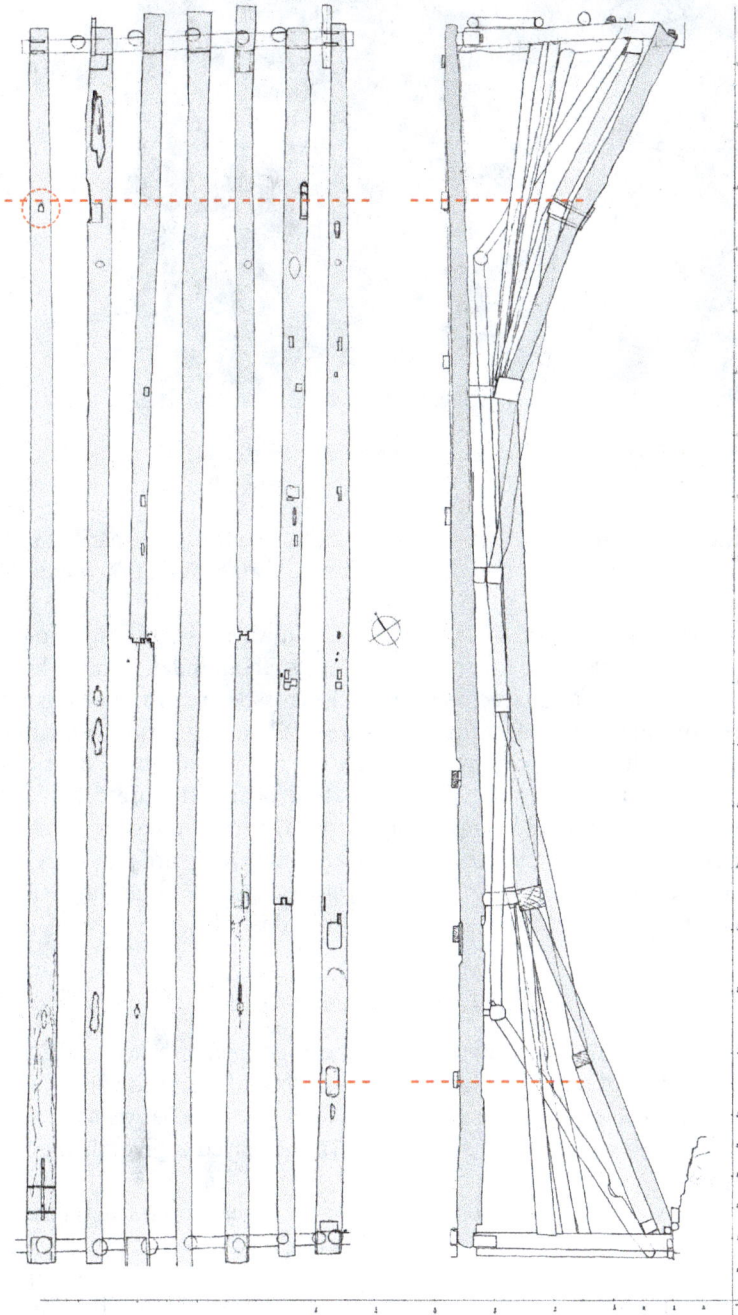

Figure 5.12 Positions of the chopping marks on the deck beams, Rulong Bridge.

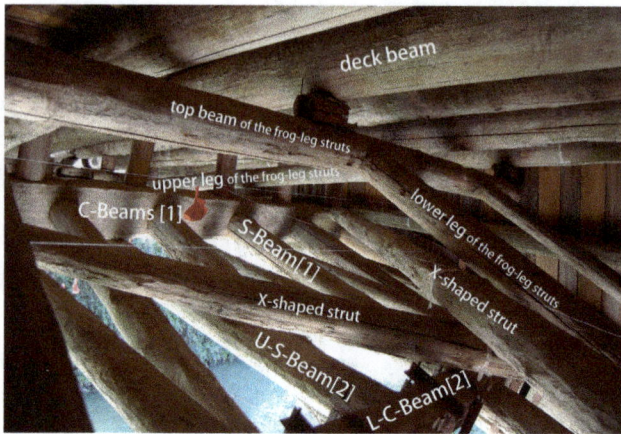

Figure 5.13 "Frog-leg struts" of the Rulong Bridge. Northern side.

in Figure 5.12, marked by a red circle). At this position, the mark is not an axe-mark like the others, but rather apparently an indentation made by a hard edge.

The crosswise top beam of the "frog-leg" is always a small round log, and therefore it is not possible for this to produce such an indentation. The only elements which have a squared section are the crossbeams of the woven arch – they are normally made of hardwood as well. When we put such rectangular crossbeams at the corresponding positions, the result is a simple strut bridge with a single three-sided arch under the deck beams (Figure 5.16). It has extremely long horizontal beams (around 14 m) in the middle.

Although this would seem to be the only possible structure that works with all the above-mentioned marks, the structure looks rather unusual. We must have further evidence to prove or disprove this idea.

5.3.2 Unexpected evidence and the construction method

After 20 days of laborious survey work, piecing together every trivial measurement, bit by bit, onto the plans, in the last week of our fieldwork, we came across an unexpected find in the dark, dirty working space between the rugged beams (inside the under-deck structure). And it played a pivotal role in the restoration.

This find was revealed when the author was measuring a detailed scale of a dovetail joint at the end of an H-Beam[1] (where it connects to the C-Beam[1]), where he had to remove the centuries-old thick layer of dust covering the beams. Doing this, the author noticed a small sag in the dust near the neck of the tail on the top surface of one of the beams. After cleaning the dust off and digging, it turned that the dust had been covering a small, deep

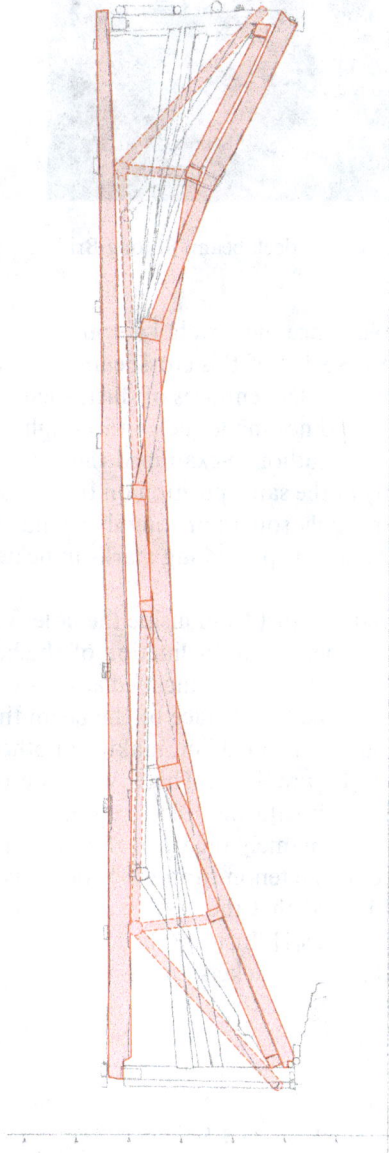

Figure 5.14 Restoration of the previous under-deck structure of Rulong Bridge, version I, the woven arch bridge.

Figure 5.15 Indentation on the deck beam, Rulong Bridge.

downward-angled hole. Since the dovetail is connected to the C-Beam[1], the hole went deep into the socket of this crossbeam (Figure 5.17).

Buried, as they were, under centuries of dust, such holes could easily have been overlooked if we had not made such a thorough examination. After the first hole was found, the author re-examined the complete structure, finding similar holes, and only at the same position on both ends of each H-Beam[1]. The openings either roughly square or roughly round in shape, with a diameter or width of ca. 2 cm. The processing marks indicate that they were made by a chisel instead of a drill.

The author removed the dust from inside the holes with a hollow bamboo stick, and measured the depth and inclination of the holes using a long iron ruler (Figure 5.17). All the holes were inclined at various angles (Figure 5.18). They went so deep from the top surface of the beam that they went through the entire dovetail along an inclined line, making another opening at the front surface of the tail (cf. Figure 4.71). There, while we could not look inside but only explore the depth with the ruler, we could feel the iron end of the ruler meet a solid surface, namely the vertical inner surface of the socket, or even the front surface of the tenon from the S-Beams[1] at the opposite end. This solid surface indicated that the holes were not used for nails to fix the H-Beams[1] to the C-Beams[1]. Instead, they must have been used before the dovetails were placed into the sockets, that is they were made and used in the construction process.

What were they used for? The diameter/width of the holes and the fact that they went right through the wooden part of the beam suggest one possibility: a rope. And if one puts a rope through such a position, it should be used to pull the logs along their axis. Otherwise, if a log is pulled in a direction perpendicular to its axis, the rope can simply be tied around the log, without having to make the effort of digging a hole.

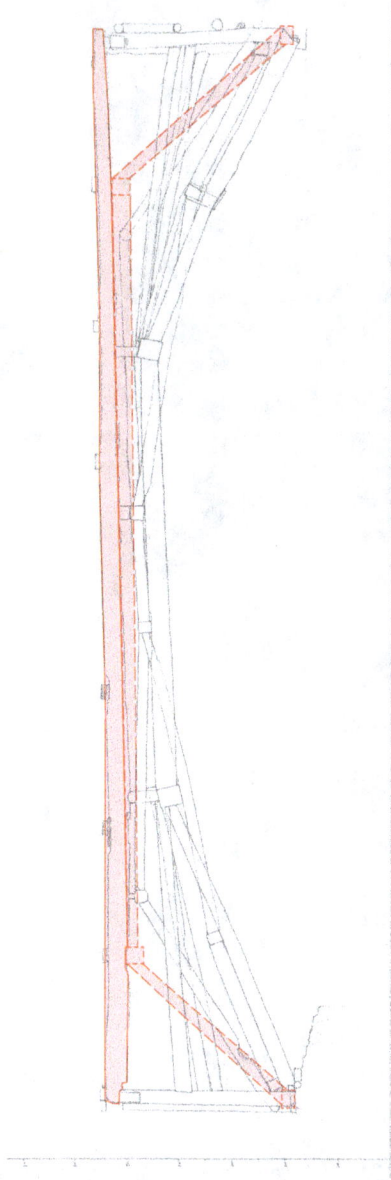

Figure 5.16 Restoration of the original under-deck structure of Rulong Bridge, version II, the strut bridge.

Figure 5.17 Holes at the ends of the H-Beams[1]. Rulong Bridge.

This assumption was further supported by two broken beam-ends: these tails were broken off with a vertical (triangular) opening along the grain from the holes to the front surfaces (Figure 5.17, c). This broken-off mode is consistent with the concept of the logs being pulled with the help of ropes, in the way described above.

This assumption, although logical, is still confusing. As we know from the last chapter, ropes were involved in the construction process in the past, especially where swing-frame-scaffolding was used. In such cases, the H-Beams[1] are pulled up from the middle of the river, by simply tying a rope around them (Figure 4.77, g). What kind of circumstances would make it necessary to pull the H-Beams[1] along their axis?

Before we found an answer, we found another kind of detail to support our pulling-by-rope hypothesis. At most beam-ends of these H-Beams[1], there are bamboo dowels or dowel holes left at the side of the beams (Figure 5.19). If someone was pulling this beam, the dowel would serve as an additional knotting position for the rope, or as handles for helping hands. Such a handle would be especially useful for moving timber when people were working on the scaffolding.

However, that still doesn't answer the question of why the carpenters would go to all the effort of chiselling holes into the beams to put a rope through them, especially when they ran the risk of damaging the beam tails.

5.3.3 Completing the causal chain

We mentioned at the beginning that the deck beams are individual large logs which go from one bank to the other, but this is a simplified expression. Readers may have already noticed from the plan drawings that actually, four of the seven rows of deck beams consist of single logs, one (the second from the east) was damaged and has been repaired using an additional piece, while the other two, the third beams from both sides (east and west side) are formed by two beams of half-span length, each ca. 10 m long. The thinner ends (tops ends) of these beams lie towards the inner part of the bridge, and meet end to end precisely at mid-span. Most interestingly, these beams have a dovetail at their top ends (Figure 5.20), and these have no construction function in their present configuration. Two of these four dovetails have holes through them.

These four beams are obviously elements from the original construction that have been recycled.

As we've seen in the previous chapter, many longitudinal members of the woven arch may have a dovetail at one end: the slanted beams may have a dovetail at the bottom/larger end, and the horizontal beams have a dovetail at both ends. Thereafter, the only members in a wooden arch bridge that have a dovetail at their thinner ends are the horizontal beams of the two arch systems (H-Beams[1] and [2]). If these 10 m-long dovetail beams were members of a previous structure, they fulfil every hypothetical requirement for the 14 m-long horizontal beams in our restoration plan above – the strut bridge version of Rulong Bridge (Figure 5.16). If correct, the larger 4 m "butt" was cut off

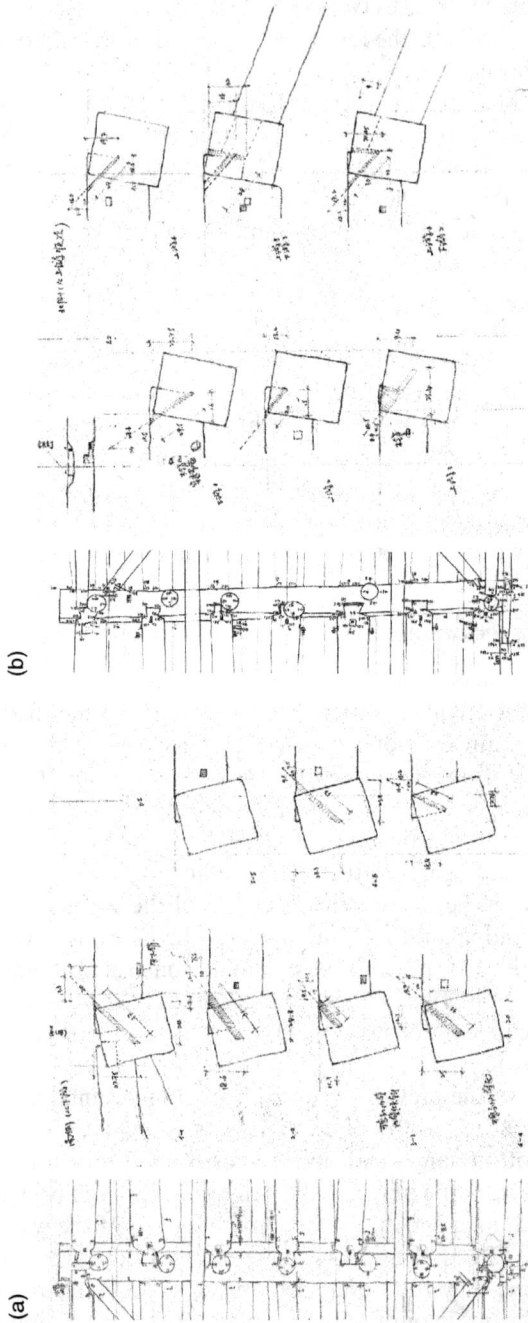

Figure 5.18 Detail of the holes in the H-Beams[1], Rulong Bridge. (a) Southern side. (b) Northern side.

Figure 5.19 An empty hole for a bamboo dowel at the sides of a beam end. Rulong Bridge.

(a) (b)

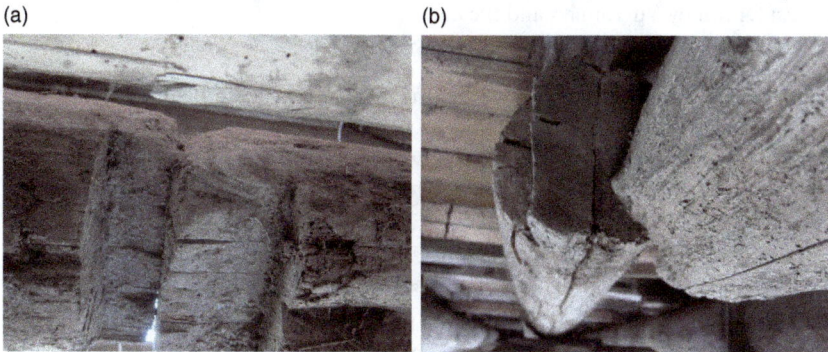

Figure 5.20 Dovetails at the beam ends of the half-span-length deck beams in the middle of the bridge. One dovetail in each pair has a hole in it. Rulong Bridge. (a) The third beam from the west/upstream. (b) The third beam from the east/downstream.

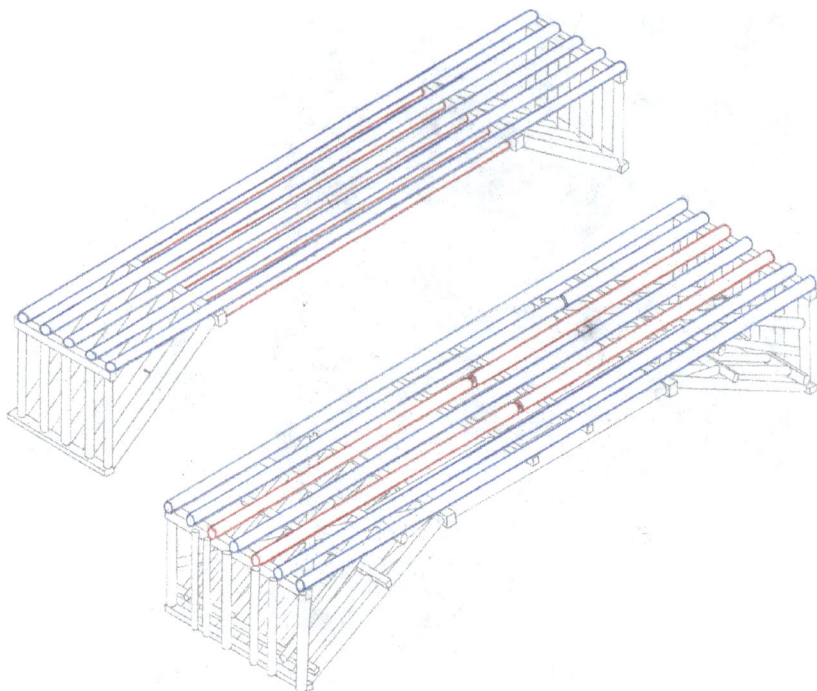

Figure 5.21 Restoration of the reconstruction of the Rulong Bridge showing the reused beams.

Above: original structure.
Below: current version of the structure.

Source: Drawn by Yu Yannan and the author.

from these 14 m-long horizontal beams, and might have been reused some-where else (not in the present bridge).

This restoration theory also works well in terms of the number of beams. There are five[2] 20 m-long and four 10 m-long deck beams in the current version of the Rulong Bridge. If we work on the assumption that all large timbers were reused in the most economical way, that all deck beams came from the original structure, and conversely, all horizontal beams from the ori-ginal structure were reused as much as possible in the reconstruction, the ori-ginal bridge would have had five deck-beams and five horizontal beams in the under-deck arch (Figure 5.21). Four of the five horizontal arch-beams were cut down from 14 m to 10 m and were arranged in pairs among the original deck beams in the rebuilt structure.

2 Four entire logs, one repaired using an additional piece.

The holes in the dovetails of these 10 m-logs indicate that these beams, when they served as 14 m-long horizontal arch-beams in the original structure, were also dragged into position using ropes. At the time, on the building site during the first construction, this action made better sense. The river channel of the Rulong Bridge is comparatively narrow and shallow (Figure 5.3). When dealing with these 14 m-long, huge, heavy logs in the first construction, it was less meaningful to move them down to the river level, in order to lift them to their position, as commonly done. At the time, it made sense to drag them from one side of the river, where they were processed, to their position mid-stream.

The processing work was obviously done on the southern bank of the river, as this side has level ground, whereas the north side is adjacent to the hill (Figure 5.2).

The logs were, as always, alternately placed, that is if the first had the thicker end to the north, the next had the thinner end to the north. Thereafter, when dragging these beams, half the beams have the rope holes on the smaller end, while the other half have the rope holes on the larger ends. This explains why there are only two holes found on the four 10 m-beams nowadays. The other two holes of the other two beams must have been on the thicker end and were cut off during the reconstruction.

5.4 Conclusions

After putting all these pieces of evidence together, the puzzle reveals the complete picture. We are able to draw the following conclusions:

The Rulong Bridge was reconstructed; the former, original structure was a strut bridge with a layer of deck beams and one three-sided arch. The rebuilt structure is the present woven arch bridge, composed of one three-sided arch and one five-sided arch.

The original structure had five 20 m-long continuous deck beams. The three-sided arch (most probably) also had five horizontal beams 14 m in length. All deck beams and four of the five arch-beams were reused in the reconstruction. The 14 m-long arch-beams were cut down to 10 m and arranged in pairs among the deck beams.

One reason for the reconstruction might have been to strengthen the structure. The load-bearing capacity of the original structure (namely the strut bridge with a 20 m span) might have been ok for a while, but it might have become weakened over time.

In the construction of both the original and the present structure, the horizontal beams of the three-sided arch were pulled along their axis by rope into their position from the southern riverbank. For this step, holes were chiselled into the beam-ends for the ropes to pass through.

This construction method was practical for the first project, since the 14 m-long middle beams were huge and heavy and cumbersome to transport by hand. However, in the reconstruction, the same method would seem to be less necessary, since the middle beams had been cut down to 10 m. Please note that the part that had been cut off, although is only 4 m in length, had the larger diameter, therefore, a simple calculation would show us that the remaining 10 m-long timber was less than half of its original weight and it would have been possible to transport it by purely manual means.

Thus, in the second construction, the need for dragging the timber using ropes, taking the trouble to chisel out the holes – thereby incurring the risk of damaging the beams – is questionable. If, when the work could have been carried out more easily, but was instead done in the more laborious old way that was used in the previous construction, we have reasons to believe that these two constructions were built by a mixed group of carpenters. That is to say, some carpenters working on the second construction (probably including the leader) had already participated in the original construction. This means that the two construction projects were carried out within the career-span of one carpenter, which is usually no more than 60 years.[3]

In the reconstruction, the number of deck beams was expanded from five to seven. Therefore, it's highly possible that the bridge corridor was completely rebuilt in the course of this project. The commander-pillars, if they had reached into the corridor in the first construction, must have been cut to half-height during the reconstruction.

The construction year inscribed in ink on the corridor beam (here, the year 1625), usually marks the completion of the corridor, and thus, in this case, it should be the date of the reconstruction project. This means that the original construction could date to as early as the latter half of the sixteenth century.

To verify our theory about the reconstruction, several samples taken from the Rulong Bridge were subjected to dendrochronological analysis in 2013. Dendrochronology is a scientific method of dating tree rings to the year they were formed. Thus, the rings of a wooden element sample tell us the years the tree was growing and esp. the year when it was felled. The felling year must be a timepoint before the timber's first use on a building project, and normally not too much earlier than that, which helps us to estimate the construction time period.

Given the fragile condition of the timber material of the old beams, it was possible to take only a few samples (those in good enough condition and with a high-enough number of tree rings) to the lab of the Institute of Archaeology, Chinese Academy of Social Sciences, for dendrochronological analysis. Although the resulting data were not enough to make a definite pronouncement as to the timeline of the construction, they could be used to confirm the theories outlined above.

Another unfortunate situation is that, in China, we do not have a comprehensive dendrochronology database, so we can only tell the time difference between the samples (relative dating) according to the overlapping time

---

3 In the mountains, carpenters are/were trained in their teenage years, and could still be working in their 70s.

Table 5.1 Relative dendrochronological dating of Rulong Bridge beams

Member	Tree rings	
	From	*To (tree felled)*
Wood samples from three-sided arch of the woven arch		
3rd H-Beam	1600	1618
5th S-Beam	1609	1623
3rd S-Beam	1466	1503
Northern C-Beam	1434	1600
Southern C-Beam	1427	**1625**
Wood samples from the deck beams		
1st deck beam	1502	1531
4th deck beam	1500	**1578**

(Longitudinal beams are counted from the upstream side)

periods they span, instead of the exact calendar year (which would be "absolute dating"). In order to get a general overview, we assume that the latest sample was felled in the year 1625, and we then dated the others according to their relative dates.

The (relative) dates of the wood samples are as follows (Table 5.1).

Although data are limited, it is noticeable that the deck beam samples are both from the sixteenth century, with the latest date being 1578 (relative date), while most of the woven arch samples date to the seventeenth century, with the latest year being our assumed 1625.

According to our reconstruction theory, all members of the deck beams (except the repaired part) are reused members of a former construction, and the woven arch members could either be reused members as well, or new timbers felled for use in the second construction. The dendrochronological results are highly consistent with the theory. The results also fit with some other general ideas, such as that the logs for the longitudinal beams (H- and S-Beams) might have been reused (the 3rd S-Beam), but the squared crossbeams – since they show many joint cuts – had to be made specifically for each individual project.

The time gap between the latest deck beam and the latest woven arch member could also be the time gap between two constructions. The 47-year time gap would fit precisely into the career-span of a carpenter who witnessed or even participated in the construction of the original strut bridge version of Rulong Bridge. At the time, the construction method – pulling the huge logs by rope to put them in place to build a three-sided arch with exaggerated proportions – probably left such a strong impression that he applied the same method half a century later to the second version of the Rulong Bridge, which this time, was a woven arch bridge.

6   Technique and craftsmen

Pedigree of the bridge carpenters and the diffusions of the technique

6.1 Our sources of information about the bridge carpenter families

Three types of sources provide us with information about the builders of the bridges. The first is the ink inscriptions on the corridor beams inside the bridges. There, if preserved, we can find the name and place of origin of the craftsmen, as well as other basic information about the project (project date, organizers and the many sponsors). Most of the known traditional bridge carpenters have been discovered by scholars tracing this kind of inscription. However, only some of the surviving bridges still carry this information today – most of them have vanished in the mists of time.

Another important source is the bridge project contracts between the master carpenters and the project directors, kept in the hands of the descendants of the two sides. The most influential family (the Xiajian Zhang family, see below) has handed down and preserved some thirty bridge-building contracts, which forms the absolute majority of the surviving bridge contracts.

Not all bridge carpenter families were able to preserve their contracts; most of these precious documents were lost over time, especially during the Cultural Revolution. Gong Difa, former director of the Shouning County Museum, searched for and collected such contracts and the relevant documentation in the MZ area. This collection is now housed in the Shouning Museum.

The third source of information is the oral knowledge and histories of the carpenters themselves. This is also limited, as the oldest woven arch bridge in this area is nearly 400 years old (Rulong Bridge, built in 1625), while the longest surviving family tradition of this profession 'only' spans about 250 years (the first-known project being built in 1767). Only two families have traditions of longer than a century. The author interviewed most of the known family bridge carpenters (those with a family tradition of at least two generations: nine people from five families) and many new carpenters who have entered the profession in recent years.

Information on the bridge carpenters from these three sources has been collected by local cultural workers and was published under the title "Investigation of the wooden arch bridges in the Ningde area" by the Ningde

Municipal Culture & Publication Bureau in 2006.[1] Gong Difa later published a revised collection of information on the bridge carpenters and their works in a single-author monograph (2013). Building on the information provided by the local cultural workers, and integrating the new finds of the author of this book, it was possible to produce an even more comprehensive version of the list. To avoid reproducing quantities of data here in the text, the results are given in diagram form, combining the technical pedigree and the works of the two leading carpenter families (Figure 6.22).

As mentioned in Chapter 4, only a very limited number of carpenter families/groups were engaged in a family-tradition version of the woven-arch-bridge-building profession. There are only three families/groups of carpenters with continuous bridge-building traditions spanning more than three generations, and of these, only two have traditions that span more than three generations before the modern times. All the families/groups are listed below:

- The Zhang family of Xiajian Masters from Xiukeng Village (an area which was formerly referred to as "Xiajian" 下荐), Zhouning County (formerly Ningde County), Fujian Province. Their first-known woven arch bridge was built in 1767, and their last bridge built in a traditional society[2] was in 1975. Over the course of some 210 years, seven (or eight[3]) generations of the family worked as bridge carpenters before the modernization in

1 Ningde is a prefecture-level city in Fujian, a region composed of a city-governed district, six counties, and two county-level cities. This area can be considered the core area of wooden arch bridges in Fujian. All surviving bridge-carpenter families with traditions spanning more than three generations live in this area today.
2 Although the time point of Chinese modernization is considered much earlier, in many aspects of the rural life in the mountains, especially in those relating to civilian constructions, development lagged behind and relatively traditional ways were followed. Regarding the bridge building activities, we set the dividing line between the traditional and modern times at 1980, when the MZ bridges first received attention in academic circes. Actually, from the latter half of the 1970s to the beginning of the twenty-first century, hardly any woven arch bridges were built in the MZ mountains.
3 If we calculate according to the family tree, seven generations of the Zhang Family have worked as bridge carpenters. However, most of the local cultural workers consider that there are eight generations of this family, based on a speculated mentorship (i.e., one member is thought to have mentored another and hence is counted as belonging to two generations) rather than actual bloodline (Ningde 2006; Gong 2013; Zhou et al. 2011).
  Both classification methods are problematic. If classification is according to speculated mentorship – as most authors do – two brothers or cousins might be classified as belonging to two different generations if the younger worked with and learned from the elder. At the same time, however, an uncle and a nephew might be listed as belonging to the same generation if they worked together under the grandfather's leadership. When classification is by bloodline, the problem is that in a large Chinese family, there might be an age difference of half a century between cousins listed as belonging to the same generation when actually they span two generations (and in some cases, a nephew might in fact be older than his uncle).
  In this book, the author classifies according to the family tree.

the mountain regions. Up to the 1970s, some 57~61[4] bridges were built or reconstructed by this family: 25 bridges validated by ink inscriptions, five according to oral information provided by family members, and the remaining 30 or so bridges are validated by contracts and other forms of written documentation).[56]

- The Xu-Zheng families[30] of the Kengdi (坑底) Masters from the Xiaodong (小东) and Dongshanlou (东山楼) villages, Township of Kengdi, Shouning County, Fujian Province. Their first-known bridge was built in 1801, and the last bridge built in the traditional society way was in 1967. In the course of approximately 170 years, six generations of this family-line have worked as bridge carpenters before modern times. Up to the 1970s, 23 bridges are known to have been built or reconstructed by this family (10 validated by ink inscriptions and 13 by oral information provided by family members).

- The Huang family of the Changqiao Masters from the town of Changqiao (长桥), Pingnan County, Fujian Province. Their first-known bridge dates to 1904, and three generations worked on building bridges before modern times. Up to the 1970s, seven bridges are known to have been built or reconstructed by this family (three validated by ink inscriptions and four by oral information from family members).

   The data from the above-mentioned sources[7] were used to generate maps to illustrate the working areas of the three bridge carpenter families (Figure 6.1, Figure 6.2). These maps provide a visual sphere of influence of the carpenter groups/families, and show that the Xiajian Masters were the dominant bridge carpenter group in the entire MZ area until the beginning of the modern era. Their working area spans almost the whole of the wooden arch-bridge area in Southeast China. And whereas they were able to get building projects in the core working areas of the other carpenter groups, the reverse was not true. The ruling status of the Xiajian Masters is evidenced by the Yangmeizhou Bridge: As briefly mentioned in the previous chapter (Sec.4.5.3.1.2), it was a case where the Xiajian Masters took over a project from the Kengdi Masters when the latter wasn't capable. This story is described in detail later on in this chapter.

---

4 Depending on whether the reconstruction of any one bridge was counted as a new bridge or not.

5 Four of the bridges are validated both by ink inscription and by contract.

6 The assessment includes about ten contracts that provide no indication as to the location of the bridge, or even its name.

7 The data puts the Xiajian Masters at an advantage, because bridges that are no longer extant today but which are certified by bridge contracts held by this family are included in the count, whereas the contracts of the other families have been lost over time, especially during the Cultural Revolution. Nevertheless, this does not alter the fact that the Xiajian Masters are the dominant force in the world of bridge carpenters.

Figure 6.1 Practical area of the leading bridge-building families (Zhang family from Xiajian, Xu-Zheng families from Kengdi and Huang family from Changqiao).

The map uses current counties as units. Circles denote the construction projects in the home county of the bridge carpenters. The directional arrows denote construction projects in other counties. The thickness of the lines corresponds to the number of bridge projects.*

*Number of the counties and their boundaries in this area have changed many times over the last five centuries.

Source: Based on Google Maps.

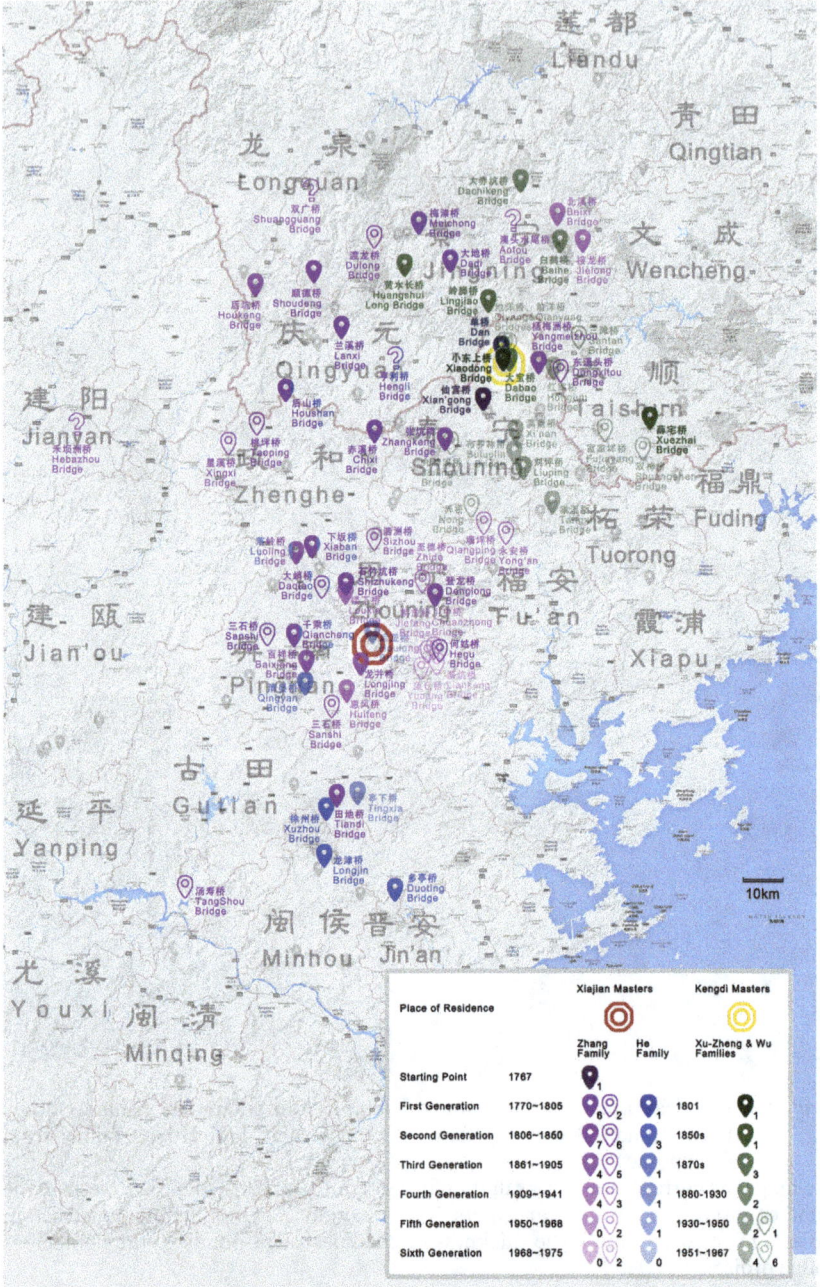

Figure 6.2 Map of the known bridge works of the Xiajian and Kengdi Masters. Solid symbols indicate bridges with the most solid certifications (exist today, or although has been destroyed in relatively recent years yet has been studied/witnessed by modern scholars). Hollow symbols indicate bridges with less solid "credentials" (do not exist, proven only by contract or oral information). Works of the Xiajian He families (see the section below) are also included on this map.

Source: Based on Google Maps.

The traceable carpentry tradition for this kind of bridge in the MZ mountains is much shorter than the construction history itself. The oldest preserved bridge structure dates to 1625 (Rulong Bridge, see Chapter 5), but does not bear the names of its builders – and the oldest literature on bridges with confirmed evidence of a woven arch structure dates as far back as half a century earlier than that.[60] The first bridge with its builders' names preserved dates to as late as 1767, and is the first work of the first generation of the Zhang family from Xiajian. In the two centuries before a known carpenter (group), bridges stood or vanished in silence, their creators remaining anonymous. Though it might be impossible to find the names lost in the mists of time, their methods of working and the techniques they passed down may yet be revealed by studying the working methods of the known carpenter groups.

Before we start our stories of the bridge-building families, a special note on the spelling rule for names used in this chapter is given.

In a traditional Chinese family, the common form of a member's name is formed by putting the family name at the beginning, the "generation name" in the middle, and the given name last. The middle generation name is prearranged[8] and recorded in the family's genealogy book. Thus, people can tell a family member's generational placement within his family simply from his name.

According to modern official spelling rules for Chinese names, the (middle) "generation name" and the (last) given name should be spelt together as one word. If we take the name Zhang Changzhi as an example, we see that Zhang (张) is his family name, Chang (昌) is actually the generation name, and Zhi (智) is his given name. His brothers are named Zhang Changyun, Zhang Changtai, etc. However, since the modern official way of spelling indicates the generation information less distinctly, for ease of reference, in this chapter – since the specific generation of the carpenter families is a crucial factor in our discussion – when we refer to members from these specific families, the middle name and the last name will be *hyphenated* rather than written as one word. Thus, names would be written: Master Zhang Chang-Zhi and his brother Zhang Chang-Yun, Zhang Chang-Tai, etc. The names of people outside these families are spelt in the official way.

6.2 Stories of the bridge carpenter families

6.2.1 *Xian'gong Bridge – the beginnings of the Xiajian Masters*

6.2.1.1 *Zhang Xin-You*

The Xiajian Zhang family lives in Xiukeng Village, in what is today Zhouning County (formerly Ningde County). In the Jiajing period (1522–1566) of the Ming Dynasty, two cousins of the Zhang family moved here from Pingnan

---

8 The generational middle name sometimes comes from a well-known text (sentence/paragraph) with propitious meanings. Each generation takes a character from the text, in turn.

Figure 6.3 Xiukeng Village, Zhouning County, Fujian Province.
Photographed in January 2019. The construction site below Xiukeng Village is for a tunnel for a new railway.

County. Since the lower arable land by the stream was already occupied, the Zhang cousins took up residence on the hill where relatively less arable land was to be found, forming what would later become Xiukeng Village (Figure 6.3). When the first generation of bridge builders emerged, the family had resided in the village for seven generations[9] (Figure 6.9). All the inhabitants of the entire village have the same family name: Zhang.

The family legend, according to the current descendant, is that Zhang Xin-You (张新祐), the family's first bridge carpenter, was formerly a master of *kung fu*, and as he was travelling in the neighbouring counties, he happened to meet a bridge master and saved him out of some troubles. They became firm friends and subsequently went back to Zhang Xin-You's home village and worked together building bridges ever since.[10]

This legend is not very reliable as a family history. Fujian province is the location of the Southern Shaolin Monastery, a legendary *kung fu* school, and

---

9 Genealogy book of the Zhang Family.
10 Zhang Chang-Zhi, in an interview with the author on 16 January 2019 in Xiukeng Village, Fujian Province. This story is also recounted in Zhou et al. (2011, 193–4).

Figure 6.4 Xian'gong Bridge (仙宫桥) in the city of Aoyang, Shouning, Fujian. Built in 1767.

therefore legends involving *kung fu* abound throughout this mountain region. However, this legend regarding Zhang Xin-You does provide us with meaningful information, namely, that this family's bridge-building technique came from a source external to the family, during Zhang Xin-You's generation, which confirms information from other sources.

In the centre of modern Shouning's county capital, Aoyang, just outside a minor city gate of the former walled city of the Qing Dynasty, is the Xian'gong Bridge (Figure 6.4) across the Chanxi Brook. Although the formal name of this bridge is Yudai Bridge (lit. "jade belt bridge"), it is commonly known as Xian'gong Bridge (lit. "the bridge by the Temple of the Immortal Miss Ma") because of its location. Immortal Miss Ma is a popular local deity in this area and her temple at the bridge-side was a rather popular spot, with many poems written describing the flourishing of the temple as well as the beauty of the bridge at its side. The temple was demolished in the 1950s, but the bridge is still very much a bustling public space today (Figure 6.5, Figure 6.6).

Like most bridges in the mountains, Xian'gong Bridge has been destroyed and rebuilt a number of times. The current structure was built in the 32nd year of the Qianlong Period (1767). In the richly-painted roof structure, a beam bears the three names of master carpenters of the project: Master Li Xiu-Yi (李秀壹), Wu Sheng-Gui (吴圣贵), and Zhang Xin-You (Figure 6.7). This inscription makes the Xian'gong Bridge the first-known work of Zhang Xin-You, and marks the

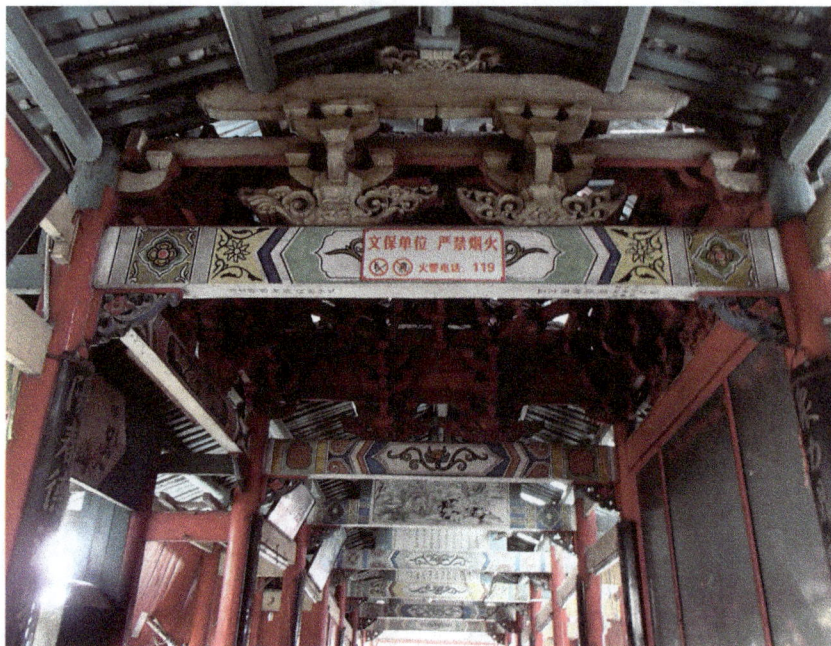

Figure 6.5 Xian'gong Bridge. Decorated roof structure of the bridge corridor.

beginning of the bridge-building career of the Zhang family of Xiajian – and incidentally, makes the Xian'gong bridge the oldest MZ bridge to be inscribed with the names of known builders.

A rather remarkable fact of the inscription is that the name of Zhang Xin-You is written after the other two masters. The name of Master Li Xiu-Yi is written separately, and in larger characters, and also records that he donated 100 Wen (money), indicating his status as the chief master of the project.[11,12] The names Wu Sheng-Gui and Zhang Xin-You are written side by side in a smaller size and Wu Sheng-Gui's name comes before that of Zhang Xin-You.[13] This indicates that Zhang Xin-You was the youngest or the least high-ranking of the three. The two other masters – especially the project-master Li

11 As mentioned, many project masters work on a construction project. We name the one who has the most responsible position as the chief master, and who is always listed first in the list of carpenter masters in the inscription on the bridge or in the contract.

12 Normally the chief master makes a donation to the project on behalf of his team. The author witnessed chief master Wu Fuyong donating after his ritual at the *dong*-festival for the Shengshuitang Bridge site in December 2012.

13 Wu Sheng-Gui is at the right-hand side and Li Xiu-Yi is at the left. The traditional Chinese in written from the right to the left.

Figure 6.6 Xian'gong Bridge. The Bridge is used as a public space for the locals. Part of the corridor has been repurposed and is now a shop.

Xiu-Yi – most probably were tutors to Zhang Xin-You, and could have been the source of the technique of the Xiajian Zhang family.

6.2.1.2 *Who is Li Xiu-Yi?*

No information is given on the origin of the masters of Xian'gong Bridge. We investigated the early bridge of the Zhang family to help us find some clues.

Zhang Xin-You built at least eight bridges during his life, and on the other seven bridges, his name usually came first on the list of project masters, working mostly with family members, including his sons, a cousin, and his nephews. On several occasions, there were also members of other families on Zhang Xin-You's team, as follows:

- A master Wu Sheng-Zeng (吴圣增), listed as the fourth and last of the master carpenters who worked on the Chixi Bridge project in Zhenghe, Fujian in 1790.[14]

---

14 The first three names were of the Zhang family. Inscription of Chixi Bridge. According to Gong (2013, 179).

Figure 6.7 Ink inscription on a beam inside the Xian'gong Bridge.

The inscription visible on the upper part of the photo reads: "Master Li Xiu-Yi donated money 100 Wen, and the inscription on the lower part, going in the opposite direction, reads: "Master Wu Sheng-Gui, (Master) Zhang Xin-You."*

*Words at two ends of this purlin are written in opposite directions, that at the bottom of the photo, appear to us upside-down. When reversed to its correct direction, the name Zhang Xin-You is written on the left side and Wu Sheng-Gui on the right side. By contrast to the modern writing system, in traditional Chinese writing, words are written from right to left, so the name Zhang Xin-You is written after the name Wu Sheng-Gui.

- A master Li Zheng-Man (李正满), whose name appears on three bridges. On the Lanxi Bridge in Qingyuan, Zhejiang (1790, Figure 6.35), he was the second name listed immediately after Zhang Xin-You and before other members of the Zhang family; he was also the second carpenter listed on the Houshan Bridge (后山桥) in Zhenghe, Fujian (1799), immediately after Zhang Xin-You;[15] and on the Meichong Bridge in Jingning, Zhejiang (1802), Li Zheng-Man was the Chief master carpenter, his name

15 According to the inscription of the Houshan Bridge.

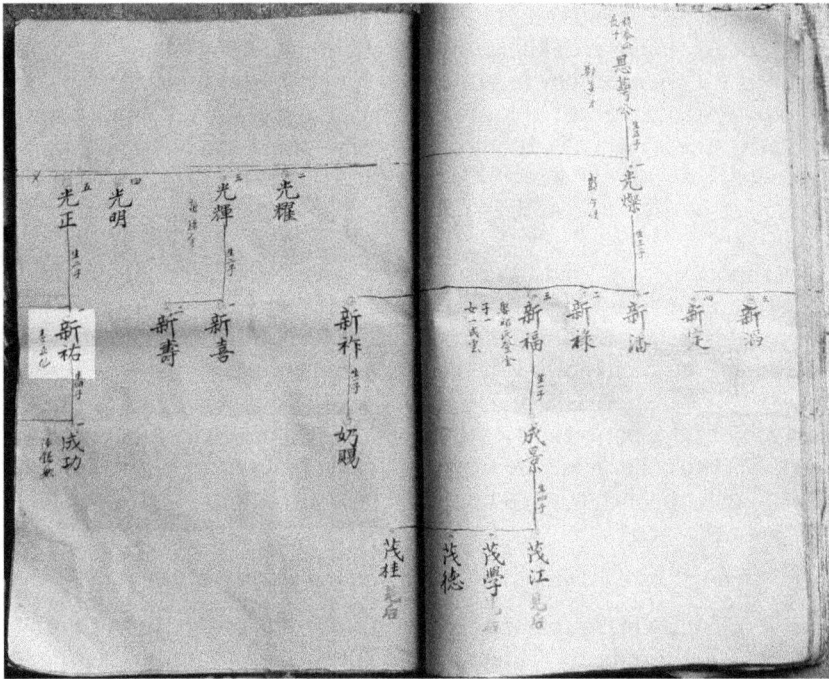

Figure 6.8 Family tree in the genealogy book of the Zhang family in Xiukeng Village. Preserved by Master Zhang Chang-Zhi. The names of Zhang Xin-You and his wife Li Zheng-Feng are marked out.

comes first on a long list of members of the Zhang family and other families (Tang 2000, 473), including Zhang Xin-You himself.[16]

- A group of masters from the He (何) family, from apparently two or three generations. (More on this family later.)

Wu Sheng-Zeng and Li Zheng-Man, who were core members of the Xiajian Masters' team. Judging from their shared generation name, Wu Sheng-Zeng was probably a brother or cousin of the master Wu Sheng-Gui who worked on the (very first) Xian'gong Bridge, and these two families worked together for at least a generation. Although we do not have the name of their village of origin, it is almost certainly around the Xiajian area.

Even less is known about the identity of Li Zheng-Man. After noticing that he bears the same family and generation name as Zhang Xin-You's wife, Li Zheng-Feng (Figure 6.8), we visited the largest branches of the ancestral

16 Zhang Xin-You was already quite old at the time, though not too old to be in charge, as he was chief master on another bridge project the following year.

halls of the Li family in Zhouning County to look through the genealogy books, and finally found these two siblings in the genealogy book in Wangsu (王宿) Village, some seven kilometres away[17] from Xiukeng village. Li Zheng-Man was the youngest boy in his family. For some reason he accompanied his second-eldest sister to her husband's village, and embarked on the bridge-carpentry career with his brother-in-law. He never married and had no descendants. As a result, he was the only carpenter master among the identified Xiajian Masters to bear the family name "Li"– apart from the mysterious Li Xiu-Yi.

Disappointingly, we did not find a Li Xiu-Yi in any genealogy book we looked through. The Li family into which Li Zheng-Man was born had no history of carpentry tradition, and thus, Li Xiu-Yi's origins remain unclear.

Although he is unknown to modern researchers, Master Li Xiu-Yi must have had a reputation as a bridge carpenter at the time in order to win the Xian'gong Bridge project – as it was not located in his county (or at least, the county his two colleagues were from), and, judging from its prestigious location and rich ornamentation, it must have been an important project.

6.2.1.3 The rise and decline of the Zhang family's bridge-building career

However Zhang Xin-You came to be introduced to this career, he successfully started a "family industry" (Figure 6.9, Figure 6.22): One or two cousin(s)[18] in his generation, three of his four sons and at least four nephews made themselves a name as bridge carpenters. This flourish in the population of the bridge carpenter family lasted three generations. At least seven members from the Cheng-generation, six of the Mao-generation, and six of the Xue-generation left traces of their works and their position on the family tree. Remember that the entire Xiukeng Village originated with two cousins, two family branches, and bridge carpenters appeared in both branches. However, the famous Zhang Xue-Chang (the chief carpenter master of the Yangmeizhou Bridge), working at the beginning of the twentieth century, was from another branch than Zhang Xin-You. During the 1940s, when the family was in its sixth generation (the Ming generation), the bridge-building career went into decline as a result of the Sino-Japanese War and the Chinese Civil War. There was no known bridge carpenter for an entire generation (the Shi generation). After the establishment of the People's Republic (1949), the family experienced a brief revival, in which two new generations worked together. Their last works (before the "UNESCO heritage era") was in the 1970s. According to the

---

17 Note: All distances given in this chapter are as they appear on the map, i.e. calculated along a straight line. However, in a mountainous area, the accessible path(s) between two points might span two or three times the distance shown on the map.

18 Apart from the Zhang Xin-Hui mentioned in Figure 6.9 there was another name, Zhang Fo-Ji (张佛济), who possibly belonged to the same generation, but his name was not found in the genealogy book.

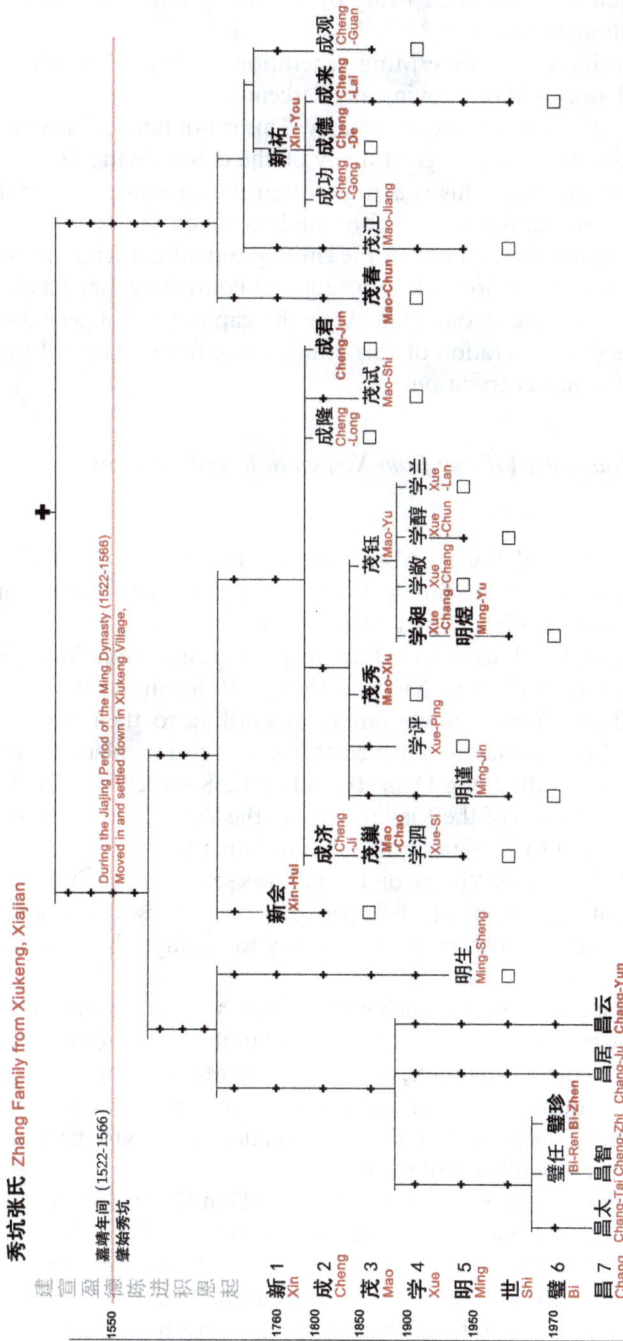

Figure 6.9 Family tree of the Xiajian Masters in the Zhang Family of Xiukeng.

According to the genealogy book of the Zhang family. A cross indicates member(s) existing in the same generation. Only the names of the bridge masters are given in the diagram. Names written in **bold** have worked on a bridge project as the chief master carpenter or as the contractor. The squares below family lines mark the dead-end of a family branch (when no child was born in that generation). In a single generation, the age difference between members could span decades. The timeline at the right-hand side indicates the active time of the most successful member(s) of a given generation.

carpenters of the current generation, in the times when there were no bridge projects, the carpenters made their living by delivering timber by waterway (rafting timbers along the river).

In pre-modern times (up to the Ming-generation), a large proportion of carpenters passed away without leaving any descendants (Figure 6.9, marked by squares); many of them did not even marry. This is not unusual if we look at the collective situation, since a great many of the entire Zhang family did not marry or bear children. This was quite often the case in poor families in traditional Chinese rural society before modern times. However, what is noticeable in this case is that not one of the entire group of carpenter masters whose name we know and who worked during that flourishing time has left a direct descendant traceable to our time. After the gap in the Shi generation, the seventh and eighth generation of this family came from a family branch without a carpenter master tradition.

6.2.2 Bridges in Xiaodong Village: from Xiajian to Kengdi Masters

6.2.2.1 The Xiajian He families

In addition to the Li and Wu members who started their bridge-building career together with the Zhang family, there was another noticeable family name among the early Xiajian Masters: the He family/families.

Members of the He family came from three neighbouring villages: Yangping, Houlong and Meidu, all close to Xiukeng Village. Residents of these three villages came initially from a single family. According to their genealogy book, one of the family's ancestors moved to the area from a place closer to the coast at the end of the Song Dynasty and in 1258 settled in Shoudong Village, only ca. 4 km north of the Xiajian area. In the Yuan Dynasty, some of the descendants moved to Xiajian, settled first in Yangping Village, and then Houlong Village. The Meidu Village on the hill was settled much later. After centuries of separate development, their pedigree became less homogeneous than the Zhang family's, which makes it less easy to distinguish between the different generations simply using names.

The He family settled the area much earlier than the Zhang family, taking the better lands by the stream, leaving the Zhang family to settle only high up on the hill. However, these two families are closely connected by marriage, ever since the first generation of the Zhang family, when they moved to the area.

Therefore, it is no surprise that the two families/family groups worked together in the bridge-building profession.

The earliest known work of the He family is the Dan Bridge (Figure 6.11), a small-scale (15 m clear span) wooden arch bridge, built in 1792, in Xiaodong Village (Shouning County), the home village of the leading group of the Kengdi Masters, who, to the best of our knowledge, had not yet started their bridge carpenter career at the time. On a wooden board hanging inside the bridge corridor, among information on the project, its project directors and

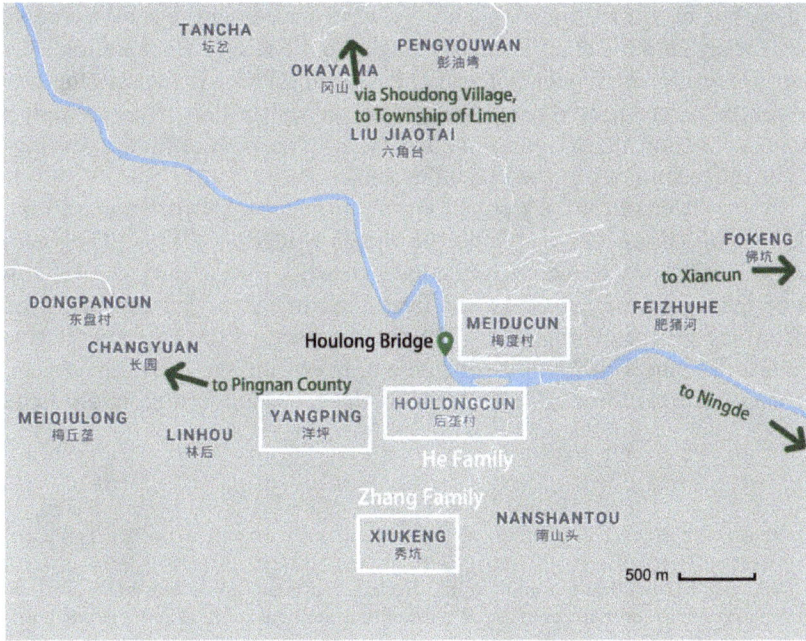

Figure 6.10 Location of the Zhang and the He families in the Xiajian area.
Source: Based on Google Maps, cf. Figure 6.3.

Figure 6.11 The Dan Bridge and the environment. The recent structure was rebuilt in 1939.

the list of donors, are two names of the bridge masters, both from the He family.[19] The construction of this bridge was overseen entirely by members of this family; and the chief master belonged to the Ren (仁) generation. The tablet gives only the names without their origin. This important clue was overlooked by previous researchers. When we realize that these carpenter masters were from the Xiajian area, we get a better picture of the development of the technology in the MZ mountains.

Five years later, in 1797, a group of He family members,[20] working under the leadership of Zhang Xin-You, built the Hengli Bridge (亨利桥) in Qingyuan County. This event was recorded in a local genealogy book (Gong 2013, 10). One of these carpenter masters was from the same Ren (仁) generation of the Houlong Village. Similarly, in 1799, two members of the He family[21] were listed after Zhang Xin-You, his son and a nephew, and the above-mentioned Li Zheng-Man, as the bridge builders of the Houshan Bridge in Zhenghe County.

19 He Ren-Yan (何仁衍) and He Zheng-De (何政德), according to Gong (2013, 205). This board is lost today.

Although, admittedly, we cannot be 100% certain that these two carpenters are from the He families of the three villages, there is compelling evidence that this is a very strong possibility, as follows:

(1) Origin of the Dan bridge carpenters: Although the Dan Bridge gave no indication as to the place of origin of the two carpenters, another bridge (Duoting Bridge, built in 1811) lists the same He Ren-Yan and another Ren-generation brother/cousin as the carpenter masters and as being from Ningde County, which at the time was the affiliated county of the Xiajian Masters.

(2) In Houlong Village, many members of the same generation and with similar names share the generation name Ren (仁), and they fit the time scope of the Master He Ren-Yan. Here, it is necessary to mention that this "Ren" is not the formal generation name predetermined by the genealogy book, but rather a nickname (used in everyday life) which appeared in a family branch in this specific village. The official generation name in the genealogy book of this generation is Yuan (元).

(3) Although the exact name He Ren-Yan does not appear in the He genealogy book, we were able to find a He Yuan-Yan (元衍) from the above-mentioned Ren/Yuan generation in Houlong Village. According to the naming system of this family, he might have been called He Ren-Yan in everyday speech. His age fits the active period of the master of the same name.

(4) Even if the master He Ren-Yan is not the He Yuan-Yan we found, we cannot eliminate the possibility that he could be another member from this Ren/Yuan generation in Houlong Village, due to the complicated naming system and naming situation there.

(5) Many members of the Houlong Ren generation intermarried with the Xiukeng Zhang family.

(6) Other members of the Ren generation worked with the Xiukeng Zhang family.

(7) To be very rigorous, even if we cannot be certain that the two carpenters of the Dan bridge project in 1792 were from the Xiajian area, it is certain that they are from the same county with which the Xiajian area is associated.

20 He Kai-Fa (何开发), He Ren-Ming (何仁明), and He Zi-Ming (何子明).

21 He Kai-Ji (何开极) and He Kai-Fa (何开发). He Kai-Fa is the same carpenter who built the Hengli Bridge.

Members of the He family worked alongside the Zhang family from the first generation of bridge carpenters right down to the most recent bridge.[22] In the first two generations, members of the He family were listed in a quarter of the Zhang family's works.[23] When they worked with the Zhang family, as a rule, their names were listed after the Zhang-surname masters.[24] Even in projects listed as being constructed entirely by members with the Zhang family surname, where no He family names are listed, there is still reason to believe that members of this secondary family might have participated in less significant positions (e.g. as common carpenters) whose names do not appear on the contract or in the inscription.

The He family carpenters also worked independently, that is they acted as chief carpenter masters themselves, without the leadership of the Zhang family. The Dan Bridge mentioned above is the earliest example of this. Their independent works appear throughout the generations, with a total of eight known works. The chief master of the Dan Bridge also built another bridge, the Duoting Bridge (多亭桥), in 1811. This bridge is located in present-day Jin'an District, which belongs to Fuzhou City, located almost at the southern end of the Xiajian Masters sphere of practice. The southern area is also the specific working area of the He family (Figure 6.2). Their early works are mostly small-scale. The Dan Bridge, as mentioned, has a span of about 15 m, and the Duoting Bridge about 20 m. Their capability for building larger bridges, up to 30 m span developed later (between the 1840s and the 1960s). Their last such project was a large bridge over a gorge – one of the most splendid MZ bridge sites: The Houlong Bridge (Figure 4.76), built in 1964, is located just outside the carpenters' village (Houlong Village). Members of the Zhang family also participated in this project.[25]

6.2.2.2 The first known bridge work of the Kengdi Masters

The town of Kengdi (坑底 lit. "bottom of a pit"), is located in a small basin at the pointed northern end of Shouning County, enclosed on three sides by counties of Zhejiang. This area is a transportation node, leading to these different counties: Taishun to the east, Jingning to the north, and Qingyuan to the west.

The Xiaodong stream flows alongside the main road passing through Kengdi. Today, the area has three woven arch bridges within comfortable walking distance of each other (Figure 6.12).

---

22 The Regensburg-China-Bridge, built in Germany in 2015 and directed by the author, see Chapter 4.
23 Seven of the 28 known projects of the Zhang family built between the 1770s and the 1840s.
24 There is only one known instance of the He master being listed as the first bridge master, before the two Zhang brothers (or cousins), and that is in the contract for the Luoling Bridge (落岭桥) project, in Zhenghe, Fujian, in 1883.
25 Zhang Chang-Zhi, in an interview with the author in December 2013.

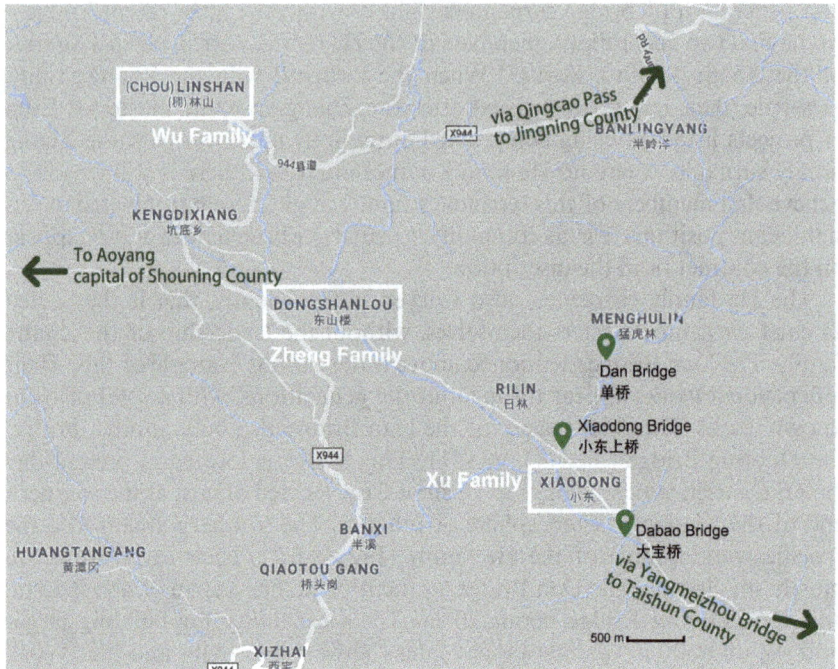

Figure 6.12 Location of the villages of the Kengdi Master families and of local bridges.

Source: Based on Google Maps and Tianditu • Fujian Maps.

There were bridges across the Xiaodong stream at least from about 1700,[26] probably at the same location of today's Dabao Bridge.[27] Then, in 1792, the Dan Bridge was built across a small brook that flows into the Xiaodong Stream. For some reason, people decided to have a woven arch bridge despite its small scale, and bridge masters from the Xiajian He family were asked to come for this specific purpose from a place of a few days walking distance.[28] Nine years later, in 1801, the Xiaodong Bridge (Figure 6.13, Figure 6.14) was built in a similar form and scale (16 m clear span), only ca. 600 metres downstream across the same brook, at the point where it flows into the Xiaodong stream. This time, though, it was built by local carpenter masters.

26 According to a local genealogy book, a member of the Wu family in Choulinshan Village managed a bridge in Xiaodong Village during his lifetime. He was born in 1671.
27 The Dabao Bridge, or its former structure, was known as the "Xiaodong Bridge" in local literature. After a second bridge was built, the Dabao Bridge was then known as the "Xiaodong Downstream Bridge" to distinguish it from the "Xiaodong Upstream Bridge".
28 Ca. 77 km straight line distance, 160 km walking route.

Figure 6.13 Xiaodong (Upstream) Bridge (小东上桥), Xiaodong Village, Shouning, Fujian.

Figure 6.14 Xiaodong Bridge. Modest in appearance, the corridor of the Xiaodong Bridge is rich in decorative members.

Figure 6.15 Ink inscription in Xiaodong Bridge bearing information about the bridge masters (written from right to left, following Chinese tradition):

Top: deputy (副), principal (都);

Middle: master carpenter (绳墨);

Below:

On the left from top to bottom: Wu (吴), Guang (光), Fu (福);

On the right from top to bottom: Xu (徐), Zhao (兆), Yu (裕).

The Xiaodong Bridge is the first known bridge built by the Kengdi Masters, as witnessed by the carpenters' names inscribed on the corridor beams: Chief Master: Xu Zhao-Yu (徐兆裕), Deputy Master: Wu Guang-Fu (吴光福) (Figure 6.15). These two names, coming from two families in local villages, marked the beginning of the two-century family tradition of the Kengdi Masters.

Unfortunately, we cannot compare the technical features of these two bridges directly. Both of them were largely rebuilt in the late 1930s. In 1937, after the reconstruction of the Yangmeizhou Bridge – whose history has been mentioned in previous chapters and will be discussed in even greater detail later on – the residual money was used to repair the three bridges around Xiaodong Village. The Xiaodong Bridge was repaired in 1938, many decorative members of the corridor are now lost, and the under-deck structure was rebuilt by replacing and reusing old building members (Figure 4.100, a). The Dan Bridge was entirely rebuilt in 1939, this time by a master descended from the same Xu family which built the Xiaodong Bridge. Thanks to the inscribed wooden board in the corridor, we have some information regarding a former construction. The Dabao Bridge was also repaired in the same year.

About the two "pioneer" Kengdi Masters who built the Xiaodong Bridge, little is known, as they did not build other bridges or leave us any other records. The better-known story of the Kengdi Masters started in the next generation of these families. After half a century devoid of information of any kind, a most dramatic bridge construction story was recorded and left to us today – that of the Xuezhai Bridge.

6.2.3 Xuezhai Bridge: the failure and success of the Kengdi Masters

6.2.3.1 Xu Yuan-Liang

The Xuezhai Bridge, located in the Town of Sankui, Taishun County, Zhejiang Province, was built in 1857.

The name "Xuezhai" literally means "Xue family home." The history of its construction is recorded in the genealogy book of the Xue family. The bridge is located directly in front of the ancestral hall of the Xue family. A bridge was initially built on this spot in 1512. After it was destroyed by flooding in 1579, the Xue family built primitive bridges as a temporary measure. In 1856, the construction of a new bridge was finally initiated. At the beginning, the carpenter master in charge of the project was a man named Wu Guang-Qian (吴光谦). However, construction was constantly being disturbed by a villager from a certain Zhang family, who believed that the bridge would have a negative effect on the *feng shui* (and therefore welfare) of his family. In August, when the construction was going on, the Zhang family gathered and threatened to destroy the bridge, whereas the Xue family crowded around riverbank to guard the construction process. Both sides gathered together hundreds of people, who stood facing each other, ready to fight. The under-deck structure was constructed in this atmosphere of confrontation and tension. On the day of completion, when the scaffolding was removed, suddenly "the bridge became a dragon on the waves" (化作长龙卧波矣, i.e. collapsed). Carried away by anger, the villagers first accused Master Wu Guang-Qian of getting the design wrong (Figure 6.16). Later, after they had calmed down, they dismissed the charge against the master, and turned against Zhang, saying that his disruptive tactics had caused the project to be carried out in great haste. The more the villagers blamed Zhang, the more determined they were to rebuild the bridge. This time, they called in a Kengdi master, Master Xu Yuan-Liang (徐元良).[29]

The record in the genealogy book conforms to the ink inscription on the corridor beam on the second construction. There, the chief carpenter master's name

29 The report on the reconstruction of the Jinxi Bridge (重修锦溪桥记略), from the genealogy book of the Xue Family of Jinxi. Preserved by the villagers' committee of Xuewai Village, Taishun County, Zhejiang Province.

Figure 6.16 The record of the Xuezhai Bridge construction in the genealogy book of the Xue family.

Preserved in the village committee office, Xuewai Village, Taishun County. Highlighted characters are the blame on the "clumsy carpenter Wu Guang-Qian."

is given as "Xu Yuan-Liang from Xiaodong," with another beam listing the five names of the deputy masters – the carpenter Xu Bin-Gui (徐斌桂), another member of the Xu family, is listed first and the following names including a carpenter from a Chen family, two from a Zheng family, and the last from the local Xue family. The reconstructed wooden arch bridge has a clear span of about 27.5 m, with a bridge deck that is higher and steeper than bridges elsewhere, which is a typical aesthetic feature of Taishun County. The Xuezhai Bridge was in good condition (Figure 6.17) until September 2016, when it was destroyed by flooding. It was later reconstructed by Taishun local bridge masters.

Not surprisingly, the three Xu masters of the two bridges mentioned so far are from the same family. Actually, they are the grandfather (Xu Zhao-Yu), the father (Xu Yuan-Liang) and the son (Xu Bin-Gui) (Figure 6.18). Neither man

Figure 6.17 Xuezhai Bridge (薛宅桥), Taishun, Zhejiang.

It was built in 1857 by the Kengdi Master Xu Yuan-Liang using mature technology and was in good condition at the time when this photo was taken, in 2012.

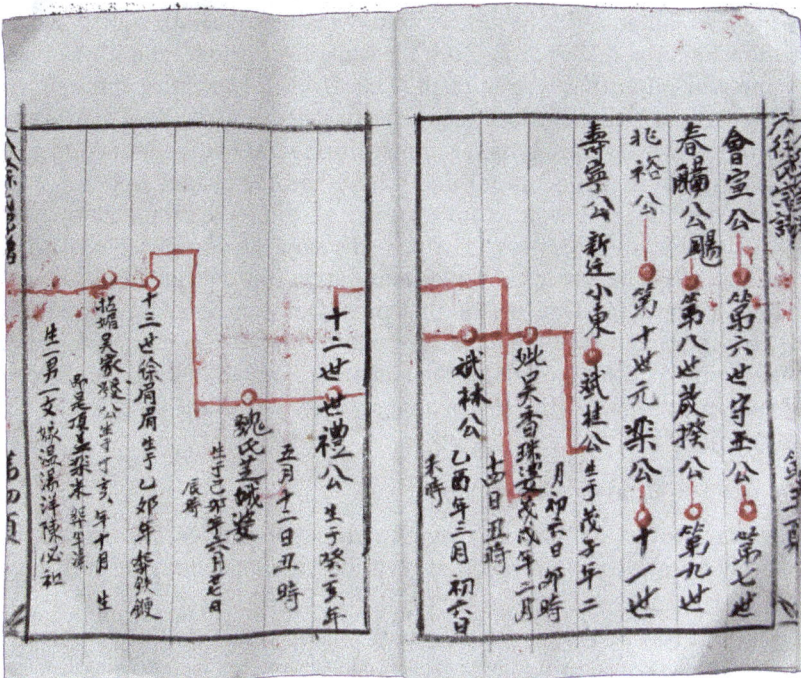

Figure 6.18 Genealogy book of the Xu family in Xiaodong Village, preserved by a descendant of that family in Xiaodong Village, Xu Qiguang (徐启光)*.

The ninth ancestor is Xu Zhao-Yu, the tenth is Xu Yuan-Liang and the eleventh Xu Bin-Gui.

*He's great-grandfather was the brother of Xu Bin-Gui. This branch of the family were not carpenters. The genealogy book was not kept very carefully, since there are only a few descendants of the entire branch left today.

from the elder generations left any other bridges, only Xu Bin-Gui did. In his life, Xu Bin-Gui built at least three bridges, in three different counties, including Jingning, Qingyuan, and his home county of Shouning. His last known work was a hometown project, the Dabao Bridge in 1878 (Figure 6.42), downstream of Xiaodong Village. With a clear span of 31.9 m, it has a rather similar appearance to the Xuezhai Bridge of the Taishui taste, with a steep, high bridge deck.

Although there is a large Xu family in Kengdi area, the carpenter masters came from a separated branch which was not prosperous. Xu Bin-Gui was the only son of Xu Zhao-Yu; he had three sons who took up a career in bridge-building, but only one grandson inherited this technical tradition. This grandson, Xu Ze-Chang (徐泽长) did not marry, and had no son of his own; he worked with a younger cousin, Zheng Hui-Fu (郑惠福) from a neighbouring village. This family bridge-building technique was thus transferred from the Xu family of Xiaodong Village to the Zheng family of Dongshanlou (东山楼) Village (which is still inside the Kengdi area). This family line is commonly known as the Xu-Zheng families[30] of the Kengdi Masters (Figure 6.19).

Probably not a pure coincidence, the first known work of Xu Ze-Chang was the reconstruction of the Dan Bridge,[31] a bridge originally constructed by the He family masters from Xiajian. As mentioned before, the first known bridge by the Xu family was a rather similar small bridge in the same village, only a few hundred metres away from the same Dan Bridge, and it was built only nine years after the erection of the Dan Bridge. Thereafter, through the Dan Bridge, two opportunities of technique transfer from the Xiajian Masters to the Kengdi Masters are possible: either during the first construction, the Kengdi Masters (first generation) might have had the chance to observe or even take part in the construction process, or, after the construction was completed, they had a century in which to study the completed structure, especially in terms of repair or even dismantling it (third generation).

It was not only the Dan Bridge that provided such possibilities of technique inspiration between these two carpenter groups. Two years before the reconstruction of the Dan Bridge, the construction of Yangmeizhou Bridge (1937) played another crucial role in a more direct way. More on this later.

6.2.3.2 Wu Guang-Qian

Let's go back to the Xuezhai Bridge. The article on the Xuezhai Bridge (in Xue family's genealogy book) did not mention anything about the carpenter

---

30 Although the Xu and Zheng families should be counted as two families in a sense and according to English grammar, Chinese scholars normally consider them as a single bridge-carpenter family (line) ("徐郑世家") in the sense of technique inheritance (Ningde 2006; Gong 2013; Zhou et al. 2011).

31 According to Gong Difa, Xu Ze-Chang also built another bridge, the Feiyun Bridge (飞云桥) in Shouning in 1938. However, there is no written confirmation of this, only the memory of and oral information provided by a descendant of the Xu-Zheng families (Gong 2013, 21). However, according to the author's on-site investigation, the technical features of this bridge do not point to the Kengdi Masters.

坑底匠人
Kengdi Masters

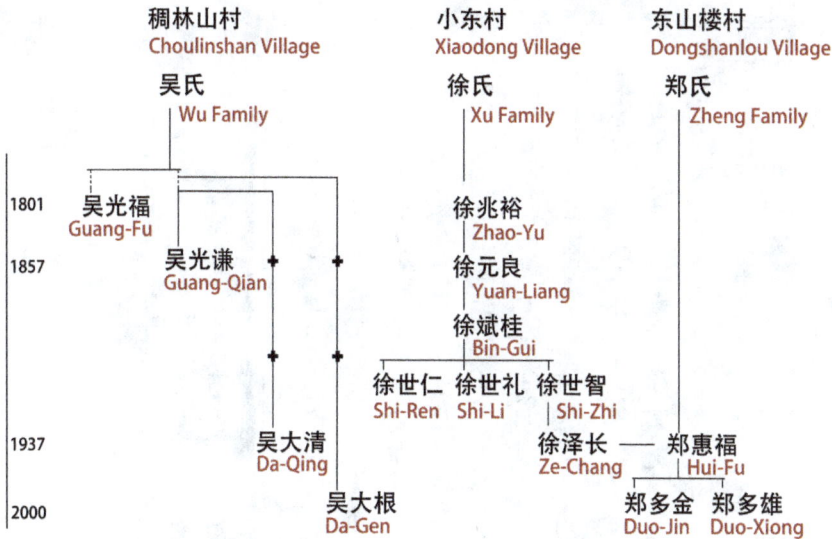

Figure 6.19 Technical pedigree of the Kengdi Masters.

master Wu Guang-Qian from the first unsuccessful construction, except his name. It is most likely that he was also a Kengdi Master, given his shared "generation name" with Wu Guang-Fu, the builder of the Xiaodong Bridge. Following this clue, the author visited the ancestral hall of the Wu family in Choulinshan Village, a neighbouring village of Xiaodong, and not only found Wu Guang-Qian's name in the genealogy book of this family, but also, surprisingly, an entire page of a biography of his life (Figure 6.20). What is surprising is that, whereas it is not uncommon to find biographies of notable members of a family in genealogy books, it is rather unusual to find the biography of bridge carpenters. In fact, this is the only biography of a bridge carpenter master found by this author in the MZ area.

In traditional rural life, every family has its own genealogy book. The continuation of a genealogy book is always a serious enterprise for a family. The wealthier and more prosperous families, usually with a tradition of education, have more elaborate genealogy books. Poorer peasant families may have only the names of three or four generations of ancestors written in their books.

According to the genealogy book of the Wu family in Choulinshan Village, this family branch has a long tradition of education. Among other branches of the Wu family in Kengdi, the Choulinshan branch was one of the most prosperous and prestigious branches in the local society. Their elevated status began at the end of the Ming Dynasty, around the 1640s, when all of rural society in southeast China was under threat by bandits.One ancestor of this

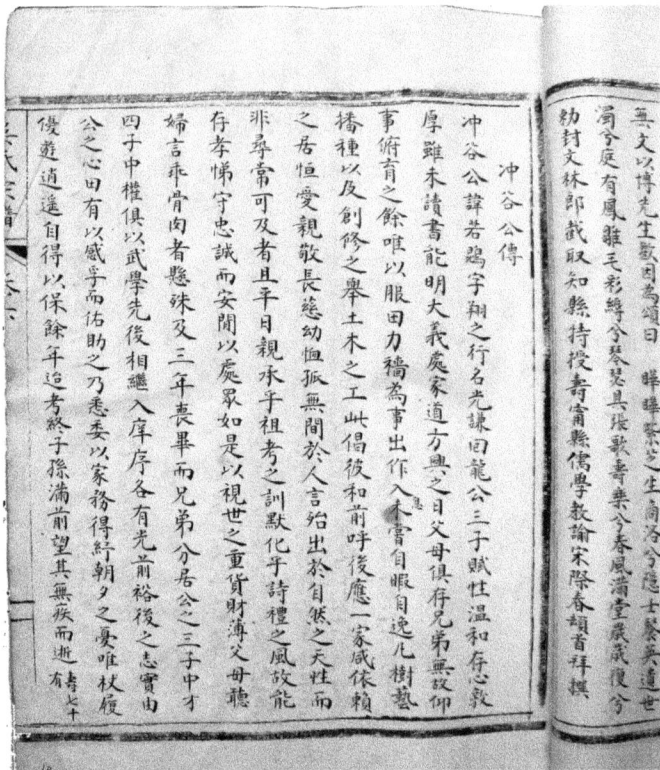

Figure 6.20 Biography of Wu Guang-Qian in the genealogy book of the Wu family. Preserved in the ancestor hall of the Wu family, Choulinshan Village, Shouning County.

branch kept calm and brave and (unlike many others) refused to flee, thus was able to protect all the members of his family. The next generation took care not only of their own family, but also of local affairs. This celebrated member of the second generation organized the construction of two bridges,[32] among many other public facilities. Since they were living in a calm time in history again, the family was able to accumulate wealth via commerce, and to put a great deal of effort into education. The "cultivation-and-education tradition"[33] reached its peak with Wu Guang-Qian's grandfather, who passed the highest imperial examination and received the title of *Jinshi*, which is a

32 One was in Xiaodong Village, probably a former structure of the Dabao Bridge mentioned above, and the other being a former structure of the Yangmeizhou Bridge, on which more later.

33 Way of life and education in traditional Chinese peasant society. In such families, people studied classics and prepared for official examinations in their spare time, or supported the education of a few specific (talented) members, while others worked on farming.

huge honour in traditional Chinese rural society. Guang-Qian's father was not as successful, although he still passed the imperial examination at the county level, and worked as a tutor for many years. There are four brothers in the Guang generation: Guang-Qian was born in 1803 and was the third boy. His two older brothers received education at government schools. Guang-Qian had no interest in studying literature, so he took over the running of the household, and was good at cultivation as well as construction. It seems most of his construction work involved building houses. Some of his nephews kept on the path of the study of literature and taking examinations, while two of his sons passed the imperial examination in martial arts at the county level.

Guang-Qian was in his fifties when he built the Xuezhai Bridge. Since the construction was not a success, naturally, it was not mentioned in his biography. In fact, his biography did not mention his bridge-building career at all. None of Guang-Qian's direct descendants was involved in bridge-building – although a great-nephew (a grandson of Guang-Qian's older brother) played a major role in the story of another legendary bridge project eighty years later, the Yangmeizhou Bridge.

6.2.4 Yangmeizhou Bridge – an encounter between the two bridge-building groups

On the road from Kengdi to Taishun, 17 km to the east of the town of Kengdi, there is a bridge across the Houxi Brook called the Yangmeizhou Bridge (named after the village nearby).

Although the Yangmeizhou Bridge was mentioned several times in Chapter 4, the focus there was on its technical features and the construction process. Here, we will reframe its story in the context of its construction history and the carpenters involved.

According to the inscription on the bridge corridor, the Yangmeizhou Bridge was first established in 1791. However, it is likely that there was a bridge built at the same or a nearby spot even earlier than that. As mentioned briefly above, a member of the Wu family in Choulinshan Village in Kengdi (the great-great-grandfather of Wu Guang-Qian, born in 1671) was reported in his biography to have organized the construction of the Yangmeizhou Bridge (and a bridge in Xiaodong).[34] If this report is reliable, the construction of these two bridges would have taken place in the late seventeenth or earlier eighteenth century.

Since the 1791 version, this bridge has been repaired and reconstructed several times. The first reconstruction took place exactly half a century later (1841), with project directors from Taishun County. The carpenter master in charge was a Xiajian Master of the Zhang family, and the work included dismantling the old structure and building a new one.[35] Then, sometime between

---

34 Biography of (Wu) Chun-Ting, Genealogy book of the Wu family in Choulinshan Village.
35 Contract for the Yangmeizhou Bridge in 1841. Preserved in Shouning Museum.

1851 and 1861 (Emperor Xianfeng Period), the bridge was again rebuilt, by unknown carpenters, with unknown organizers.[36] In 1869, the bridge underwent a reconstruction, organized by a Wu family member in Kengdi,[37] from another branch of the Kengdi Wu family (i.e. not the bridge-builder branch).[38] In 1937, the bridge was in urgent need of repair/reconstruction again. This time, the reconstruction project was documented in various ways, including a contract, an inscription, and descriptions given to the author by multiple eyewitnesses in the course of his fieldwork.

The inscription on the corridor beams records the project directors as being from the Kengdi local area and from the county-capital of Shouning. The majority of the funding came from local villages in the Kengdi area, and from villages in other parts of Shouning County, as well as neighbouring counties including Taishun, Qingyuan and Jingning. The project was entrusted to Master Wu Da-Qing (吳大清) from Choulinshan Village in Kengdi. He was a grandnephew of Master Wu Guang-Qian, and a cousin of Master Wu Da-Gen (吳大根) mentioned in Chapter 4, who was still active at the time of the author's research. (The age difference between these cousins is some sixty years).

Wu Da-Qing was born in 1899 and was 38 years old when he was put in charge of the project. He was a little younger than the above-mentioned Kengdi Masters of the Xu-Zheng families, as Xu Ze-Chang was born in 1892, and Zheng Hui-Fu in 1895. All three of the masters were at their best age for carpentry. Somehow, Wu Da-Qing won the project instead of the other team.

There were stories told by members of the Wu family of Kengdi Village to the author when he was looking through the genealogy books. According to one legend, Wu Da-Qing won the project because he was a relative of the director's. Although this is possible, we must bear in mind that almost all of the inhabitants of the Kengdi villages are more or less related, and both surnames – Wu and Xu – appear among the names of the directors of the project. Judging from the name and the village of origin, the Wu-family director does not seem to be a close relative of Wu Da-Qing. The same goes for the Xu-family project director.

According to another story, there was someone else who was competent, who came from elsewhere, but Wu Da-Qing was able to scare him off by quoting the complicated local building rules relating to the bridge corridor. However, this kind of tale is typical of craftsmen.

Although Wu Da-Qing was the chief master of the bridge, other Kengdi Masters also participated in the construction process. Zheng Hui-Fu even

---

36 Contract for the Yangmeizhou Bridge in 1937. Preserved in Shouning Museum.
37 Inscription on a corridor beam, and the biography of Wu Guang-Gu (吳光穀) in the genealogy book of the Wu family in Kengdi.
38 They are only distant related. This Wu branch lives in Kengdi Village, while the bridge builder branch lives in Choulinshan Village nearby.

brought his nine-year-old son Zheng Duo-Jin (郑多金) to the building site.[39] This boy took on the family tradition and became a bridge-building master in later decades.

There was another boy at the construction site at the time: 12-year-old Dong Zhiji (董直机). He lived in the township of Lingbei in Taishun County, and at the time of the bridge construction, he happened to be around Yangmeizhou Village visiting relatives. Full of curiosity, he joined the carpenter group and voluntarily assisted in fetching and carrying. Wu Da-Qing was impressed by this smart and diligent boy, he received this boy as an apprentice on the site – half seriously and half in jest. He taught the boy some basic rules and skills of bridge-building – without going into the core technical knowledge (more on this later). At that time, Dong Zhiji did not know any carpentry: he kept this knowledge earnestly in mind for the rest of his life. The nominal master-and-apprentice bid farewell to each other when the project was finished, and never saw each other again (partly because of the wars lasting from 1937 to 1949). Five years later, Dong Zhiji officially apprenticed to another carpenter and started learning woodworking. His master was a house-building carpenter, and so was Dong Zhiji, for most of his life. However, the memory and dream of (woven arch) bridge-building stayed with him; he never forgot what he saw and learned at the Yangmeizhou Bridge site. When he was finally "discovered" in the twenty-first century by local cultural workers as a bridgebuilder, he got the opportunity to build his first woven arch bridge in 2004 at the age of 79. From then on, he became known to the public and built another bridge in 2008. He passed away in 2018.

The title of "National Intangible Cultural Heritage Inheritor" was conferred on Dong Zhiji in 2009. In 2011, Zheng Duo-Jin received the same title. During an interview with this author, in January 2019, 91-year-old Zheng Duo-Jin recalled that he remembered the other little boy at the Yangmeizhou Bridge construction site, saying "that boy was three or four years older than me."

When Wu Da-Qing arrived at the Yangmeizhou bridge site, he was overwhelmed, at first. Although he was capable of bridge-building, this bridge was to span a river pool more than 20 metres deep, not to mention the fact that it would need to have a clear span of ca. 33 metres, catalogued under the large-scale bridges of this type. Neither Wu Da-Qing nor, apparently, his colleagues, had the confidence to take on this construction project. The project committee had to turn to the Xiajian Masters – again – as the bridge committee had done in the 1841 reconstruction of the bridge. This time, the Xiajian Masters were called only for the under-deck structure. Wu Da-Qing remained in charge of the bridge corridor and kept his status as a chief master.

39 Zheng Duo-Jin, interview with the author on 15 January 2019 in Dongshanlou Village, Fujian Province.

The leading master came from Xiajian and was called Zhang Xue-Chang (张学昶), a fourth-generation bridge builder in his family. He arrived with huge troops of carpenters. The technical construction process is described in detail in Chapter 4 (Sec.4.5.3.1.2 and 4.5.3.2.2). The key was a method using swing-frame-scaffolding, a rather skilful but dangerous construction technology requiring well-trained and systematically-organized carpenters, who were divided into three groups: one went down at the water, one went up on the scaffolding, and one group left on the riverbank. In addition to having an in-depth knowledge of their craft and technological processes, the carpenters also needed to be good climbers and swimmers. Dong Zhiji recalled that the carpenters who worked on the most dangerous part were offered with two silver coins (in a red envelope) and a bowl of liquor before they stepped onto the dangers scaffolding (Introduction, scene seven).

Zhang Xue-Chang and his carpenters left the construction site as soon as the under-deck structure was finished. Wu Da-Qing led his colleagues in finishing the rest of the work. On a corridor beam, Zhang Xue-Chang's and Wu Da-Qing's names appeared as the carpenter masters, side by side. Interestingly, to keep a balance in this embarrassing situation of how the labour was divided between the two groups of masters, although Zhang Xue-Chang's name came first, his title was given simply as "carpenter," whereas Wu Da-Qing's title, although his name was listed second, was given as "chief master" (Figure 6.21).

The swing-frame-scaffolding construction method of the Xiajian Masters made a strong impression on the Kengdi Masters. What they witnessed and experienced was passed down to the next generation. On the construction site in 2013, Master Wu Da-Gen described and criticized the construction method to the author quite openly. He had some reasonable doubts and criticism on some specific technical details that he had heard about from different sources including his elder cousin Master Wu Da-Qin, and father-and-son of the Zheng family, Zheng Hui-Fu and Zheng Duo-Jin, who are also related to him by marriage.[40]

6.2.5 *The bridge carpenter families: how did the carpenters form groups?*

When we discussed the family-tradition bridge-building carpenters in the previous chapters and at the beginning of this chapter, we identified the Xiajian Masters with the Zhang family, and the Kengdi Masters with the Xu-Zheng families, out of convenience and based on the current shared knowledge of most authors on this topic. However, by now, readers have probably realized that each group of carpenters was composed of several families from the local area (Figure 6.22).

---

40 Zheng Hui-Fu had another son named Zheng Duo-Xiong, the younger brother of Zheng Duo-Jin. Wu Da-Qin and Zheng Duo-Xiong married two sisters. In Wu Da-Qin's narration, he learned bridge building technology directly from Zheng Hui-Fu.

Figure 6.21 Inscription giving the names of the carpenter masters of the Yangmeizhou
Bridge.
(The photo was taken in August 2017, using infrared photography.)
Right-hand side (ranked first): Carpenter Zhang Xue-Chang* from Xiaokeng**
Village in Ningde County.
*On the beam, the name was mistakenly written as "Zhang He-Chang" (张鹤昶),
 because "He" (鹤) is pronounced the same as "Xue" in the local dialect.
**Another common name for Xiukeng Village.
Left-hand side (ranked second) Chief Master Wu Da-Qing from Choulinshan Village
in Shouning County.
Bottom: With skilful technique and craft.

The Xiajian Masters began with three master carpenters with different
family names: one called Li, one called Wu and one called Zhang. In their
first known project, the Zhang family ranked last on this list. Also, when the
other two families gradually left their bridge-building careers, although the
Zhang family undoubtedly took on a dominant position, it was still closely
connected to the other families, through cooperating on projects, and through
marriage. Among these connections, the He family was the most important.

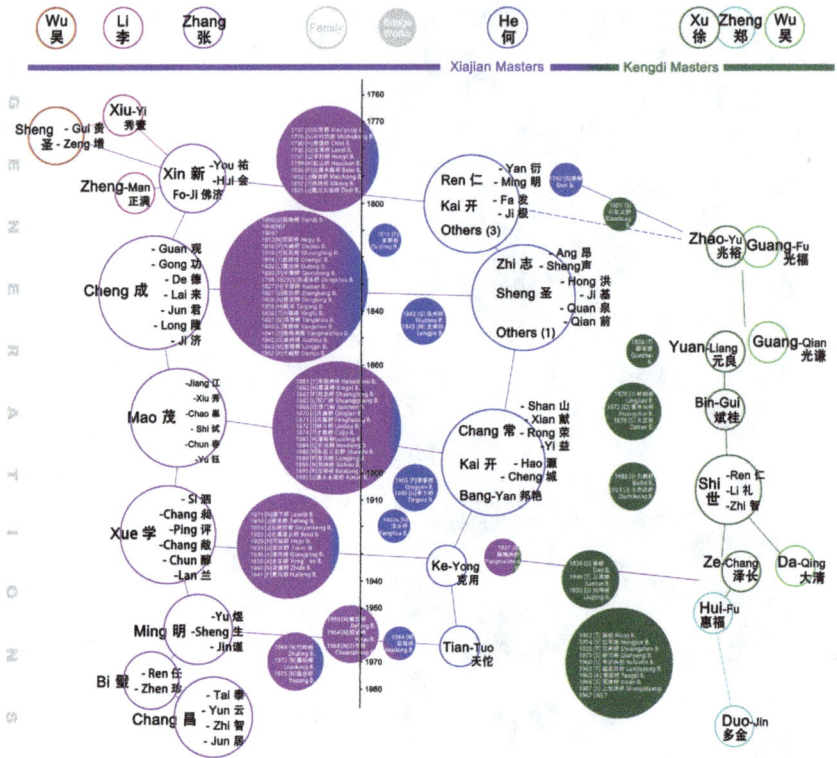

Figure 6.22 Technical pedigree and bridgeworks of the Xiajian and Kengdi Masters. Abbreviation of location: A – Fu'an; F – Jin'an; G – Gutian; H – Zhenghe; J – Jingning; N – Zhouning and Jiaocheng; M – Minhou; P – Pingnan; Q – Qingyuan; S – Shouning; T – Taishun; W – Wencheng; Y – Jianyang.

The He family lived in neighbouring villages and appeared on the scene a generation after the Zhang family. The two families worked together from time to time and mostly under the leadership of Zhang family members. Although the He family also built bridges on their own during the two centuries, these bridges (especially the early examples) are mostly small-scale constructions. All these clues indicate that the He family learned their bridge-building techniques from the Zhang family through observation and cooperation, and that they became experienced enough to handle small-scale projects by themselves, although their technique could not compete with their Zhang surnamed colleagues. Theirs was a relationship of strategic cooperation. As demonstrated by the 1937 Yangmeizhou Bridge project, the Xiajian Masters comprised a huge troop of associated craftsmen under the leadership of the Zhang family. Even though we do not have a list of the names of all the carpenters involved, it is reasonable to assume that members of the He family were involved, as well as members of many other families from various villages in the Xiajian area.

Whereas the Xiajian Masters had a clearly-identifiable core consisting of a single family, the Kengdi Masters were the result of the cooperative efforts of two main families. The leading families were the Xu family (which later became the Zheng family, due to a lack of descendants with the Xu name), and the subordinate Wu family, these two families working together from their first known project to the most recent constructions:

In the leading family, the Xu family exhibits a clear tradition of bridge carpenter craftsmanship: most of the family members took up this career, the skills were handed down from father to son (without other assorted relatives such as cousins or nephews) from Xu Zhao-Yu to Xu Ze-Chang, for five generations (Figure 6.19). This compact family team had one main obvious disadvantage, and that was that it could easily be "interrupted" if one "link" in the chain broke, which is exactly what happened in the fifth generation. If the carpenter master had no son of his own, the family career had to be continued under another family name, so as not to be abandoned outright.

The vulnerability of the Xu family is a result of the family status in the locality. The carpenter family moved to the Kengdi area relatively recently. In their somewhat crudely-compiled genealogy book (Figure 6.18) the early history of their family branch is unclear, and it mentions the Xu Bin-Gui brothers as the ancestors who moved from elsewhere to Xiaodong Village. However, we know that Xu Bin-Gui's grandfather, Xu Zhao-Yu, had already built the Xiaodong Bridge. So it's more likely to be an inaccurate entry in this (in itself blurrily narrated) genealogy book: that is the ancestor who first settled in the Kengdi area was probably Xu Zhao-Yu (if not even earlier), rather than his grandsons. He moved in alone and set up home in the marginal area outside the villages of the larger families in the area, and worked as a craftsman, which was a good survival method for people who have little arable land to cultivate.

The Wu Masters, on the other hand, came from a long-established, large family, which, surprisingly, had a tradition of education. Cultivation and education formed the backbone of the family. Those members of the Wu family who preferred to earn their living by becoming craftsmen (considered an inferior career choice) did so, but they did not feel obliged to pass on their craftsmanship to their descendants. The Wu family carpenters were also less successful compared with their Xu-Zheng colleagues. Of the four known members (Figure 6.19), the second, Wu Guang-Qian, failed to build the Xuezhai Bridge and the third, Wu Da-Qing, was unable to build the Yangmeizhou Bridge and had to ask for outside help from the Xiajian Masters. The first Wu Guang-Fu and the last Wu Da-Gen worked as deputy masters on projects, assisting the colleagues of the Xu and Zheng families. All in all, the bridge-building capabilities of the Wu family carpenters seem to come not from their own family line, but rather from their cooperation with the leading Xu/Zheng families.

Of all these families, the dominant group, the Xiajian Zhang Masters, provided the most stable model for technique inheritance. This group of carpenters had a large crew. Multiple branches of the family were involved, with brothers and cousins, fathers and uncles, sons and nephews working together in different permutations. There was not a continuous direct blood-line in this family: the longest spanned only three generations. Actually, as mentioned, it was not unusual for bridge carpenters to be childless – the dangerous work was a profession usually taken up by poor people with a limited range of options, and they naturally had fewer opportunities to marry.[41] However, thanks to the dedication of this large extended family, it was possible for the technique to be handed down among several branches of the family. It is also the sheer numbers of the family involved in carpentry that kept the carpenter team supplied with members, and enabled the Zhang family to form a systematically-organized team with a sophisticated system of labour division.

Although they were and are the bridge carpenters with the longest trad-ition, neither the Kengdi Masters nor the Xiajian Masters developed their technique by themselves. The Xiajian Masters have a legend about how their ancestor embarked on the career, as their first known work showed signs of external leadership. The Kengdi Masters built their first bridge only nine years after the work of the Xiajian Masters from outside their neighbour-hood, on the same brook and on the same scale, which is not likely to be a coincidence. However, we cannot categorically state that there is a relation-ship based simply on location and timeline. In addition to written and spoken words, clues are also provided by structure and technique.

6.3 Construction features and technical evolution

6.3.1 *Distinctive features of bridge pedigree*

Among all the sources we've mentioned regarding any bridge's carpenter master, oral information is listed as the least reliable. For example carpenters may exaggerate their role in a project, so if we listen only to the stories of the Zheng descendants of the Kengdi Masters, we might include the Yangmeizhou Bridge in the list of their works.[42] However, all the technical details of its under-deck structure exhibit features of the Xiajian Masters, and our specu-lation is confirmed by the inscription, bridge contract and narrations from other sources. For the same reason, we can also remove another bridge from the list of works by the Xu-Zheng families as provided by other authors,[31]

---

41 Poor families were unable to pay a 'bride price' for their sons.
42 In Zhou et al. (2011, 48), the name of the chief master of Yangmeizhou Bridge was given as Zheng Hui-Fu from Kengdi. However, according to our research, although he was part of the building crew from the construction in 1937 – but only as a common carpenter assisting, working mainly on the corridor. See Sec.6.2.4 above.

since the bridge was included on the list based on descendants' statements, but it exhibits features that are rather different from other works by this family.

In their silent way, the structures can also tell us about their builders. In Chapter 4, we mentioned some of the crucial technical features which are tightly linked with specific carpenter families/groups, including the proportion of the arch members (Sec.4.3.2.3, design [Rule III]), the angle/inclination of the crossbeams (Sec.4.3.2.4, design [Rule V]), and the form of specific wood joints (Sec.4.6.2). If we bear these typical features in mind, we can distinguish between the bridges made by the two largest groups of carpenters (Figure 6.23).

These features somehow identify a group, because they belong to the core knowledge of bridge design (the professional secrets) from the time before modern draughtsmanship, and thus were shared only amongst the core members of a specific carpenter group.

Some of the features are rather characteristic, remaining unchanged throughout the family history. For example joint type is one of the most reliable ways for us to differentiate between a bridge built by the Xiajian Masters and one built by the Kengdi Masters, since the form of the joint is a result of the working process and construction method, and this hardly changes in the technique handed down within a particular carpenter family/group. This is not difficult to understand: changes in joints at specific points will cause a chain reaction in the working process, involving the cooperation of the entire team. If such a change appears, it must be a radical change or even a technical revolution. This indeed did happen, but very rarely – the *choudu*/crossbeam-ramming technique mentioned in Chapter 4 being the best example of this. (A detailed analysis of its development is given below.)

Some other features are less durable over time. They might change within a generation or be changed by specific carpenters, with or without reason. The proportion of the slanted beams between the first and second system of the woven arch (design [Rule III]) is an example of this. As a rule, the ratio used by Kengdi Masters (ca. 1/2) is slightly lower than that used by the first generations of the Xiajian Masters (ca. 2/3). In some of the earlier works, they look similar and cannot be distinguished from one another simply by this single feature. However, this proportion of the Kengdi Masters was radically reduced to a much lower ratio – down to ca. 1/3 in the works of the latest generation.[43] Such changes are a simple question of design, and have less influence on the construction site, that is do not involve the cooperation of the whole carpenter group, and are therefore introduced easily.

The inclination of the topmost crossbeams (U-C-Beams[2]) is another useful but less reliable feature. On the one hand, it is a stable identifying feature, since the angles of the crossbeams are among the most "experience-dependent" and

---

43 In Zheng Duo-Xiong and Wu Da-Gen's works in recent years.

mortise & tenon

x = ca. 1/2 (earlier g.)
    1/3~1/2 (later g.)

Kengdi Masters

horizontal

Inclination of the top crossbeam
(U-C-Beam[2])

Location of the middle crossbeam
(L-C-Beam[2])

inclined

Joint form at the F-Beam[2]

Sign of *choudu* method   "piercing form"

Inclination of the C-Beams[1]

Inclination of the S-Beams[1]   inclined
                                 (but flatter than the S-Beams)   5/10 – 6/10 (when clear span 18–40m)

Horizontal Proportion of the First System   x = ca. 0.9–1.35

x = ca. 2/3

Xiajian Masters

Xiajian vs Kengdi

Scope of the "Quintessential Form"

Some Shared Features

dovetail

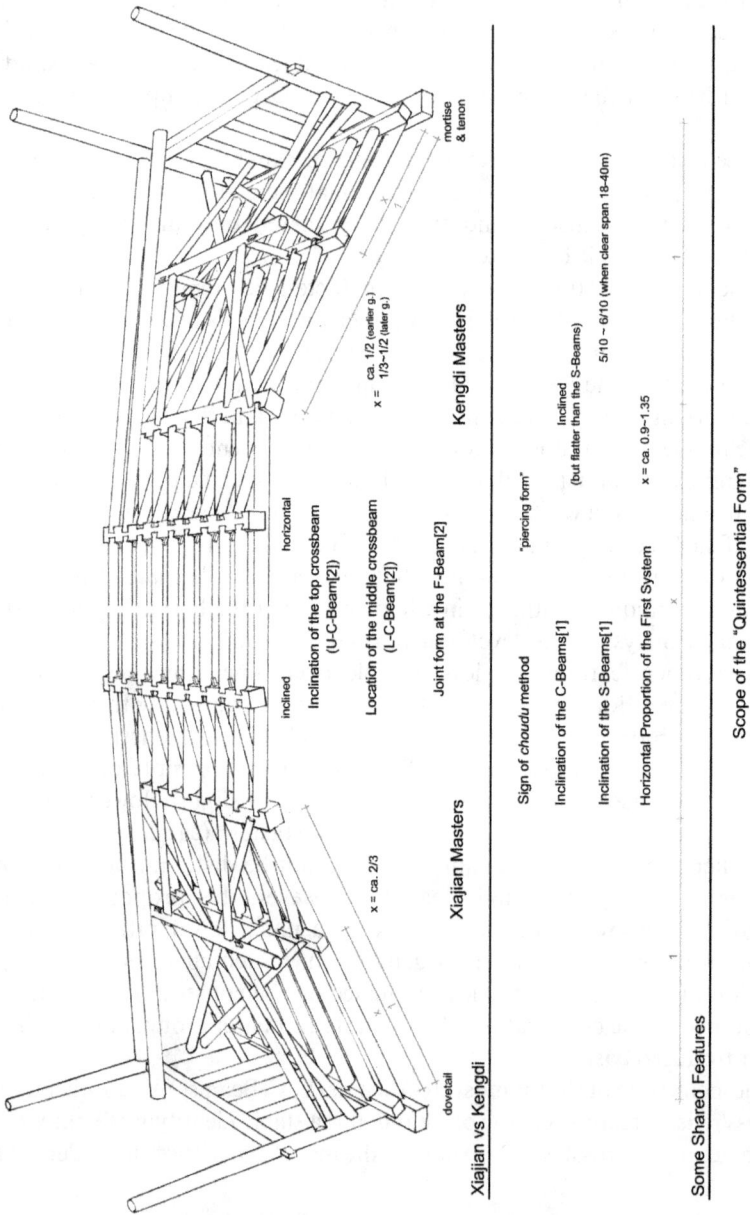

Figure 6.23 Xiajian Masters and Kengdi Masters: differences and similarities in the technical features of typical bridge design.

therefore most jealously-guarded secrets of the carpenter families' construction knowledge. On the other hand, the inclination is a less reliable factor for observation and measurement and furthermore, it can be influenced by structure deformation. Therefore, this factor is used only as an auxiliary reference point for our study of the technique pedigree.

It must be said that it is not possible to securely attribute a bridge to a specific group of builders using only the features listed above if the list of potential builders includes more than the Xiajian and Kengdi Masters. For example bridges built by the Changqiao Masters (Huang family) share quite similar features to bridges built by the Kengdi Masters. Some bridges not built by these families also share similar visual features with bridges built by them; and carpenters who do not have a family tradition learn from their more experienced colleagues, especially when they get a chance to join the projects of these influential families.

To be more specific, if we define a scope of "quintessential form" (Figure 6.23) of MZ bridges, being the result of the design methods used by the three most influential families, we cannot conclude that every bridge that fits this form was built by one of these three families.

However, from another point of view, it is much safer to exclude the possibility that an anonymous bridge was built by one of these families, if the bridge has features that are obviously different from the "quintessential form." Such features include but are not limited to: uncommon proportions of the woven arch members (extremely long or short beams, if the bridge span is not extremely small), less reasonable angles of the crossbeams, uncommon joint forms, etc. Examples will be analysed in the following sections.

6.3.2 Choudu: *a sign of the family-tradition bridge carpenters*

The role of the *choudu*/crossbeam-ramming step was emphasized in the discussion of the construction methods in previous chapters as being the most crucial technique for MZ-bridge construction: for pushing the woven arch tight and thus make it stable. It's a key process emphasized by each group of the family-tradition carpenters. Therefore, it is important to know whether a bridge was built using this technique or not. Naturally, this information is not recorded in written form. However, as long as the bridge is still standing, we can still find the answer from examining the structure.

6.3.2.1 Piercing form

If a woven arch structure was built using the *choudou* step, the "weaving-beams" (U-S-Beams[2]) must pierce through the topmost crossbeams (U-C-Beams[2]). This form is called "piercing form" (Figure 6.24, left; Figure 6.25, a) in this book and considered to be the "mature form." In this form, the H-Beams[2] do not meet the U-S-Beams[2] head to head in the crossbeam (namely, the H- and S-Beams[2] are not in a line), as the longitudinal beams in the first system. Instead, the H-Beams[2] lie above the H-Beams[1], and are located between the slanted beams of their own system (S-Beams[2]). Thereafter, two particular signs appearing together could be considered to indicate that the *choudu*/crossbeams-ramming process has been applied:

(1) The number and position of the H-Beams[2]: the H-Beams[2] being one more in number than the S-Beams[2] and being laid between the S-Beams[2]. Let's put it in a more visual way: if the first system has nine groups of beams (namely, nine slanted beams at each side and nine horizontal beams), the second system has eight groups of slanted beams (eight upper and eight lower at each side), located in between the beams of the first system, but nine horizontal beams, which do not match in number and position of the other (slanted) beams of their own system, but match that of the first system. In this form, the H-Beams[2] are hidden behind the H-Beams[1] seen from below.
(2) The joint of the weaving-beams: the S-Beams[2] pierce the U-C-Beams[2]. The tenon ends piecing out of the crossbeam (or cut at the surface of the crossbeam) is visible from below.

All the bridges of the family-tradition carpenters were built in this form (except for a very limited number of early examples, which will be discussed in more detail further on).

6.3.2.2 Matching form

Before the *choudu* method was developed, it was natural and intuitive to have the H-Beams[2] match in number and position with the slanted beams in the same system (Figure 6.24, right; Figure 6.25, b) (e.g. the first system has nine

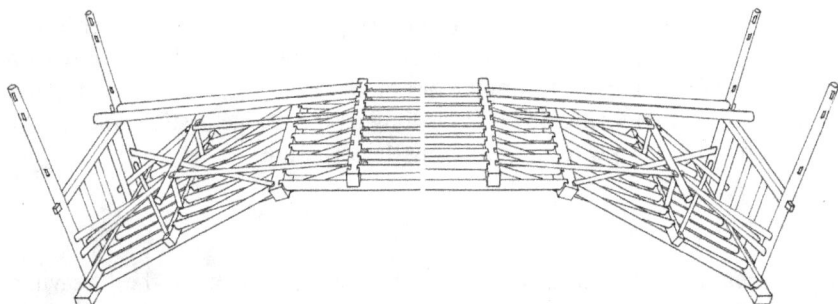

Figure 6.24 Sketch of the woven arch structure of the "piercing form" ("mature form"; left) and the "matching form" ("immature form"; right).

Figure 6.25 Under-deck structure "piercing form" ("mature form") and the "matching form" ("immature form"). (a) Longjin Bridge (龙津桥), Pingnan, Fujian. "Piercing form." (b) Zhangkeng Bridge (张坑桥), Shouning, Fujian. "Matching form."

groups of beams, and the second system has eight groups, horizontal and slanted, and all these longitudinal beams are aligned with each other). The H- and S-Beams meet each other in (the joint cuttings of) the crossbeams (hereinafter referred to as "matching form" in this book). With this form, it is impossible to apply the *choudu* / crossbeam-rammer process during construction. Therefore, it is a reliable indicator that the bridge was built by carpenters without or have little career tradition. Hereinafter, this form will be known as the "immature form."

Almost all bridge carpenters start with the "immature" form at the beginning of their bridge-building career – unless they had the chance to learn the advanced technique from a senior. A most representative example of this "intuitive" choice of form is the Tongle Bridge, built by Master Dong Zhiji. We told Dong Zhiji's story above: he worked on the construction site of the Yangmeizhou Bridge in 1937 at the age of 12. There, a Kengdi Master, Wu Da-Qing, taught him some basic bridge-building methods, and Dong Zhiji witnessed the working process of the Xiajian Masters' team. Only in 2004 did Dong Zhiji get his first chance to build a woven arch bridge in his home town: the Tongle Bridge in Lingbei, Taishun (Figure 6.26). The bridge was built in the "immature" "matching form" and quite understandably, considering it was a professional secret, Wu Da-Qing did not teach the boy this crucial step of *choudu* at all. It is also reasonable to assume that, as a little boy, Dong Zhiji

Figure 6.26 Tongle Bridge (同乐桥), Taishun, Zhejiang. Built in 2004 by Master Dong
     Zhiji. "Matching form."

wouldn't have been able to recognize the function and significance of such a
detailed measure even if he had observed it being implemented. So, after two-
thirds of a century, even though he tried his best to remember everything he
learned from the Yangmeizhou bridge site, Dong Zhiji built his own bridge in
a non-family-tradition way.

   Ironically, another carpenter who was awarded the title "Provincial
Intangible Cultural Heritage Inheritor" from the same county (Taishun),
Master Zeng Jiakuai (曾家快), also builds bridges in the same way. His biography tells us clearly that, originally a skilled young carpenter, he started his
bridge-building career by observing historical and new bridges. After finishing
a couple of small-scale bridges, he was noticed by local culture workers, and
almost "forced" by them to acknowledge Dong Zhiji as his master in 2011.[44,45]
This is not surprising, as all his bridges – whether they were built before or
after his "apprenticeship," are built in the "matching form" (Figure 6.27).

6.3.2.3 Interlacing form

While a bridge that uses the "matching form" is almost certain to be built by
non-family-tradition carpenters. However, a bridge with a "piercing form" could
have been built by non-family-tradition carpenters, if they had the opportunity
to work with and learn from family-tradition colleagues during construction.

   Non-family-tradition carpenters have various kinds of possibilities for
learning. Some of them start by joining a project and working with the
experienced masters. For observant individuals, it's a good chance to assimilate

44 Interview with Zeng Jiakuai and culture worker Ji Haibo on the Xiujian Village Bridge con-
     struction site in August 2011.
45 Having "apprenticeship" status was actually an advantage for his application to obtain the
     "Intangible Cultural Heritage Inheritor" title, as a part of the local cultural politics.

Figure 6.27 A new bridge by Zeng Jiakuai in Xiujian Village, Taishun, Zhejiang. "Matching form" (The dovetail sockets in the middle are left for the H-Beams[2]. They are placed opposite (matching) the inclined beams). (Photographed in August 2011, during construction)

Figure 6.28 Huilan Bridge (回澜桥), Shouning, Fujian. "Interlacing form."

the crucial technique. Others learn from examining existing bridges. In this case, even if they studied a perfectly-constructed mature structure form, it's less likely that they would figure out the *choudu* technique by themselves.

The Huilan Bridge (Figure 6.28) in Shouning County is an example of this kind, built in 1964 by a local carpenter without a family tradition. He said he learned the construction technique from studying an old bridge in his village which burnt down in1979 (Ningde 2006, 60). The bridge he built following the old example takes its form feature from the "mature-form," that is the H- and S-Beams[2] are interlaced with each other. However, the weaving-beams do not go through the crossbeams, and this indicates that *choudu* step was not executed. The form was simply copied without learning the most crucial method. This form will be referred to as the "interlacing form" in this book.

This bridge also exhibits other technical features that show this carpenter's lack of experience. The most notable aspect is that the C-Beams[1] are placed horizontally without an angle, leaving a huge gap between the

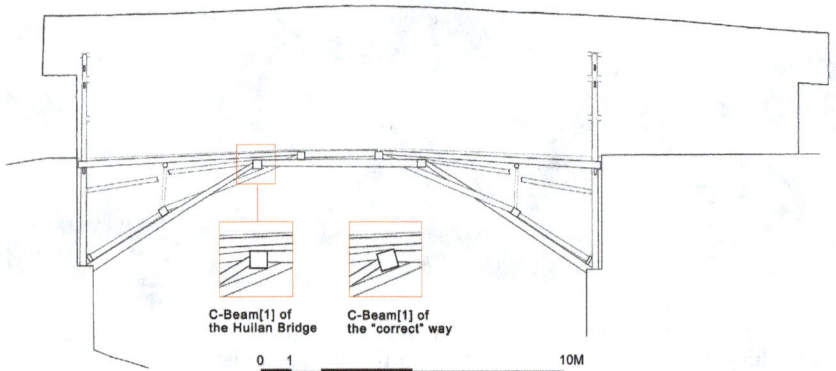

C-Beam[1] of
the Huilan Bridge

C-Beam[1] of
the "correct" way

0   1   10M

Figure 6.29 Longitudinal section of the Huilan Bridge, focusing on the form of the
       C-Beam[1].

weaving-beams underneath them (Figure 6.29). This indicates that the carpenter could not figure out the design of the angle for the crossbeams. The
horizontally-laid C-Beam[1] is a waste of material and destabilizing for the
structure (the woven structure can't be woven tightly afterwards). It appears
on some other bridges, and is a reliable indicator of an inexperienced bridge
carpenter.

Here further clarification is necessary. Even though these carpenters built
bridges with an "immature form," this fact is important only for the cat-
egorization of the bridge-building technique-pedigree, and does not necessarily reflect on the experience or skill on woodworking of the carpenters
themselves. We have seen above that, carpenters who have been awarded the
"Inheritor" title have built bridges with an "immature form," and later on we
will see that even the most powerful carpenter family went through a stage of
building "immature form" bridges. Below is another special example, where
bridges with an "immature form" prove to be an ingenious structural work.

This example, the Shunde Bridge, in Longquan County, Zhejiang, was
briefly mentioned in Chapter 4 (Figure 4.37). According to a bridge contract
preserved by the Xiajian Zhang family (Ningde 2006, 106–7), a bridge in this
village had been built in 1840 by the Xiajian Zhang brothers.[46] However, the
most recent version of this bridge was built in 1915, most probably at the
same location. This current structure was built in "interlacing form" structure
by unknown carpenters who were probably locals. Apparently, they did not
seek the assistance of the Xiajian Masters again, as the villagers had 75 years
before, but instead, they had simply studied the old structure. Since they
hadn't had the chance to witness their predecessors' working methods, they

46 Zhang Cheng-Jun (张成君) and Zhang Cheng-Ji (张成济).

Figure 6.30 Shunde Bridge (順德桥), Yangshun Village, Longquan, Zhejiang. Built in 1915 by unknown carpenters in "interlacing form."

copied the structural form without applying the *choudu* method. And yet, this produced a fabulous result: seen from below, the woven arch structure is very impressive, with naturally-curved timber being used where the beams (U-S-Beam[2]) are interwoven, and forming an extremely neat and beautiful structure (Figure 6.30).

6.3.3 The evolution of choudu in the Zhang Family of the Xiajian Masters

Even the family with the longest tradition starts with an immature period. The early works of the Xiajian Masters witness the development of the *choudu* technique in this carpenter family.

The first known work of the Xiajian Masters, the 1767 Xian'gong Bridge (Figure 6.4) has a clear span of 23 m and was built in this "immature" "matching form" (Figure 6.31). We already know that according to family legend, in the first generation of this family tradition, Zhang Xin-You worked as the third carpenter master on this project, under the leadership of a mysterious Li Xiu-Yi, and with a colleague called Wu Sheng-Gui. It appears that it was during this project that the bridge-building technique was passed onto the Xiajian Zhang family.

Figure 6.31 Xian'gong Bridge, Shouning, Fujian, built in 1767. The earliest known bridge of the Xiajian Masters. "Matching form."

The origin of this technique is untraceable. Considering the fact that Li Xiu-Yi was able to win such an important project, he might well have had experience in buildings bridges of this kind. The structure of Xian'gong Bridge might reflect a norm of that time.

In any case, during the time of the Xian'gong Bridge, there is no evidence that *choudu* existed. Xian'gong Bridge is the third-oldest surviving woven arch bridge today. The two bridges that came before it (Figure 5.5; Figure 6.32, Figure 6.33) and those (five in all, e.g. Figure 6.34) built in the two and a half decades after it were built in the "immature form," including the second oldest-surviving bridge of the Xiajian Masters, the 1790 Chixi Bridge in Zhenghe, Fujian[47]. This bridge was built under the leadership of Zhang Xin-You, together with a cousin, a nephew, and others. Its scale is similar to the Xian'gong Bridge (ca. 24 m). The *choudu* technique had not been invented at the time.

47 An even earlier bridge by the Zhang family was built in 1779 by Zhang Xin-You and his cousin Xin-Hui – the Shizhukeng Bridge (石竹坑桥) in Zhouning, Fujian. Its clear span was ca. 30 m. It was demolished in 1978 (Gong 2013, 10) and no clues are left regarding construction details.

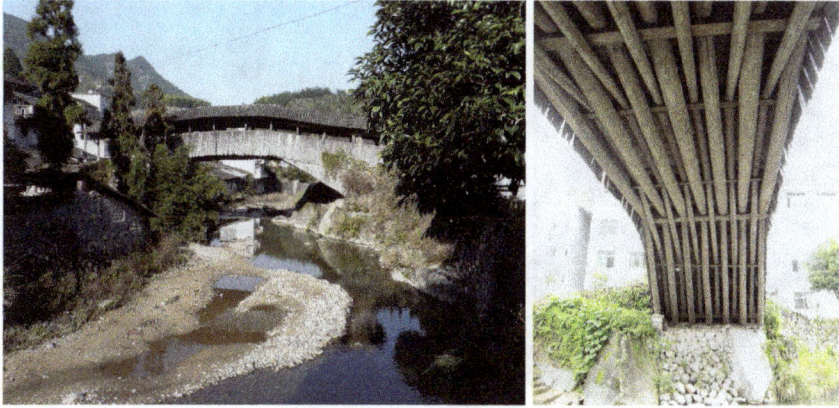

Figure 6.32 Dongkeng Xia Bridge (东坑下桥, Dongkeng Downstream Bridge), Jingning, Zhejiang. Built in 1698, unknown carpenters. "Matching form."

Figure 6.33 Dongkeng Xia Bridge. The immature technique has caused the woven arch to become loose, namely the L-C-Beam[2] has partly departed from the S-Beams[1].

Figure 6.34 Niao Bridge (袅桥), Qingyuan, Zhejiang. Built in 1769, unknown carpenters. "Matching form."

It was only four years later, in 1794, that the "piercing form" emerged and so it is highly possible that this was the corresponding *choudu* technique, as a solution to a large-scale challenge in the form of the 35 m-span Lanxi Bridge (Figure 6.35). The chief carpenter master was still Zhang Xin-You, while his brother-in-law Li Zheng-Man and a nephew as the deputy masters. Today, this is the earliest existing bridge built in the "piercing form." This bridge, together with three other bridges built in the same structural form under Zhang Xin-You's leadership,[48] witnessed the development of the *choudu* technique in the Xiajian Zhang family, a technical innovation in the construction history of the MZ bridges.

Zhang Xin-You retired in the 1800s. From then on, the next generation, the Cheng (成) generation, worked independently. The most active members included two sons of Zhang Xin-You's; Zhang Chengde (张成德) and Zhang Cheng-Lai (张成来). It appeared that some members of this younger generation had questions or doubts about the newly-invented technique. They sometimes worked with the new method, and sometimes they reverted to the old way. The Zhangkeng Bridge (Figure 6.25, b) in Shouning County in 1827, for example was in the old"matching form," built by Cheng-De and Cheng-Lai. However, other works by the same brothers before and after the Zhangkeng Bridge – including the Qiancheng Bridge (千乘桥) in Pingnan in 1820, and the Denglong Bridge (登龙桥) in Zhouning in 1836, were built using the new "piercing form." Moreover, yet another bridge[49] built in the same year (1827), by their cousins, was also built in the "advanced" form. After this initial period of going back and forth between the two styles, the "piercing form," with its

---

48 Houshan Bridge in Zhenghe County, 1799; Meichong Bridge in Jingning County, 1802, Dadi Bridge in Jingning County, 1803. For one bridge, (Meichong Bridge), the Chief Master was Li Zheng-Man, with Zhang Xin-You listed second; for the others, Zhang Xin-You's name was listed first.

49 Xiaban Bridge (下坂桥) in Zhenghe, Fujian.

Figure 6.35 Lanxi Bridge (兰溪桥), Qingyuan, Zhejiang. Built in 1794 by Master Zhang Xin-You. "Piercing form."

This bridge was relocated and rebuilt in 1984 to make way for the construction of a reservoir at its original location. The arch beams were cut short and repaired during the reconstruction, while the structure remained unchanged.

integral *choudu* technique, became a firm fixture of the Zhang family's works, and was retained in the design of all subsequent bridges: an unchangeable feature of this family's works, right up to the current generation.

Since even the Zhang family descendants had gone through a period of hesitation before they accepted the new design form, it took time for other carpenters to take up the design. For example in 1811, the Xiajian He family built a bridge[50] in the "matching form." However, by the early 1840s, bridges[51] built by this family were using the new "piercing form." And as we will see later, from the 1810s onwards, even bridges built by other (unknown) carpenters showed signs of *choudu*.

To sum up, on the one hand, by the first half of the nineteenth century, the *choudu* technique had become the predominant technique in the professional bridge carpenter. On the other hand, although become common knowledge and a standard working process within the career, the technique remained a closely-guarded secret, unknown to those outside this specific career.

6.3.4 Bridges with different features

We have discussed at length how common features of MZ bridges emerged, and how they appeared in the most influential bridge carpenter families. But

50 Duoting Bridge (多亭桥) in Jin'an, Fujian.
51 Xuzhou Bridge (徐州桥) in Gutian County, Fujian, and Longjin Bridge (龙津桥) in Minhou, Fujian.

how different can bridges look when they are built by carpenters external to these families?

Well, it depends. Structures do look very alike, especially when they are built to respond to similar challenges. When faced with the issue of a large span, the natural limitations to timber length will make the woven arch form a natural and reasonable solution, which is part of the reason that all experienced carpenters choose similar proportions and inclination for the arch, which greatly influence the appearance of the bridge. For those who came from a family rooted in a long bridge-building tradition, or who learned from such a master, these proportions are part of the standard design, and also applies to smaller bridges.

However, those who were self-taught were less restricted, as they mainly built small-scale bridges, which allows a larger variation in proportion to be applied. The Jiaolong Bridge (Figure 6.36) in Zhenghe County is a good example of this. The bridge, which has a clear span of only ca. 13 m, has rather unusual proportions, with an extremely steep arch and a short horizontal

Figure 6.36 Jiaolong Bridge (蛟龙桥), Zhenghe, Fujian.
The bridge has extreme proportions. According to the inscription on the bridge corridor, this bridge was built in 1835 by the Lin brothers (or cousins) from Hexi Village, Shouning, Fujian.

beam section. It was built in 1835 by a group of carpenters from the Shang-Hexi Village in Shouning County (which today belongs to Zhouning County), only 10 km away from the construction site. The names of the carpenters are inscribed on the corridor beams. The chief master was called Lin Shi-Guang (林仕光), and this was not his first project participated.

Four years earlier, in 1831, Lin Shi-Guang had worked on another bridge project, the Liren Bridge (里仁桥) in Shouning County (Gong 2013, 8), 20 km[17] from his home. On this project, he worked only as a common carpenter. The carpenter master came from another county, some 45 km away.[17] Although this master is not related to any known bridge carpenter family, this bridge shares some common features with the works of the Kengdi Masters.

Lin Shi-Guang's work shows us how someone becomes a bridge carpenter through participation and observation. Whether or not he set out with the intention to learn (or maybe even "steal") the technique – he learned by joining a project led by an experienced bridge carpenter. Although he worked as a common carpenter, he had the opportunity to assist and observe the work of more experienced colleagues. Even if the senior carpenter had hesitated to reveal his secret knowledge (e.g. the design method), he would not have been able to conceal the actual construction steps. When the observer later realized his own project, he partly followed the method he observed, and partly came up with his own solution.

Bridges with unusual forms of this kind appear occasionally, when novice carpenters were still in their exploration period. In other cases, the unusual appearance stems from a local preference. Some bridges in Taishun County are of this type.

Compared to the usual humble appearance of the MZ bridges, some of the early bridges (before 1900) in Taishun are characterized by their large span (27–34 m), towering arch, steep deck, and in some cases, upturned flying eaves (Figure 6.37–Figure 6.41).

To achieve a steep arch, the relationship between the lateral deck beam (LH-Beams) and the woven arch has to be changed which, however,

(a)   (b)

Figure 6.37 Sister bridges Xidong and Beijian. Sixi, Taishun, Zhejiang. (a) Xidong Bridge (溪东桥), rebuilt in 1827. (b) Beijian Bridge (北涧桥), rebuilt in 1849.

Figure 6.38 Wenxing Bridge (文兴桥), Taishun, Zhejiang is known for its deformation. (Photo: left: Hu Shi; right: author)
Built in 1857, "matching form." Condition in 2009 before its repair.

Figure 6.39 Wenxing Bridge. Western side. The U-S-Beams[2] have separated from the C-Beam[1], 2009.

endangers the structural stability, as it allows the crossbeams (C-Beams[1]) to rotate (for a detailed explanation see Sec.4.3.4). In an extreme situation, this might result in a deformation such as that seen in the Wenxing Bridge (Figure 6.38, Figure 6.39)[52]. The experienced bridge carpenters in Taishun

52 The deformation of the Wenxing Bridge was caused not only by the structural form, but also by an error in the design of the length of the arch-beams, which thus failed to weave the arch tight.

Figure 6.40 Xianju Bridge (仙居桥), Taishun, Zhejiang. Built in 1801. "Matching form."

Figure 6.41 Beijian Bridge, Taishun, Zhejiang.
This bridge has a steep arch, high L-C-Beams[2] and short weaving-beams (U-S-Beams[2]), so as to weave the arch tight. In this example, the crossbeams are too high and the middle posts ("mid-legs" of the "frog-leg struts") supporting the lateral deck beams cannot stand on the crossbeam any more. They have to have additional support below.

solved the stability problem by another means: by shortening the length of the weaving-beams (U-S-Beams[2]) to restrict the possible rotation (of the C-Beams[1]) (Figure 6.40, Figure 6.41). This short length of the weaving-beams (i.e. the high position of the L-C-Beams[2]) is an identical feature to that of the unknown Taishun masters. If we compare the Xuezhai Bridge built by the Kengdi Masters in 1857 with the other Taishun examples, the Xuezhai Bridge, although built in the same steep style as the local (Taishun) taste, the proportion of its beams still follow the Kengdi norms (Figure 6.17).

Figure 6.42 Dabao Bridge (大宝桥), Xiaodong Village, Shouning, Fujian. Built in 1878 by Master Xu Bin-Gui.

Two decades after Xu senior and junior (father and son) built the Xuezhai Bridge in Taishun, the son, Master Xu Bin-Gui built the Dabao Bridge just outside his own village (Xiaodong) in 1878, in an uncommonly steep form very similar to the Taishun style (Figure 6.42, Figure 6.43). The practices of the Kengdi Maters in response to the Taishun taste are a good example, showing that the aesthetic and/or external requirements of structural form change more easily than the specific constructional features we ascribe to a certain technical pedigree.

Judging from the "pedigree features," especially the joint forms of the woven arch,[53] builders of these early Taishun bridges do not belong to any known family. No names are inscribed on corridors or appear in local literature. Furthermore, these bridges have been repaired or rebuilt multiple times. Although there were probably one or more group(s) of carpenters working in this area in the eighteenth and nineteenth century, somehow, by the mid-nineteenth century, they had vanished. In 1857, Xuezhai Bridge and Wenxing Bridge were completed at almost the same time. One bridge was built by carpenters from outside (from another county), and the other was built by inexperienced bridge carpenters, who had such a problem with deformation.

In addition to the variations due to lack of experience and regional preference, there are cases that appear to solve specific problems with unique solutions. A good example of this is the Fushou Bridge (Figure 6.44) in Shouning County. Built in 1814, with a clear span of ca. 31 m and a relatively steep arch, this structure solved the construction problem which the

---

53 For example, Beijian and Xuezhai Bridges use "duck-beak" at the F-Beam[2], a feature not seen outside Taishun. Their *choudu* is also different to that of the other families: the arch is in "piercing form", but the H-Beams[2] are unevenly distributed (Figure 6.41).

Figure 6.43 Dabao Bridge, corridor.
Built in the Taishun style, the deck of this Bridge is so steep that when we standing at one end of corridor, we cannot see anyone coming from the other end.

Taishun bridges also face, but with a special solution: adding an additional layer of horizontal beams in the middle (Figure 6.45, Figure 6.46). The newly-developed *choudu* technique has been applied, and with three layers of horizontal beams, this technique is of even greater significance in the construction process. We have no information about the carpenters, but considering the *choudu* technique and the neat structure, they must have been among the most experienced carpenters of the time.

We've realized how influential the leading families/groups of carpenters were, at least in the last two centuries before the modern era, through their capability to build huge structures and through the geographical scope of their working area. At the same time, we have to keep in mind that, of the 100 bridges or so that survive, only about one third were built by the three leading families/groups. There was always space for "non-bridge-tradition-carpenters" to take up this career, from simpler projects. At places outside the working area of the core carpenter groups, bridges were built with even more unconventional forms.

Figure 6.44 Fushou Bridge (福寿桥), Shouning, Fujian. Built in 1814. On the façade are three pairs of crossbeams, indicating the visible three layers of horizontal beams.

6.4 Marginal variant: atypical MZ bridges

The bridges discussed so far, no matter how "unusual" compared to each other, are still within the scope of the "typical" form, in that their under-deck woven arch is composed of two systems: a main system consisting of a three-sided arch, and a secondary system, consisting of a five-sided arch. Bridges of this type of structure are called "3+5 (woven arch) bridges" in the sections that follow.

With a three-sided arch as a basic form,[54] there are theoretically four possibilities for variations of the woven arch, that is 3+X, 3+2, 3+3, 3+4 and 3+5 bridges.

Amazingly, all these possible variations appear in the two provinces. However, they appear only in places outside the conventional 3+5 bridges areas. We would call the scope of the typical (3+5) bridges the core MZ-bridge area, and that of the untypical bridges, the marginal regions of the MZ area (Figure 6.47).

---

54 Theoretically, the first system need not necessarily be a three-sided arch: arches with two, four, or even more sides are also possible. However, since such forms would change the construction process fundamentally and go beyond the natural development of MZ bridges, these possibilities are not discussed in this chapter.

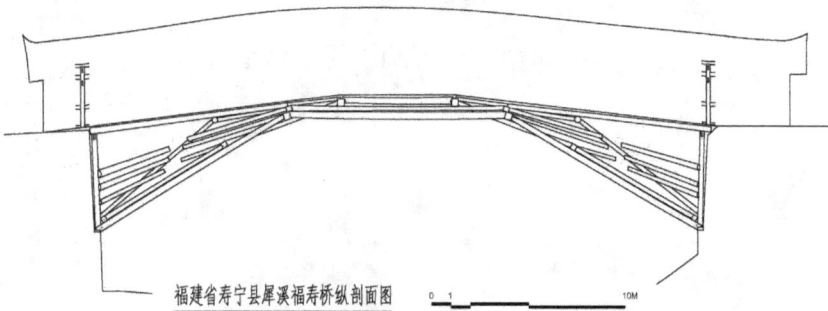

Figure 6.45 Longitudinal section of the Fushou Bridge, showing the three layers of horizontal beams.

Figure 6.46 Under-deck structure of Fushou Bridge. Three layers of horizontal beams and three pairs of X-struts that connect separately to the three levels of crossbeams.

6.4.1 "3+3"-format bridges as a deviation from "3+5"-format bridges

3+3 is a most intuitive form of variation: if we imagine cutting off the lower part of a typical 3+5 woven arch at the height of the L-C-Beams[2], the remainder of the top part is a 3+3 woven arch.

In a 3+3 bridge, the second system is also a three-sided arch, like the first system. The slanted beams of the second system (S-Beams[2]) are the weaving-beams. The weaving mechanics is crucial to form (or for us to identify) a 3+3 woven arch bridge. Bridges composed of two overlapping three-sided arches (struts) are sometimes classified by some authors as belonging to the wooden arch bridge group, but they are clearly not woven arch bridges.

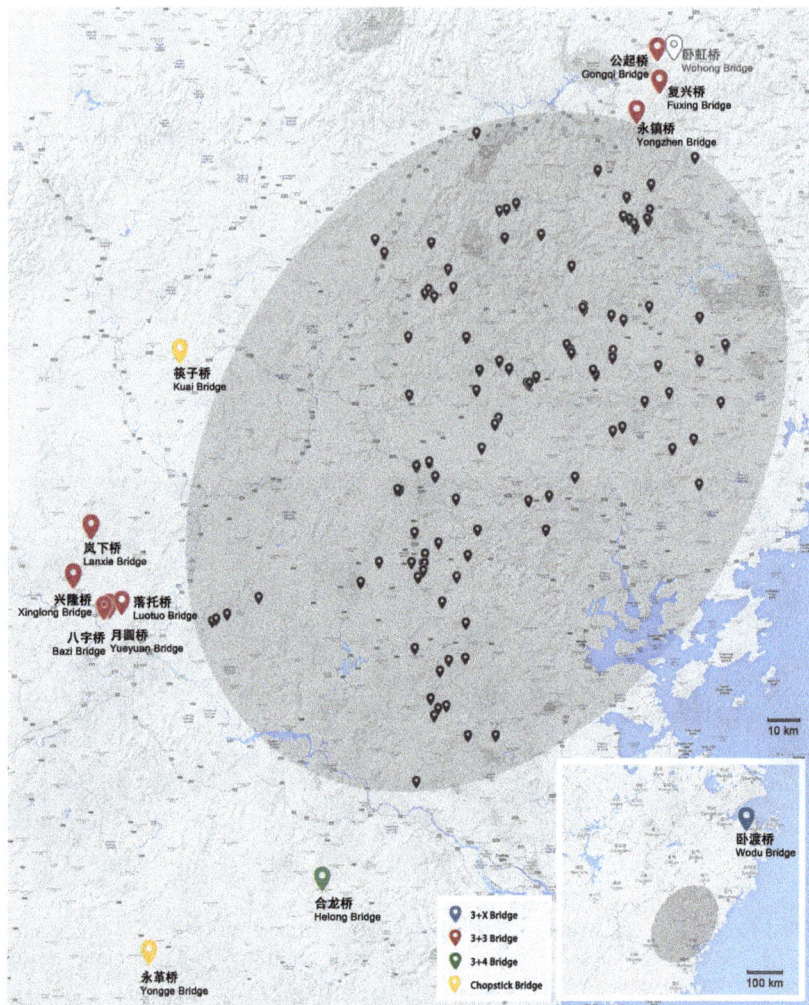

Figure 6.47 Topographical map of the atypical wooden-arch bridge area in Southeast China.

The grey translucent oval marks the area of the (core) area of the typical wooden-arch bridges. The bridges that deviate from the typical structure are marked in the following colours: red (3+3 bridges), green (3+4 bridges) blue (3+X bridges) and yellow (chopstick-bridges).

Source: Based on Google Maps.

3+3 bridges can be found at the northern (Figure 6.48) and southwestern edge (Figure 6.49) of the MZ area, outside the area of typical wooden arch bridges. All known 3+3 bridges are later constructions: the earliest being built in the 1870s, and most of them dating from the 1930s and 1940s (Su and Lu 2010).

Figure 6.48 Yongzhen Bridge (永镇桥), Jingning, Zhejiang. The bridge is located on the northern edge of the MZ area.

Figure 6.49 Yueyuan Bridge (月圆桥), Yanping, Fujian. The bridge is located at the southern edge of the MZ area.

With two segments less than a 3+5 bridge, the 3+3 bridge is the optimum choice for smaller wooden arch bridges. The known historical 3+3 bridges have spans ranging from 15 m to 22.5 m.

When designing the Regensburg Bridge for the Nepal Himalayan Pavilion in Regensburg, Germany, described in Chapter 4, the structural task – a clear span of 7.5 m – was too small for a 3+5 bridge. Therefore, the author designed a 3+3 bridge (Figure 4.32).

6.4.2 "3+2" ("3+X") bridges: the most basic woven arch form

The construction aim of the weaving mechanics of the woven arch is to restrict the possible rotation of the primary system – the three-sided arch.

This rotation occurs at the intersection between the crossbeams and the longitudinal beams, namely, at the two C-Beams[1]. Therefore, the simplest form of the woven arch has two groups of beams in the secondary system, each of which restricts one of the C-Beams[1].

Although no full-size "theoretical bridges" of this type are to be found in the MZ area, smaller-scales examples have indeed been constructed by local bridge carpenters. Master Dong Zhiji of Taishun County, whose story is well-known to us from this chapter and from Chapter 4, made such a model at home as his own creation (Figure 6.50). In this, the theoretically simplest wooden arch bridge, the three-sided arch is rather flat in inclination and serves as the deck structure system as well. The second system has only four slanted beams, two on each side, reaching from the abutment to the crossbeams of the first system of the other side, while interlocking and supporting the crossbeam of its own side.

Although no full-scale bridge was realized according to Dong Zhiji's innovative model-design inside the MZ area, there is a bridge structure constructed on similar principles ca. 200 km outside the woven arch bridge area.

The Wodu Bridge (Figure 6.51) is located in a mountain village in the township of Fenghua, Ningbo, in the northern part of Zhejiang Province. This bridge was built in 1921, and we have no information about its builders. Similar to Dong Zhiji's design, in this bridge, the slanted beams of the second system reach from the abutment to the crossbeams of the other side, interlocking and supporting the other crossbeam at their mid part. Less pared-down than Dong Zhiji's design, it has additional deck beams on the top, laid

Figure 6.50 Model of a 3+X bridge by Master Dong Zhiji from Taishun, Zhejiang.

Figure 6.51 Wodu Bridge (卧渡桥), Fenghua, Ningbo, Zhejiang.

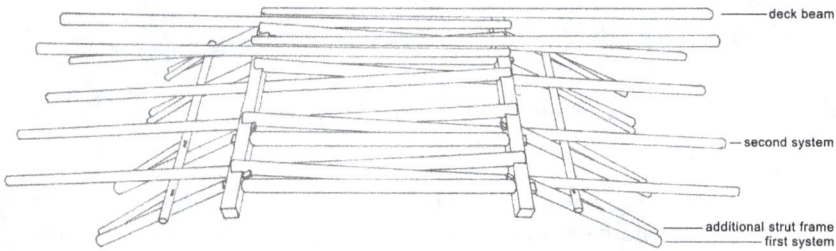

Figure 6.52 Sketch of the under-deck structure of the Wodu Bridge.

alternately from the two sides of the bridge, reaching from the abutments to the crossbeams at the other side (Figure 6.52).

6.4.3 "3+2" bridges as an unfavourable deviation from the "3+3" bridges

There is another way to achieve a 3+2 bridge, and that is by simply removing the horizontal beams (and a corresponding crossbeam) from a 3+3 bridge. However, the structural form generated in this way is unfavourable.

A bridge with an even number of segments of the second arch system is not a common choice for MZ bridge builders. As discussed in Chapter 4, the lengths of the middle segment of the second system (H-Beams[2]) are hard to calculate in advance, and are as a rule determined during the construction progress. The adjustment of the length of these middle beams in-situ makes it possible to tighten the woven structure. Therefore, without the H-Beams[2], it is difficult for the bridge carpenters in MZ-area to interweave (and control) the two arch systems tightly.

The author was able to find only one example of such a 3+2 bridge, and it is a rather new one. We came across a construction site in 2013 in the town of

Figure 6.53 Town of Dongkeng, Jingning, Zhejiang. A 3+2 bridge under construction.

Dongkeng, Jingning, Zhejiang. The bridge (Figure 6.53) was being built by a local carpenter who had not built a bridge before. In this "wrong"[55] design, the second arch system of the wooden arch takes the form of a three-sided arch, but without the middle H-Beams. Thereafter, in the second system – the "two-sided arch"– there are twocrossbeams at the top, lying close together in the middle of the upper part of the arch.

6.4.4 "3+4" bridges as a special form of MZ bridge

As discussed above, bridges with a second arch system with an even number of segments are considered an unfavourable shape for an MZ bridge. However, there is a historical 3+4 bridge example, located on the southern edge of the MZ area.

The Helong Bridge (Figure 6.54) in Minqing County, Fujian, was built in 1927. It is a double-span bridge. The larger bridge span (ca. 14.7 m) is a typical 3+5 woven arch. The smaller span (ca. 12.5 m) has a 3+4 woven arch construction. Here, the 3+4 woven arch is probably a deviation from the 3+5 woven arch (Figure 6.55).

Its joint system of the 3+4 span points to a different construction method from the typical bridges. In the second arch system, the upper beams (U-S-Beams[2]) have a tenon at their upper end, and a dovetail at the bottom, while the lower beams (L-S-Beams[2]) have a dovetail at the top and a tenon

---

55 Criticized by Kengdi Master Wu Dagen at the construction site, when he and the author walked past this bridge in December 2013.

Figure 6.54 Helong Bridge (合龙桥), Minqing, Fujian.

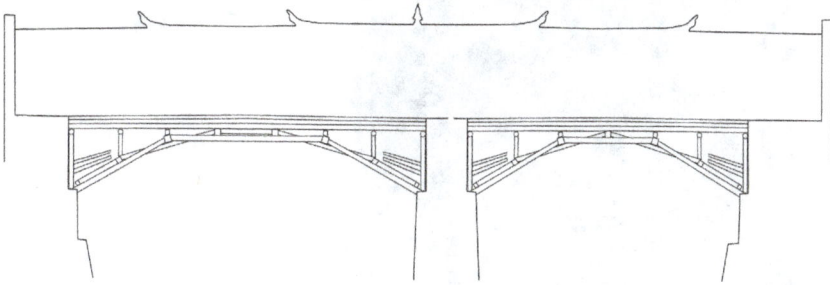

Figure 6.55 Schematic longitudinal section of Helong Bridge.

at the bottom (Figure 6.56). This indicates that the lower beams have been installed after the upper beams, the reverse of the process used in a typical 3+5 bridge. As a matter of fact, in a 3+4 woven arch, it is not possible to install the upper slanted beams before the lower ones. The construction at the arch feet confirms this construction process (Figure 6.57).

6.4.5 "Chopstick-bridges" as a rudimental form

Two of the most striking examples, which gave a strongly experimental impression, distinct from all other forms, may provide us with some ideas about how woven arch bridges came into being.

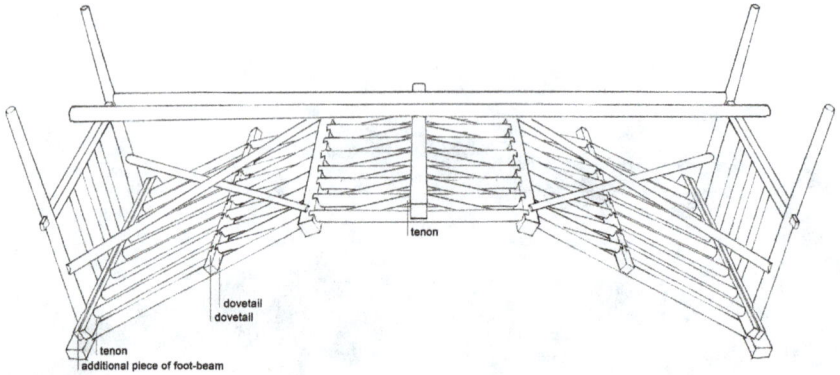

Figure 6.56 Schematic drawing of the joint form of the 3+4 woven arch of Helong Bridge.

Figure 6.57 Arch feet of the Helong Bridge. Two parts of feet beams are pressed together. The exterior one needs to be knocked in at the very end, to make the entire arch tight.

Figure 6.58 Yongge Bridge (永革桥), Dehua, Fujian.

The first case, the Yongge Bridge (Figure 6.58), is located outside of the MZ bridges area to the south Quanzhou, Fujian (marked on the map in Figure 6.47 by lower of the two yellow pointers). This bridge was built in 1948, and has a span of about 20 m.

Since the bridge is comparatively young, modern scholars were able to trace the master carpenter, and hear the story of the bridge construction from his own lips (Su and Lu 2010).

The bridge project was directed by a former official by the name of Guo Zhenhua (郭振华), who previously worked in Shouning County, which is, as we know, the core area of the woven arch bridges. After he retired, he returned to his hometown, and initiated a bridge project, appointing a local carpenter, Huang Yizhu (黄以柱) to this project. However, this area, much further south in Fujian Province, had no woven arch bridge tradition. The carpenter (Huang Yizhu) had no idea how to build such a bridge. Therefore, Guo Zhenhua taught him the method, and the bridge was constructed as a result of this teaching.

But wait – how could an official instruct a carpenter to build a bridge? Well, the bridge construction gives us a clue.

Some scholars (Su and Lu 2010) claim that the under-deck arch (Figure 6.59) of the Yongge Bridge is a 3+4 arch. However, these so-called two systems are not as distinct as the common MZ bridges. The longitudinal and transverse beams are connected to each other not by means of wooden joints, but by iron bolts (Figure 6.60). Thus, the woven arch of the bridge can

Figure 6.59 Under-deck structure of the Yongge Bridge.

Figure 6.60 Iron bolts in the joints of Yongge Bridge.

be seen as an expanded version of a "chopstick-bridge" (cf. Figure 6.61 and Figure 6.62, also see Figure 3.3). Chopstick-bridge is a very popular game in the MZ bridge area: every child knows how to play it, and it is widely believed by the carpenters and the villagers that the game incorporates the construction principle of their bridges.

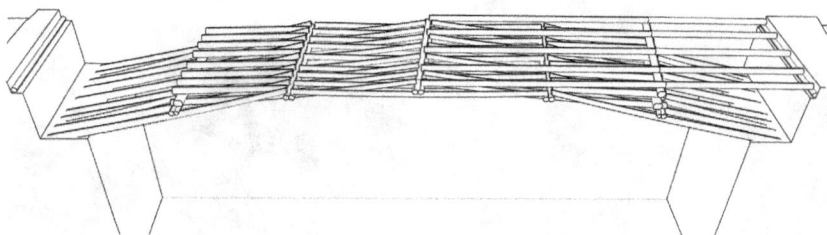

Figure 6.61 Sketch of the under-deck structure of the Yongge Bridge.

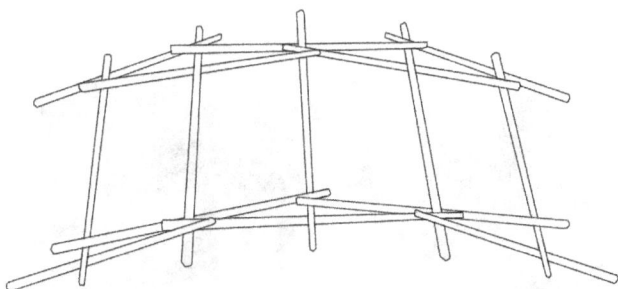

Figure 6.62 The chopstick-bridge model corresponding to the Yongge Bridge.

If we take that idea as a basis, then it's entirely plausible for an official to be able to teach a carpenter to build a bridge, using some 15 chopsticks to demonstrate the construction principle of the structure. Without ever having seen a "real" woven arch bridge, the carpenter erected the bridge according to the principles of the chopstick game. This may also be why, instead of traditional wooden joints, the Yongge Bridge has iron bolts connecting the beams, when iron bolts only began to be imported into Chinese construction in the late nineteenth century.

The second example even more clearly indicates its origins in the "chopstick-bridge" game. The Kuaizi Bridge (Figure 6.63) is located in Ankou Village, Jianyang, Fujian, on the western margins of the MZ-bridge area (Figure 6.47). Built in 1964, the under-deck structure is an almost identical reproduction of the most common form of the "chopstick-bridge-game" played by the locals (Figure 6.64). Beams are fixed with simple joints only at the most indispensable positions, otherwise they just simply hold together. As if that weren't enough, the name of the bridge, "Kuaizi Bridge" is exactly "Chopstick-bridge" in Chinese.

These two examples are clear and almost direct imitations of the chopstick-bridge game: they are almost a one-to-one realization of the game with logs

Figure 6.63 Kuaizi Bridge (筷子桥), Jianyang, Fujian. (Photos: Ying Jiakang)

Figure 6.64 A local showing us the "chopstick-bridge-game" in the corridor of the Lanxia Bridge (岚下桥), Shunchang, Fujian, a 3+3 bridge.

instead of sticks. Furthermore, even though these two bridges appear fairly late in the chronicle, they incorporate within themselves a possible rudimental prototype of the woven arch bridge: the chopstick-bridge-game. Going along this route, the problem encountered in the structural experiment – namely the main difference between a real bridge and an expanded version of the chopstick-bridge game – is the joints.

6.5 Technique inheritance and dissemination

6.5.1 Bridge-building technology and carpenter-group composition

From our long journey, which went from modern bridge-building sites to the old structures of MZ bridges, and from the knowledge shared by the carpenter masters, to their family stories and legends, we can now come to some conclusions about the work pattern of the MZ bridge builders.

Building woven arch bridges is a special profession, which uses methods and techniques that are a closely-guarded secret among certain groups of carpenters. Only very few carpenter families/groups are able to build large-scale bridges in difficult environments. Their capability is informed by a tried and tested body of experiential knowledge, techniques and skill sets that have been passed down through generations of carpenters, and by a systematically-organized carpenter alliance with a clear division of labour.

On the other hand, bridge-building techniques, though kept secret within the profession, were not entirely a mystery to those who do not belong to the alliance. The techniques spread in direct or indirect ways. New bridge carpenters appeared all the time, building smaller-scale bridges here and there. However, no matter how many and how successful these newly-emerging carpenters were, none of them was able to shake the dominant status of the leading family or group.

To understand how the bridge-building techniques were protected, transmitted and spread, we will classify them, and the carpenters involved, into five levels (Figure 6.65):

- *Core knowledge*: This is the most secret knowledge, which determines the success of bridge construction, mostly relating to the design rules of the woven arch (the set of empirical rules for determining the scale and details of structural members, including proportions, angles and other crucial parameters). The core knowledge is shared only between the *core members*, who are the leading carpenter masters[56] in a given bridge-building family or multiple-family group. Since this knowledge is applied mostly during the design process and thus invisible to those who do not belong to this group, it is the most protected professional secret within a carpenter family or group and only handed down via teaching selected members of the offspring and those associated/affiliated with the family or family group (e.g. sons-in-law, husbands of sisters, etc.).
- *Core technique*: We use the term "technique" to refer to the visible construction methods, as distinct from the "knowledge," as described above. The core technique is the part that is visible but, thanks to its high technical threshold and difficulty of execution, is hard to copy (or "steal"). Building bridges over gorges or over deep water with the simple swing-frame-scaffolding is the most representative example of this core technique. The entire carpenter family/group who work together (and most of the time over many generations) share the core technique. They are the *allied members*. They commonly come from the same or neighbouring

56 Carpenter masters who have the capability to act as chief or deputy masters of construction projects.

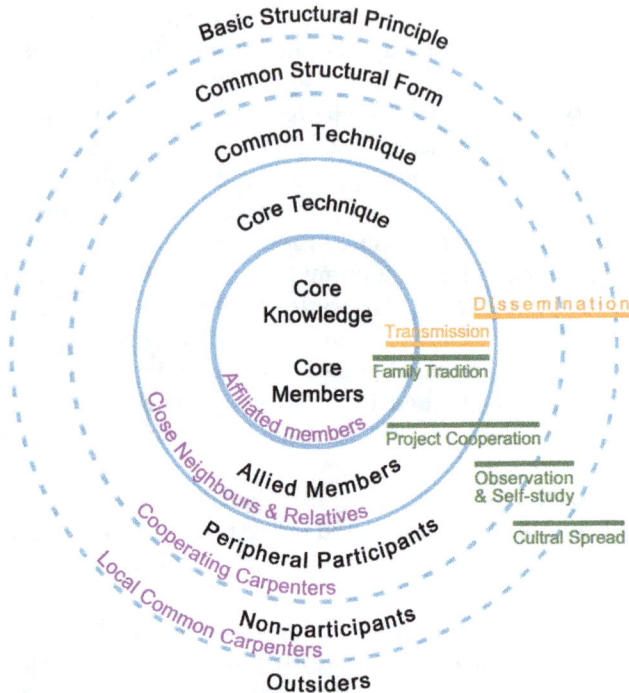

Figure 6.65 Diagram of the composition of the bridge-building technology and bridge carpenters.

village(s), and are more or less closely related to each other. Their organization and division of labour are also included in the core technique. Unlike the core knowledge, the core technique is not usually taught using words but rather by showing what to do.

- *Common technique*: The basic technique includes the normal working processes and construction methods, is visible on the construction site, and can be carried out by anyone who can do common carpentry. The *choudu* technique is between this level and the previous higher classified one (more on the classification of *choudu* will be explained later). In the case of most bridge projects, a group of professional bridge carpenters from outside take charge of the under-deck structure, and some local common carpenters work as assistants and (sometimes) take charge of building the bridge corridor. The assistants who do not belong to the group of bridge professionals are *peripheral participants* in the project. Although they are not specifically invited to learn the bridge-building technique from the bridge carpenter family, neither can they be forbidden to do so. Smart and observant local carpenters are able to learn the basic technique and equip themselves with bridge-building skills for taking on smaller and simpler projects later on.

- **Common structural form**: as soon as a bridge is built, it is exposed to view throughout its lifetime. All the visible structural and construction forms and features belong to this level. For a **local common carpenter** who is familiar with common carpentry and construction methods, it is not difficult to figure out a set of construction solutions by himself and to make an imitation, even without contact to the specialist carpenters. In this case, the local carpenter will miss out some of the key methods and techniques (e.g. the *choudu* technique) and/or neglect some minor technical features (e.g. joint form), and produce a structure with different construction features.

- **Basic structural principle**: The woven arch principle, which is to compose the timber members using interlocking principle (chopstick-bridge-game) to achieve a larger bridge span. Those who have not seen an MZ bridge with their own eyes, but have heard of the structural principle, may "invent" a structure with the woven arch mechanics in an entirely different structure form. The above-described Yongge Bridge and Kuaizi Bridge are the best examples of this case.

Before we move on, it would be appropriate to give a brief explanation of the place of *choudu* inside this catalogue. The *choudu* technique could be classified as belonging both to the basic technique and to the core technique. The *choudu* step is definitely shared operationally by all the members of the construction crew, and is visible to all the peripheral participants as well. However, without an indication or explanation by the specialist carpenters, its structural function and significance remains hidden, and might not always be well understood or learned by anyone witnessing it. For example as we have seen from Master Dong Zhiji's story, he was unable to learn the *choudu* method on the Yangmeizhou Bridge construction site (this is not meant as a criticism or to be derogatory to him, as he was just a child at the time).

The *choudu* technique is also a good example for explaining the difference between common technique and common structural form. As described, the visible result of the *choudu* working step, the "piercing form" or the "interlacing form," is a structural form that can be copied by non-participant carpenters, but the treatment for tightening the arch by means of the *choudu* method is a technique.

Using this construction model for bridge-building technology helps to give us a clearer understanding of the different types and working/cooperating mode of MZ bridgebuilders. There are basically three types of bridge carpenter:

(1) Family and family association

Bridge construction in the MZ Mountains, especially for large-scale structures, always involves dozens of craftsmen in the woodworking phrase. As a result, the most mature carpenter groups are an alliance of two or more families from neighbouring villages[57] and work together over more than one generation. Within the group, there might be a leading family that always acts as the "core" of the project. The family not only holds the key professional knowledge, but also acts as the contractor and representative of the entire group (and thus signing their names to the contracts and on the corridor beams). The later generations of the Xiajian Zhang family and the Kengdi Xu family played the role of the single-family core of their group. In addition, the allied families might also share the core position during particular periods, for example such as happened with the Li and He families (Xiajian Masters) and the Wu family (Kengdi Masters). The prosperity of the carpenter groups is founded and largely depends on (in addition to external conditions) the prosperity of the families themselves. In times of family hardship or in the case of families without issue, the inherited technique might die out (as was the case with the anonymous early bridge builders) or passed on to an allied family (as was the case with the Xu-Zheng families of the Kengdi Masters).

The real bond of the allied families is engendered not only through their geographical proximity, but even more importantly, by the affinity that arises as a result of this proximity. Only close members can be entrusted with the crucial professional secrets.

The marginal families and other members of the allied group work as simple builders. They are able to master the detailed construction technique, but are not involved in the design and decision-making process. Most of the time, they are not privy to the key knowledge. They work under the leadership of the core members. In the most developed carpenter group, namely the later generations of the Xiajian masters, they are organized into a well-trained group with a clear division of labour, including work at water level or over the water, height work, scaffolding work and common carpentry work.

(2) Hands-on training through teamwork

While the core family keep the design method secret, they cannot hide the construction steps from the others. Everyone that works with the family gets the chance to have on-the-job training. This happens inside the associations of carpenters, where clever members of marginal families acquire the confidence to then take on small-scale projects independently (e.g. the Xiajian He family). This also happens frequently in projects when specialist

---

57 Most these mountain villages were one-family villages, where the entire village shares the same surname and the same ancestors.

bridge carpenters are invited to join a project from outside, and have to work together with the local carpenters, and the latter thus had the opportunity to get to know about the bridge-building technique.

It is also common for a carpenter without bridge-building experience to be able to shape his career according to his own wishes. For example he may travel to join a group of bridge builders and stay until he gets enough experience (as with the example of Master Lin Shi-Guang from Zhenghe, who built the Jiaolong Bridge, Sec.6.3.4 above). In some special cases, he may take on a project in his hometown and invite a master carpenter to be his instructor.[58] In this case, the master carpenter from outside does not take full charge of the project, but only gives advice when necessary. After receiving the instruction, the student goes on to establish his own "lineage."

(3) Autodidacticism

In some cases, due to circumstances or local requirements, clever and ambitious carpenters take over a construction task without any former bridge-building experience, simply through observing and examining the old bridges around them. Such stories are occasionally heard (e.g. the story of the Huilan Bridge, see Sec.6.3.2.3 above). Also in the 1960s, there was even a case where a local carpenter built a medium-scale (28 m span) bridge over a gorge (the Longtan Bridge 龙滩桥 in Zhenghe County) (Gong 2013, 182). Occasionally, a self-taught carpenter will start a new career after the first attempt. For a modern example see Zeng Jiakuai in Taishun County (Sec.6.3.2.2 above). However, in most cases, the project in or around their own village is a one-off job, and is the only bridge they build.

6.5.2 *Technique, families, and locality: rethinking the Xiajian and Kengdi Masters*

We have emphasized in the timeline, that the Kengdi Masters built their first work in their own village, nine years after they had invited a group of bridge-building masters from outside to build a very similar bridge only ca. 600 metres upstream. The invited masters most probably belonged to the He family, from Houlong Village in Xiajian, a marginal family of the Xiajian Masters. During the first project, the future Kengdi Masters had the opportunity to participate or at least observe the construction process – and even if they didn't, they had plenty of time to study the finished structure and to prepare to embark on this career.

The technique of these two groups of masters tells the same story, in an even more direct way: the technique of the Xiajian and Kengdi Masters are highly consistent with each other, the only differences being in details. Most

---

58 Wei (韦) family in Pingnan County. Interview with Master Wei Shunling in December 2012.

importantly, they follow an almost identical method in terms of design, with differences only in limited parameters. In terms of their technical features, these fit well in the third category of technology constitution (Figure 6.65), that is where the junior group has most likely worked as peripheral participants on one or more bridge projects and learned from their senior colleagues. On sites where the secret parameters were concealed from them, they developed their own ways and rules in their later practice.

Given that these two groups of carpenters have the longest-standing bridge-building tradition, and are closely related to each other, it is interesting to compare these two groups of craftsmen, keeping the following question in mind: Why were only two groups of carpenters successful in passing down the technique, given that so many others attempted to do this? And why is it the Xiajian Masters, and not the Kengdi or any other later families or groups, who occupy an incomparable and dominant place among the MZ bridgebuilders?

It's very interesting to note that in both cases, the leading carpenter family was indeed the family with an inferior position in the locality. For example the Zhang family moved into the Xiajian area three centuries later than the He families, when the better lands along the riverside and on the sunny side of the hill had already been taken over by the He family villages. The Zhang family had to make do with cultivating land on the shady side of the hill. The Xu family moved into Kengdi even later and was never able to expand to become a large family. This is not only the result of the inferior status of the craftsmen: if we remember how dangerous the activities connected with bridge construction could be, we will be less surprised by this "natural selection" among the local carpenters.

Following this line of thinking – that a less-developed settlement coupled with inferior social standing – results in a better and more stable inheritance of bridge-building technique, it's easy to see that it's not a coincidence that the Xiajian Masters were much more successful in bridge-building than the Kengdi Masters.

Both groups live in the central mountain area, but a major difference between the Xiajian and Kengdi areas is that Kengdi is located at a traffic node between four neighbouring counties and two provinces, while Xiajian is hidden away in an area far from main roads.

Residents in Kengdi had comparatively more arable land, better communication with the outside world, more abundant external resources and chances to accumulate wealth via commerce. Thus, larger families could afford to begin a cultivation-education tradition (as the Wu family of Choulinshan Village). In such circumstances, people were confronted with a choice as to what they did with their lives, between (higher) education, farming and various crafts, such as the different life-choices made by Master Wu Guang-Qian and his brothers.

By contrast, in Xiajian there were no such resources. Although the arable land was sufficient to provide staple foodstuffs for local consumption, there was no surplus to sell, to generate money for buying other goods. The local economy relied greatly on timber transportation. Although hard to reach by road overland, there was a big stream running through the Xiajian area, the Houlong Brooke, which was suitable for timber rafting. As recalled by the living bridge carpenters from the Zhang family, until the end of the twentieth century, during the winter, when there were no bridge-building projects, the family relied on timber and bamboo rafting. The *Shanmu* ("China fir") timber was rafted from the upstream mountains down to the downstream distributing centre. Since the stream was rather narrow in places, tree trunks were not tied together to make rafts (as they were in other places), but floated in the stream in single file, one after the other. People running the rafts stood on the middle of the trunk, holding a bamboo pole in their hands to keep their balance, and rushed down the stream standing on the trunks, even at places which had a waterfall higher than a metre.[59] This was a dangerous career and required cooperative teamwork as well as superior swimming skills. And the skill required in timber felling and transportation is similar to some of the core technique (namely, construction with the swing-frame-scaffolding) of bridge-building! (Figure 6.66)

The rugged natural environment, together with the lack of economic resources, might be two of the decisive reasons that pushed the Xiajian Masters (especially the Zhang family) to embark on the path of developing dangerous and sophisticated craftsmanship, and to occupy an invincible position in this profession.

6.5.3 Tracing the source of the technology

Although the Xiajian masters have an uninterrupted tradition of 200 years of bridge-building, the longest of all the existing groups, they did not start the MZ bridge-building tradition. The actual history of the MZ bridges is far longer than that of the known carpenter families. The oldest preserved structure dates to 1625 (Rulong Bridge, Chapter 5), and literature with credible information on such bridges was available as early as 1579.[60] There must have been capable carpenter groups that lived long before the family groups

---

59 Zhang Chang-Zhi, in an interview with the author on 16 January 2019 in Xiukeng Village, Fujian Province. Zhang Chang-Zhi boasted (at the age of 71), that even now, if given a tree trunk in the middle of the stream, he could still stand on it and control it. Such timber rafting skill is a performance in a tourism town (Town of Huotong) downstream of the Houlong Brooke today.

60 Many authors, esp. the local scholars, claim that the woven arch bridge tradition in the MZ Mountains began as far back as the Song Dynasty (from ca. twelfth century). The author of this book does not agree with this statement. The literatures they cite mention bridges built in Song times or even earlier, but they are not necessarily woven arch bridges. The earliest literature on the local woven arch bridges found by this author, is from local chronicles. The earliest ones were written in 1579 and 1588, mentioning a large-scale bridge over a cliffy gorge in Jingning County. (万历七年（1579）《栝苍汇纪》："东向为矿坑岭，达北溪至白鹤射桥。又南为大漈岭。"万历十六年（1588）《景宁县志》："射桥，在六都白鹤，跨悬崖几十丈最高。")

Figure 6.66 Drawings of timber transportation in the mountains (of Sichuan
Province) from the *Collected Essays on Timber-Felling in the Western
Region* (西槎汇草), Jiajing reign (1522–66) of Ming Dynasty.

a. Sending the log off a cliff with ropes.

b. Pulling the logs over a gorge with traditional cranes.

Source: Library of Congress.*

*www.wdl.org/en/item/4696/. accessed: 4/5/2020.

we are aware of, and the technique was passed down to the "visible hands of our times" by these anonymous predecessors.

If we are trying to determine when the technique originated, there is no solid material evidence for us to turn to: we can only look into social history for support.

It is possible that the MZ mountain area was sparsely settled long before records started. However, the first real tide of immigrants arrived in the late Tang Dynasty (late ninth century), and the second tide at the end of the North Song Dynasty (in the 1130s). These early immigrants were refugees escaping the chaos of wars or fleeing from northern invaders. They hid in the mountains, lived on agriculture and whatever they could find in the forest, with few requirements for contact with the outside world.

However, life in the mountains was not peaceful, like a Shangri-La. There was always traffic between the mountains and the Empire – a flow of silver. The MZ mountains were rich in silver and other metal deposits, on both the Zhejiang and Fujian sides, and had a large overlap with our bridge area. Government mining activities started under the Song Dynasty and were unsustainable exploitation of local societies. Mining was not a path to wealth – it was the gate to hell. A large portion of workers were prisoners, and they were treated like slaves. If the mining activities did not produce the required results, high taxes were imposed on the local society (Tang 2002; Tang 2011).

The situation became especially ugly in the first century of the Ming Dynasty (1368–1644). After a series of plagues destroyed many residential areas in the mountains, the government repopulated some areas with prisoners, and the pressure on mining continued, relentless. From the early fifteenth century, the conflict between the governmental and private (which was illegal) mining escalated, becoming violent, resulting in a decade-long conflict between government forces and the rebelling miners – or rather bandits. The deep mountains provided the best hiding places for the bandits. In the 1450s, a government administration was first established in some of the most remote mountain areas, to bring the unstable situation under control once and for all. The counties of Taishun, Jingning, and Shouning – the central counties of the MZ-bridges area, were all founded at that time. However, it took centuries for civilization to really take hold in this remote mountain area. Zhouning County, the heart of our bridge area and home to the Xiajian Masters, was founded as recently as the 1940s![61]

It was only with the arrival of peace, and after the gradual abandonment and prohibition of silver mining in the fifteenth to sixteenth centuries that the society of the MZ mountain area began to regenerate. In the fifteenth to sixteenth centuries, the MZ mountains underwent their first-ever economic development, through planting forests, commercial products such as indigo, mushrooms, tea,

---

61 An detailed study of the history of the MZ mountains, their social environment, and the bridge-building families, is given in a slightly expanded Chinese version of this book.

and oil (Xu 2014). The upturn of the economy brought with it the first wave of trade with the outside world. This resulted in an increasing demand for roads and bridges, together with an increase in social wealth to afford them, and finally to the bridge-building boom in the centuries that followed, and the professional bridge carpenter families and groups. There was a huge increase in the number of bridges recorded in local chronicles between the late Ming and the late Qing dynasties.[62] The technology for the woven arch bridges is very probably an achievement that occurred during that same social process.

Although from the mists of time we have hardly any information as to how exactly the technique first emerged and what the first stages of its evolution were, what we can observe today may still contain clues to some hidden elements of the past. The different levels of knowledge and craftsmanship shown in our diagram (Figure 6.65) may also represent the different stages of the technique's development. The route from being an outsider to being a core member of the bridge carpenters' group might be similar to the route from an early-stage experiment to the mature "high-tech" version of the bridge technology, a generations-long evolution which developed in the mountains. And even after so many centuries, this technology was still being experimented with, innovated upon, handed down and spread in exactly the same manner, among various local carpenters until very recently.

References

Gong, Difa
龚迪发. 2013. 福建木拱桥调查报告. 北京：科学出版社.
Ningde
宁德市文化与出版局.2006.宁德市虹梁式木构廊桥屋桥考古调查与研究.北京：科学出版社.
Su, Xudong and Liu Yan
苏旭东,刘妍.2010."双三节苗"木拱桥——木拱桥发展体系中的重要形式. 华中建筑 (10)，39–42.
Su, Xudong and Lu Zeqi
苏旭东,陆则起.2010.永革桥的结构和技艺流播.2010年古桥研究与保护国际学术研讨会.
Tang, Huancheng
唐寰澄. 2000. 中国科学技术史·桥梁卷. 北京：科学出版社: 461–92.
Tang, Lizong
唐立宗. 2002.在"盗区"与"政区"之间. 台北：国立台湾大学出版中心.
唐立宗. 2011.坑冶竞利：明代矿政、矿盗与地方社会. 台北：国立政治大学历史系.
Xu, Xiaowang
徐晓望. 2014.明清东南山区社会经济转型：以闽浙赣边为中心. 北京：中国文史出版社.
Zhou, Fengfang, Lu Zeqi and Su Xudong
周芬芳, 陆则起, 苏旭东. 2011. 中国木拱桥传统营造技艺. 杭州：浙江人民出版社.

---

62 Taking Jingning County as an example: in 1588, 1778, and 1873 chronicles, bridge numbers recorded were 24, 42, and 99, respectively.

Part III

Conclusion

7 Rethinking the histories of woven arch bridges

7.1 Universal uniqueness

The emergence of woven arch bridges in human construction history is a story both extraordinary and universal. Featuring an identical beam-weaving mechanism, the Da Vinci bridges, the Huntington Moon Bridge, the Rainbow Bridge, and the MZ bridges share some striking similarities (Figure 7.1).

This characteristic is so unusual that on the one hand, many of the bridges have been considered to be one-of-a-kind examples in architectural histories (Tang 2010; Hirahara 2013) and on the other hand, as soon as the existence of similar examples in other areas of the world is made known to them, scholars will tend to assume a technique and/or culture spread between these different parts of the world; while China, which provides the earliest examples, is considered as the origin of the technique.[1]

These two widespread assumptions, the uniqueness or the Chinese influence, are both incorrect. Woven arch constructions could be and very often were maverick examples of structural phenomena in their local history, while still having their roots in the locality and having developed from their own cultures. They occurred in parallel in different cultures and historical periods.

The influence of the earliest known woven arch structure, the Rainbow Bridge over the Bian River in China, did not reach very far at all. Apart from the historical literature in the eleventh and twelfth century cited in Chapter 3, and another scroll-painting from the same period (more on this later more), there is no evidence of another such bridge ever being found outside the North China cultural area (centred in Henan, Shanxi and Shandong provinces), nor was it handed down to later dynasties. It was very unlikely that this influence reached as far as Da Vinci's time, or to Europe. Even the idea that the much later Chinese examples found in other areas (MZ area or Gansu) had their "origins" in the Rainbow Bridge, as many other scholars have suggested

---

1 Thönnissen (2015) in his research on the history of reciprocal frame structure, suggests that Da Vinci would have been inspired in his design by the woven arch structures in the Orient and the idea of the Rainbow Bridge in China (Thönnissen, 46). Simily opinion is also hold by Italian scholars (27). Liu Jie (2017) suggests that the Huntington Moon Bridge was influenced by Chinese paintings (438–40).

Figure 7.1 Identical form of woven arch bridges in different cultures.
a. Leonardo da Vinci's woven structure, 1480s Italy (Ch. 1. from: Codex Madrid.I, folio 45. *Source*: ©Biblioteca Nacional de España.
b. The Huntington Moon Bridge, 1913, Japan/US. (Ch. 2).
c. Rainbow Bridge, Qingming Scroll, 11th century, North China. (Ch. 3).
d. Kuaizi Bridge (lit. "Chopstick Bridge"), Ankou Village, Jianyang, Fujian, Southeast China. (Ch. 6).
e. Lanxia Bridge, Shunchang, Fujian, Southeast China. (Ch. 6).
f. In the Corridor of the Lanxia Bridge, the chopstick game shown by a local. (Ch. 6).

(Chapter 3), is also doubtful, due to the difference in construction, the centuries between them, the geographical gap, and lack of evidence.

Isolated or related? In what way and at what level? Before we hurry to form an assumption, we have to build up an overall understanding of the entire catalogue of woven arch structures in the whole world.

7.2 Typology of the woven arch structures

Although other examples with a similar construction principle emerged in the history of various cultures, there is a paucity of literature and physical material for them, so this section includes all known woven arch structures built before modern times, to enable us to compose an overall vision of this structural type.

At the basic level of the typology, woven arch structures are divided into two groups, and then further into three types:

Group I. Pure woven arch structures.

The group of structures takes the woven arch as the main body. The purest and most 'extreme' example of this group can be seen in the "chopstick-bridge-game." Indeed, other bridges in this group can be considered as a realization of the chopstick-bridge-game translated into real bridge scale.

Group II. A combination of the woven arch principle and other structural forms.

The bridges in this group occur almost exclusively in China. They could be further differentiated by the other structural form involved.

7.2.1 Type A. Enlarged chopstick-bridge

This type covers the same scope as Group I, with only the chopstick game itself excluded. It's evident that the Rainbow Bridge, the Huntington Moon Bridge, and the Da Vinci bridges are all included in this group. We can add a couple of new examples to this catalogue:

7.2.1.1 A "rural" Rainbow Bridge

The Rainbow Bridge in the Qingming Scroll is not the only graphical representation of its kind in the Song Dynasty. In recent years a different woven arch bridge – one that appeared at almost the same time – was first noticed by architectural scholars.[2] It is a small rural bridge, depicted in the bottom left-hand corner of the "Landscape Scroll in the Autumn" (*Jiangshan Qiuse Tu,* 江山秋色图, "Landscape Scroll" for short, Figure 7.2) by Zhao Boju (赵伯驹, 1120–1182), a painter of the Southern Song Dynasty (1127–1279)[3], while the Qingming Scroll (beginning of the twelfth century) is earlier, and from the Northern Song Dynasty. Both painters served as official artists at the court of the Song Government. The Landscape Scroll was done only a few decades after the Qingming Scroll. Judging from its autumn mountain view and building form, the landscape is obviously northern China. This indicates that

---

2 The author's attention was drawn to this new source by Dr Liu Diyu (Tongji University, Shanghai, China), an expert on the study of the Qingming scroll. Also see: Liu (2014, 295).
3 After the Song government moved from the north to the south.

Figure 7.2 Jiangshan Qiuse Tu and the rural Rainbow Bridge.
Source: ©Palace Museum, Beijing, China.

the painter's knowledge of the bridge might also have come from sources in Northern times, or even the same source of the Qingming Scroll.

The Rainbow Bridge from the Qingming Scroll ("Qingming Rainbow Bridge" for short) and this later rural bridge, although greatly different in scale, are very similar in structure. They both have two groups (rows) of longitudinal beams, one with an odd number of beams and one with an even number, and their crossbeams are both clamped together without wooden joints. All beams are squared in section. The only difference is that the Qingming Rainbow Bridge has two more arch sections, and thus is much larger in span. This can be demonstrated by showing two steps of the chopstick game: the easier one could be expanded to form the larger one, from a three-segment chopstick-bridge, to a five-segment bridge (Figure 3.3, b and d).

7.2.1.2 The Garyu Bridge

The second example is found in Japan. It was an actual construction, the only woven arch bridge ever known built in Japan. However, it existed only for a short period of time and is visible today only in historical records. The Garyu Bridge (臥竜橋), literally: lying dragon bridge) is located in Shiraiwa, Sagae, Japan. A bridge bearing this name was mentioned for the first time in 1744, but was destroyed by water and rebuilt only in 1827, in a new place. This second bridge was about 40 metres long. Two sets of design drawings dated in this year have been preserved by local carpenter master families, they provide two versions of bridge design, both depict a wooden bridge composed of lateral cantilever structures and a middle woven arch.

The design preserved by the Watanabe family is designed by a carpenter master Ishikawa Migiheiji, the drawings are drawn on a surface 90 cm wide and 2.1 m long, and are a combination of plan, elevation and perspective drawings in different layers (several layers of paper cut and put together) (Figure 7.3, a).

The drawings provided by Suzuki family were designed by an ancestor of this family (Suzuki Kumagoro) (Sagae 1985, 78), arranged in a similar layout; they depict a similar bridge with a simpler woven arch structure in the middle of the span (Figure 7.3, b). From this family, an account of materials for the bridge construction dated in the same year was also shown to the researchers. The bridge is mentioned as a cantilever bridge (刎橋) in this historical literature.[4]

It's not clear today which design was carried out in 1827. After its establishment, this version of the bridge suffered damage by flood several times and was repeatedly repaired. In 1858, it was converted into a cantilever bridge (Sagae 1985, 26–7).

Ishikawa's design is represented here by a 1:50 wooden model made by the author (Figure 7.4) and which is now preserved in the Deutsches Museum, Munich, Germany. The middle part is clearly a chopstick-bridge, with inwardly placed longitudinal beams. The crossbeams are five-sided in section corresponding to the angles of the longitudinal beams. It's highly probable that the beams are fixed using iron nails rather than wooden joints (iron elements being a common element of traditional Japanese bridge construction).

Similar to the relationship between the rural and Qingming Rainbow bridge, the two versions of the woven arch structure of these two designs could also be understood as two steps of the chopstick game: while Suzuki's design takes the shape of the simplest version of a chopstick-bridge (Figure 3.3, a, cf. Figure 1.19), Ishikawa's design is a much more complicated one.

Observant readers may be aware that the Garyu Bridge was built only slightly less than a century earlier than the Huntington Moon Bridge (Chapter 2),

---

4 Information and material provided to the author by Kazuhiko Nitto in April 2017. Also see: Nitto (2004); Sagae (1985).

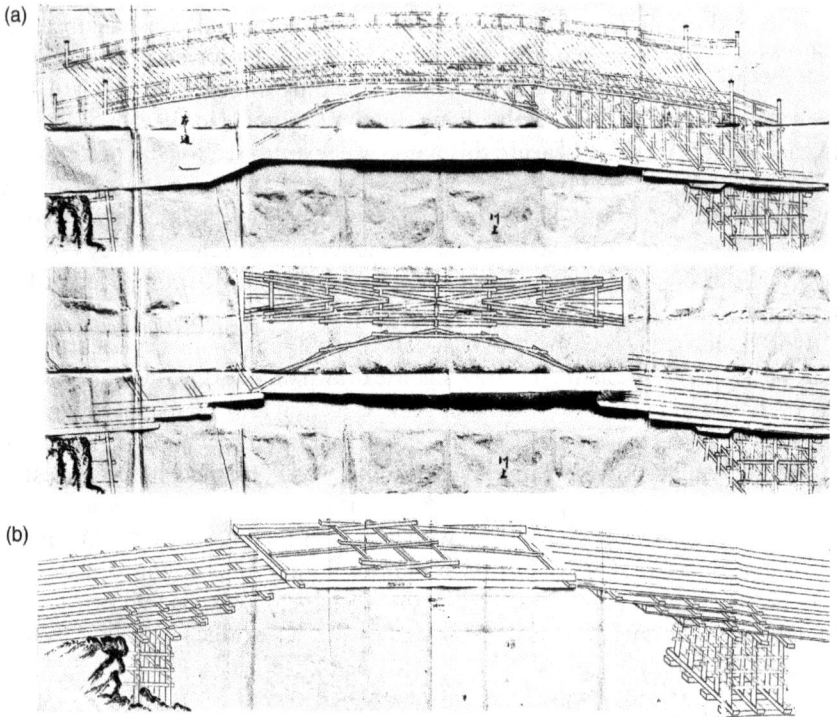

Figure 7.3 Historical drawings of the Garyu Bridge, Japan, built in 1827.
(a) Drawings designed by Ishikawa Migiheiji (石川右平次) and preserved by Watanabe (渡辺) family.
(b) Drawings designed by Suzuki Kumagoro (鈴木熊五郎) and preserved by Suzuki (鈴木) family.
Source: Photos by: Kazuhiko Nitto; combined and revised.

which was also built by a Japanese carpenter, one whose father-in-law happened to be a bridge engineer in Yokohama (Sec.2.2.1). However, it's still too early to draw the conclusion that these two bridges are related. Firstly, Yamagata and Yokohama are far apart from each other. Yokohama is located in a harbour area in east-central Honshu, while Yamagata is located in the mountainous area in the northwest. In the steel era, knowledge about a local timber bridge construction from a less-developed mountain area was not necessarily available to a modern bridge engineer from a more urbanized district. In fact, the Garyu Bridge is still seldom mentioned in Japanese architectural history today.

7.2.1.3 *The so-called "Caucasian Bridge"*

The third example comes from Germany. In his carpenter's handbook (*Das Zimmermannsbuch*) of 1895, German scholar Franz Sales Meyer refers to

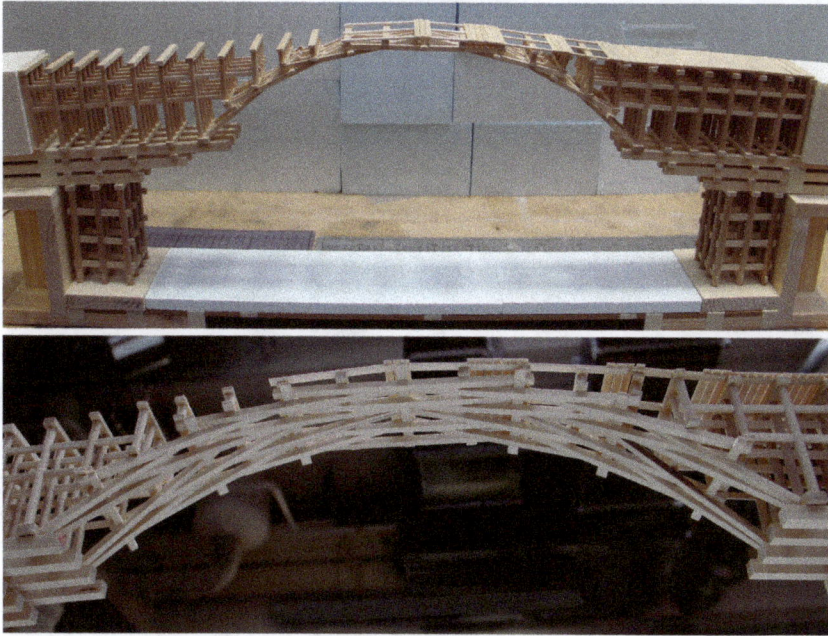

Figure 7.4 Wooden model of the Garyu Bridge according to the Ishikawa's design.
Source: Made by the author.

a bridge "the so-called Caucasian Bridge"[5] (Figure 7.5) is described in the section on bridges made of natural wood for gardens and parks, indicating its foreign origin, most probably from a mountainous area (i.e. the Caucasus), and its playful feature and function:

> Eine Stegkonstruktion, bekannt unter dem Namen 'Kaukasische Brücke'. Die Konstruktion ist gut, sieht gefällig und leicht aus und hat etwas Außergewöhnliches. Notwendige Voraussetzung ist ein unbedingt festes Widerlager, da die Belastung auf ein sogen. Knie wirkt und fast ganz als Horizontalschub zu Geltungkommt. So wie der Steg gezeichnet ist, könnte er etwa eine Breite von 4 m überspannen. Die Längshölzer des Sprengwerkes sind mit den Querhölzern möglichst solid zu verbinden und unter sich zu **verbolzen**. Der Beleg ist ein Dielenbeleg, der zweimal gebrochen ist. Auf der Bruchstelle oder besser unter derselben dürfte ein schützender Blechstreif sehr angezeigt sein. Die äußersten Geländerpfosten sind in den Boden gerammt, die drei mittleren finden ihren, allerdings nicht bedeutenden, Halt durch Befestigung an den

5 The "so-called" is from the original writing ("*sogen.*" = *sogenannt*) beneath the figure.

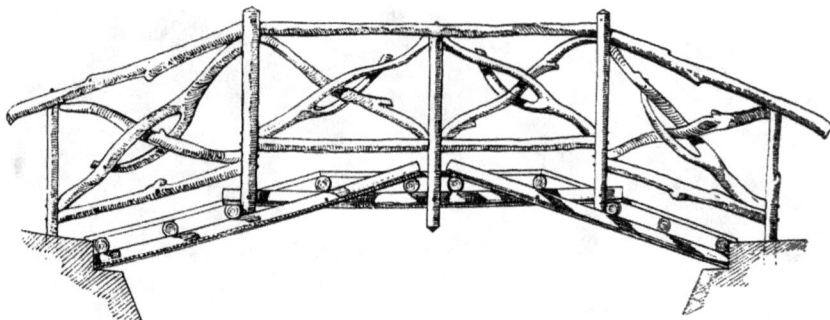

Fig. 354.
Sogen. kaukasische Brücke.

Figure 7.5 The "so-called Caucasian Bridge" from *Das Zimmermannsbuch* by Franz Sales Meyer.

Source: Meyer 1895, Fig. 354.

Trägerhölzern. Da die Träger nicht wesentlich verstärkt werden können, ohne die Konstruktion überhaupt unmöglich zu machen (weil der Mittelteil zu sehr erhöht würde), so muss auf möglichst leistungsfähiges Holz gesehen werden.

[A bridge structure, known as the 'Caucasian Bridge'. The construction is good, looks pleasing and light and has something extraordinary [about it]. The essential prerequisite is an absolutely solid abutment, since the load acts on a so-called knee and is almost completely applied as a horizontal thrust. As the depicted bridge, it could span a width of about 4 m. The longitudinal timbers of the structure should be connected to the transverse timbers as firmly as possible and be bolted together. The deck is made of boards, [the line of] which breaks twice. At the break points, or rather below, a protective sheet of metal strip should be very appropriate. The outermost railing posts are rammed into the ground, the three middle ones get their (not significant) hold by fastening on the bearing beams. Since the structural beams cannot be significantly strengthened without rendering the construction completely impossible (because the middle section would then be too high), care should be taken to select the most appropriate type of wood.]

7.2.2 Type B. Handshake bridges: improved cantilever bridges

In the 1970s (before the "discovery" of the MZ Bridges), in his *Science and Civilisation in China*, Joseph Needham (1971) categorized the Rainbow Bridge as a subtype of cantilever bridge and used the term "multi-angular soaring cantilever" (163) to describe it. A cantilever in the context of historical wooden

Figure 7.6 Cantilever of a bridge in Zagu'nao, Sichuan, Southeast China, by American sociologist Sidney D. Gamble, 1917–1919.

Source: Sidney D. Gamble Photographs. Vol.1. 1908 ‒ 1932.

bridges is composed of multiple levers of overhanging timbers which reach out from an abutment. A cantilever bridge commonly has two cantilevers from the two banks, reaching towards the middle span, and connected by a simple beam. This kind of structure is the most common type throughout Asia, from Japan to Turkey, for large-scale wooden bridges. So common in Asia and so seldom seen in Europe, timber cantilever bridges are even thought by some scholars to be a bridge type exclusive to Asia, which is not true. Cantilever bridges are also common in the mountainous areas of Norway (Figure 7.12), and bridges consisting of cantilever combined with other forms are also seen in the Alps.

Chinese examples, the largest in number and form, are particularly renowned both within China and in the western world. The earliest literature regarding this bridge type stems from the fifth century AD, and refers to a nomadic people in the Northwest,[6] near to Tibet. Several thousands of years later, the most magnificent examples in that same area shocked the earliest western explorers (Figure 7.6).

Thanks to the predominance of the cantilever bridges among Asian wooden bridges, Needham was not the only one to classify a rainbow-bridge-type structure as a cantilever bridge. When the drawings of the Garyu Bridge

---

6 Shazhouji (沙州记) by Duan Guo (段国). This kind of bridge is recorded as an achievement of the Tuyuhun (Azha) people.

Figure 7.7 Wo Bridge (Baling Bridge) in Weiyuan, Gansu Province, Northwest China.
Source: Tang 2000, Figs 5-80, 5-82.

were first discovered by Japanese researchers, they too classified it (esp. the vesion of Ishikawa's design) as a cantilever bridge (刎橋) (Nitto 2004).—They had a better reason to do so, since the Garyu Bridge is a combination of can-tilever and woven arch structures.

Such classification was never accepted by Chinese scholars, nor is it in this book, as we have numerous ways to tell the fundamental differences between a cantilever and a woven arch. Although, interestingly, these two types do have some relationship. We've seen this combination in Japan, and in the following sections we will see that they are both found in the same area in Norway. In China, there are some bridges which would be seen as a hybrid or a combin-ation of cantilever and woven arch. Their relationship with the woven arch was first noticed by Tang Huancheng (Sec.3.2.3).

The first type was found in Gansu, Northwest China: there were two of them of the same type, both named "Wo Bridge," but with different Chinese characters. One literally means "Lying Bridge"(卧桥)[7], and is located in Weiyuan (Figure 7.7; Figure 3.10); the other literally means "Handshake Bridge" (握桥) in Lanzhou.[8]

The second type is an example found in the mountains on the boundary between Hubei and Chongqing provinces, Southwest China: the Qunce Bridge (Figure 7.8; Figure 3.11).

Both these western China types have inclined cantilevers as their main structural bodies. The northwest examples involved a woven arch at the

7 This bridge also called "Baling Bridge"(灞陵桥) as mentioned in Sec.3.2.3. This bridge was rebuilt in its old form in the modern era.
8 This bridge was torn down and rebuilt in concrete.

Figure 7.8 Qunce Bridge, Enshi, Hubei, Mid-South China, 2016. (a) Landscape of the Qunce Bridge (b) Iron wedges are applied at the jointing positions of the beams. The author is captured in the photo under-deck doing the survey.

Source: Photos: Wu Chao.

very top part, the interweaving beams connecting the cantilever and the top horizontal beams, and thus form an inverse curve on the arch shape. The Southwest example simply interweaves the cantilevers and the horizontal beams together. In both examples, the woven arch principle is applied to combine the cantilever arms and strengthen their otherwise overhanging ends, just like two hands reaching out to hold each other.

7.2.3 Type C. Norwegian clamped bridge: reinforced strut bridges

The European examples of woven arch bridges that we have discussed so far – the Da Vinci bridges and the German case – are not closely connected to local bridge-building practices. The Da Vinci inventions are for use in the war following an example of the Roman time, while the German example is an imported idea and meant only as a decorative garden item. However, in another part of Europe, we do find an example with its roots firmly anchored in local building practices.

This intriguing group of bridges can be found in central Norway. Rich in mountains, valleys, and timber resources, the environment for wooden bridges is very much similar to that in China.

Before the seventeenth century, this area was only sparsely populated by the Sami, but it was heavily developed because of mining in the eighteenth to nineteenth centuries. The transport requirements for mining purposes led to the erection of a great many timber bridges with various forms: cantilever, strut, truss, and hybrid. Among them, the strut is the most common structure, alone or in combination with other forms (Figure 7.9–Figure 7.14). Since they were intended for industrial use, less care was taken with the woodworking aspects of these wooden structures: joints are roughly hewn and iron elements are applied liberally.

Figure 7.9 Historical bridge in Røros, Norway. Three-sided strut with X-shaped struts. Norway, 2017.

Among the many forms, three-sided strut is one of the most common structure type used in these bridges. Since the three-sided strut is easily subject to deformation, people had to take measurements to strengthen the joints. Nailing an additional wooden block at the joint place is the most simple, crude, and effective way (Figure 7.10, Figure 7.13). For the same reasons, an old bridge in Einunddalen (Figure 7.14) provided a fascinating example of the woven arch approach. The struts are interlocked with the horizontal beams, very much like the rural Rainbow Bridge and the "so-called Caucasian Bridge" in form. However, there is a fundamental difference in between: while we would classify the latter two examples as belonging to the chopstick-bridge group, the example in Norway shows a clear genetic relationship in terms of an evolutionary route among the local struts-structures.

7.2.4 Type D. MZ bridges: improved strut bridges

The basic construction form and the structural challenge in the MZ mountains are similar to that in Norway, namely, with the strut bridge as a basic form before the development of the woven arch structure, and by combining several layers or forms of struts to strengthen the load-bearing capacity where the bridge spans a greater distance. However, as the bridge becomes larger, no

Figure 7.10 Historical bridge in Loholet, Nord Fron, Norway.
Source: Photo: Sparby, Kaare/Anno Norsk skogmuseum.*
*SJF.1990-04737. https://digitaltmuseum.no/021016555236/gammel-og-ny-bru-over-elva-vinstra-ved-loholet-i-grenda-ruste-i-nord-fron. Accessed: 4/5/2020.

matter how many layers of supports are added, there is another construction problem: the deformation relating to the rigidity of the joints.

As we've seen in previous chapters, all woven arch bridges in the MZ area, whether they are 3+X, 3+2, 3+3, 3+4 or 3+5 bridges in form, share the same basic system of a three-sided arch. At the same time, strut bridges with a single three-sided arch being a basic and common structure form in the mountains.

Woodworking joints, in particular the dovetail, are not actually rigid connections in themselves, but rather semi-rigid, partially-rotatable nodes. A three-sided strut/arch is therefore deformable. If it is topped by a layer of continuous deck beams, for example like the earlier version of the Rulong Bridge (Figure 5.16) is, the rotation tendency of the struts will be repressed by the heavy deck beams above. In cases where the span is greater, and exceeds the length of a single tree trunk, the deck-beams cover only a part of the span, and thus cannot restrain the deformation of the structure underneath, and beam-weaving mechanics are then applied to interlock the strut joints. The Wodu Bridge (Figure 6.51, Figure 6.52) exactly expresses this idea of the 3+2 (3+X) bridge. From this (3+2 or 3+X) form onwards, up to the most mature 3+5 form bridges, the aim is the same: to stiffen and strengthen this basic three-sided arch system – as the bridge carpenters articulately explained (Sec.4.6.1.2)

Figure 7.11 Historical strut bridge in Røros, Norway. A combination of struts, three-sided strut and truss.

Source: Røros Museum.*

*RMUB.000168 https://digitaltmuseum.no/021016631215/sleggbrua-hyttelva-og-malmplassen-sett-fra-sorost-ovre-del-av-kurantgarden, accessed: 4/5/2020.

Figure 7.12 Historical bridge in Røros, Norway. A combination of cantilever and struts, 2017.

Figure 7.13 Historical strut bridge in Rondane area, Norway. Three-sided strut bridge. The joints are strengthened by wooden blocks, 2017.

7.3 From idea to technique

7.3.1 *Origins of the woven arch structures*

After our long journey exploring all the known historical woven arch structures of the world, and putting our collections into our catalogue system, it is now time to return to discussing the "origins" of the technique for each of these historical examples. Instead of looking for a birthplace and a point in time, it is of greater technical value to ask how the idea itself came about.

7.3.1.1 *To strengthen the joints*

The *praefectus fabrum* (military engineer) of Caesar's Roman Army, the Norwegian carpenters, and the early Chinese bridge builders in the MZ mountains faced the same construction problem: when working on a bridge structure and applying a three-sided arch frame, how to enhance the stability of that frame by strengthening the rigidity of the joints. A genius engineer in the Roman army came up with the idea of inserting an interlocking element (*fibulis*)between the posts and the beam, while the carpenters in Norway and China came up with the interlocking beams.

Deliberations on the joint design were a starting point for some of these woven arch structures. On the other hand, as Da Vinci's bridge designs show, the woven arch structure could be seen as the direct representation of an enlarged joint. This concept is also put forward by some modern structural design scholars. Vito Bertin (2012) suggests that a theoretical structural principle of the entire category of "leverwork" (or "reciprocal frame structure"), to which the woven arch structure belongs, is to deal with of the joints by interlocking members.

7.3.1.2 *A playful route to the structural design*

On the other hand, it's not necessary to be an engineer or a craftsman to come up with the idea of a woven arch. The simplest woven structure can

Figure 7.14 Historical bridge in Einunddale area, Norway. Photo entitled "clamped bridge in Einunddalen." ("*Bru, elv. Klemmet brua i Einunddalen*").

Source: Anno Musea i Nord-Østerdalen.*

*MINccessed: https://digitaltmuseum.no/021016387398/bru-elv-klemmet-brua-i-einunddalen, accessed: 4/5/2020.

be created with three wooden sticks (Figure 1.24, a), and with three pairs of chopsticks, you can build a "bridge" on the dining table (Figure 3.3, a). The basic principle is so simple and convenient, that it could be created in nature, by a couple of fallen tree branches, or be created by a child. It does not need to have an inventor, or a place of origin: it could simply come into being, at any time, anywhere.

As soon as a game like that appears, it could be studied and passed down through generations, gaining variety and complexity in its form, and with the right person who appears in history from time to time, with a keen curiosity and active hands, the idea would be put into practice: the enlarged chopstick-type bridge is born. Structures created in this way do not necessarily fulfil a real purpose or respond to a construction challenge in the actual world, but rather are playful items, to be enjoyed for the fun of it. All the garden structures, the Huntington Moon Bridge, the German "Caucasian Bridge," show such a self-evident playful attitude.

In professional hands, the folk game could turn into a study model. A craftsman "thinks with his hands," and we've seen in so many bridge stories that such a game or model played a significant role. Da Vinci definitely studied models, as shown by his sketches. Toichiro Kawai, the Japanese carpenter, had a model at home for his design of the Huntington Moon Bridge, as reported by his son. The Chinese scholars used chopsticks and matchsticks in their studies and lectures on the Rainbow Bridge – and the people of the MZ bridge area have played the game since their childhood...

Indeed the woven arch structure itself has such a playful nature, that it is fairly safe to deduct that wherever a bridge exists, a similar game exists or existed there too. And when we talk about a possible relationship between these bridges, the game is a much more convenient medium to transport the idea, rather than the building technique.

7.3.2 The gap between idea and technique

Not every experimental structure led to a mature bridge technique. None of the "enlarged chopstick game" type bridge structures formed an enduring tradition – they all disappeared over time. Those lucky enough to leave visual clues are usually very late examples. We have no idea how many others were lost.

To go from a playful garden structure to a real bridge, the "technique-gap" needs to be bridged. Technique is an entire set of means and methods for translating an idea into reality. For a new technique to be accepted into a regional building tradition, it must fit into existing local technical conditions.

The woven arch bridge provides many new challenges. In a 'pure' reciprocal frame structure, thanks to the mutual support of structural members (the "reciprocity"), the failure of one member will result in the failure of the entire frame. Take the chopstick-bridge as an extreme example – remove any member from it, and the entire structure collapses. Although this problem is partly solved in the MZ bridges by applying a group of beams, the accuracy required in design, processing, and construction is still the first challenge that is incomparable to any other timber constructions. The many angles involved between all structural members and the interreliance between one member and another mean that this is not an easy task for university students, or engineers, even today![9] It might well be that the technique for the Rainbow Bridge has been lost partly due to the level of difficulty.

The construction process itself is the second problem. Not all bridges can be built on land as Da Vinci's transportable war-bridges could, or in an accommodating setting like the garden bridges were. If the bridge actually needs to be assembled over water, the mutual supporting mechanism again calls for special methods and measurements to ensure security. This situation is very much like the construction of a stone arch – i.e until the very last piece is in place, the entire structure is unsafe. In this sense, a scaffolding system similar to that for masonry construction would be suitable for a woven arch bridge. However, it is precisely in situations when it is not feasible to build a stone bridge that a woven arch bridge is called for – as we've seen in the case of the busy canal of the Rainbow Bridge, and over the deep valleys of the MZ mountains. In other words, if conditions allow a stone bridge to be built, it is built, as it's stable and enduring and much more conventional in most societies. The domain that is left for woven arch bridges is precisely that of the technically-challenging area of difficult construction conditions.

The Rainbow Bridge endured for a while: the people who built it must have figured out a set of methods for doing the calculations and to manage the construction (at great cost and involving sophisticated technical measurements). Much later, in our time, when bridge engineer Tang Huancheng built the Jinze Bridge according to the technique probably used in the Song Dynasty, he had to build the bridge twice to ensure the correct design and construction: He first built the bridge on land, disassembled it, and reassembled it above the river (Tang 2010, 115) – a rather narrow and calm river. Constructions involving this level of sophistication of technique and cost might have been justifiable when it was a question of building the gateway to the capital city (Bianliang city of the North Song Dynasty), or other governmental projects, as recorded in the literature, but this technology soon disappeared in the turbulent years towards the end of the dynasty.

7.3.3 The success of the MZ bridges

The success of the MZ bridges is based precisely on solving the challenges in both design and construction. The first key fact is that they were not building an entirely new type of bridge (as was the case with the Rainbow Bridge), but a modified version of a local strut bridge tradition. The two arch systems not only provide them with suitable construction conditions even in unsatisfactory environments, but also with flexibility in design: Thanks to this two system constitution, the first arch system is practically self-supporting and can

---

9 The larger scope of the reciprocal frame structure has been incorporated into undergraduate training in many architectural schools, including Nanjing University where the author has worked. Scientific methods of design have been studied by many scholars, including Bertin (2012) and Thönesson (2015).

support the rest of the construction, so that even the simplest swing-frame-scaffolding can provide enough stability; and thanks to the even number of the arch section, the exact scale of some building members can be adjusted during the actual construction, much easing the requirements on design.

Another key feature of the MZ bridges, compared with other woven arch bridges, is the joints technique. As mentioned above, theoretically, there is no need for elaborate woodworking joints in a woven arch structure – in a finished structure – since each member restricts the other's movement. However, precisely because of this mutual restriction, until the last member is put into its place, an unfinished (unclosed) woven arch structure could be entirely unstable. Thereafter, the woodworking joints of the MZ bridges, tenons, and dovetails, give some rigidity to the joints of an unfinished woven arch, enabling step-by-step assembly. We've seen in technique discussion of the MZ bridges, how important the selection of joints is, in the entire toolkit of the bridge carpenters' technology. The survival of the MZ bridge tradition (instead of any other types and attempts), is down to the joint system inherited from the local building tradition.

7.4 Structural thoughts: dealing with the challenge of the span in different cultures

In this section, we will go behind the scenes to see how is the span-challenge dealt with in different cultures, from which the woven arch bridges emerged. Before we do that, there is a fundamental question: we all accept that the woven arch structure is something unusual, but what makes it so special?

7.4.1 Ambiguity in terms of the structural science

7.4.1.1 Compared to an arch

The first question is: is a woven arch actually an arch?

There's no doubt that it's arch-shaped and could be called arch in the common understanding of the word and by the bridge builders in China (see Sec.4.6.1.1). However, this question has led to confusion in the field of engineering science. In the scientific sense, the arch has a specific definition of its mechanical features (on the opposite of a beam): an arch is good at resisting compression, while it is weak at dealing with tensile stress and bending force. On the contrary, a beam is used to resist the bending force. An arch pushes the base forward, and if this thrust is not offset by the counterforce provided by a solid base, a masonry arch collapses. A beam does not apply thrust to the base.

In a woven arch, however, compression and tension co-exist, and the bending force plays an important role. The woven arch is formed by weaving (bending) beams, and it acts as an arch from the macrocosmic view, but as a beam from the microcosmic (element) view. Some scholars define the woven arch as a structural form that is somewhere between an arch and a beam (Tang 1987, 64).

The bending action keeps the structure together. The *choudu* technique for the MZ bridges is used precisely to increase the bending force to enhance stability. Unlike a masonry arch, in a woven arch, if the thrust is not offset by the base, the arch itself will expand, but unlike the masonry arch, the expansion stops when the force is balanced by the bending force from inside – this can be demonstrated by playing the chopstick-bridge-game on a slippery table surface (cf. Figure 4.96).

7.4.1.2 Compared to a truss

Some have also compared the woven arch bridges to wooden truss-bridges. "A truss is a single plane framework of individual structural members connected at their ends to form a series of triangles to span a large distance" (Shekhar 2005, 431). In this sense, a woven arch and a truss do have some similarities: they both consist of a number of shorter members put together to achieve a larger span. However, a truss and a woven arch are almost opposites in terms of statics features. Whereas ambiguity in the state of forces is a characteristic of the woven arch, the truss is a model of clear force division.

The modern truss is a typical product of science. Although they have been widely used in the western world since Roman times or even earlier, the scientific study of the mechanics of truss structure is a basis of modern structural science. Furthermore, like physics, modern structural engineering developed based on abstract models in which forces are studied first in theoretical, ideal conditions. An ideal truss consists of two-force members only: the tensile or compressive members, with structural stability ensured by the two-force bars arranged in a triangular formation.

Using pure and abstract models as a basis, the load-bearing capacity of a structure can be calculated quite precisely, and it is possible to construct colossal iron truss bridges across the sea. Engineers' minds are trained to work with such abstract analytical models – which is why it's a bit confusing when they first encounter the woven arch concept. Ambiguous in definition and classification, it's part of the reason that the woven arch fascinates so many people with a scientific background, including the author, in his first engineering class in 2002 (see Afterword).

7.4.2 The truss tradition in western building cultures

What were the common ways to bridge larger spans with timber in the times before modern science?

In his book *I quattro libri dell'architettura* (1570), on the next page after his study of Caesar's Rhine Bridge, Andrea Palladio took up a new theme: wooden truss bridges. He designed four types of truss bridges (Figure 7.15), of which one is said to be inspired by a German example. Based on this study, Palladio is considered as a pioneer of the modern truss design, and his influence extended as far as the early development of modern iron bridges.

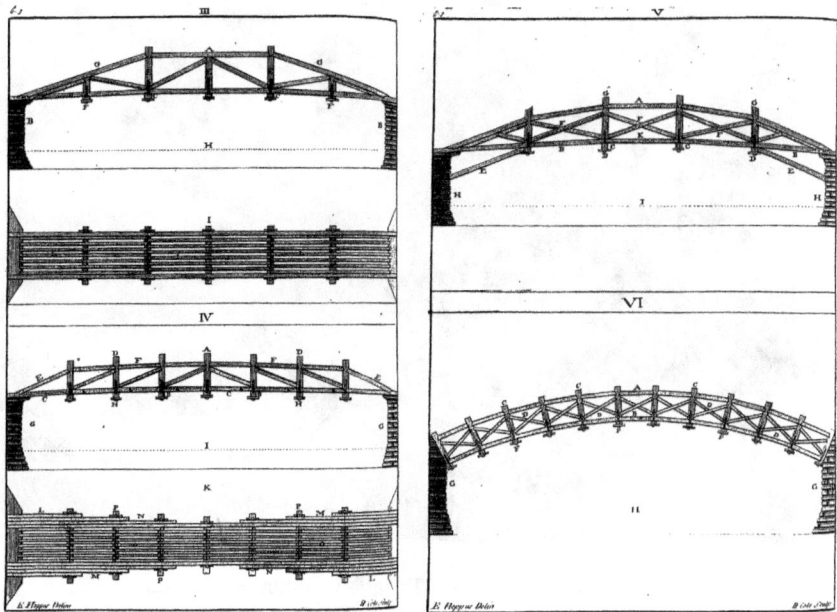

Figure 7.15 Palladio's truss design in *I quattro libri dell'architettura*.
Source: Palladio 1570.

The truss form has a long tradition in European civilization, and is used in both bridges and roofs. A famous truss bridge from Roman times is Trajan's Bridge, across the Danube, built around 105 AD and subsequently visualized in relief on the Trajan Column in Rome (Figure 7.16). A fourth-century Roman image of a truss roof on a fresco in the Old St Peter's Basilica in Rome (Figure 7.17) shows a triangular roof with a middle post which is clearly in tension (hanging post).

Members in tension are a crucial indicator of the truss form. The tensile strength inside the wood, parallel to the direction of the grain is two to three times greater than its compressive strength in all types of wood. Employing wooden components to bear tensile forces is one of the most important early achievements of the history of mechanics, and this step is expressed by the emergence of the hanging post (king post) in the truss, in both bridge and roof structure evolution.

Since bridges and roofs both face the same span challenge, even though they may develop separately, their techniques are related. From the Renaissance onwards, the relationship between them was even more distinct. In Palladio's design, at least one of his truss bridges may have been based on techniques from the practice of roof truss construction (Tampone 2003). This cross-pollination was even more evident in the works of the eighteenth-century

Figure 7.16 Trajan's Bridge, relief on Trajan's Column, Rome, Italy.
Source: Wikipedia.*
*https://en.wikipedia.org/wiki/File:072_Conrad_Cichorius,_Die_Reliefs_der_
Traianss%C3%A4ule,_Tafel_LXXII_(Ausschnitt_01).jpg, accessed: 4/5/2020.

Swiss master family Grubenmann. Their largest masterpiece include bridges
with over 60 m clear span (a legendary bridge, originally designed with a clear
span of over 120 m) and 20 m x 38 m clear roof space (Killer 1985).

The truss is no doubt the most important structure of timber construction in the western world, and the most effective timber structure before
our era of modern science. In the series of structural design competitions at
Tsinghua University in the 2000s, when the author was studying there, a hand-
made model of a wooden truss bridge, with a clear span of 1 m, was able to
bear a dynamic load of 15 kg, even though the bridge itself only weighed
some 60 g![10]

Not only is it effective in load-bearing, the truss structure is also convenient
for construction. An entire frame can be made onshore and hung into position, in both bridge and roof construction. So no matter how inhospitable
a site is, European craftsmen will never take the risks that the MZ bridge
carpenters take with their swing-frame-scaffolding.

The stability of the truss is ensured by the triangularly-arranged members,
making it easier to take control of the performance of the form, and to process the joints. Consequently, design and processing accuracy of the individual

---

10 https://news.tsinghua.edu.cn/info/1276/46081.htm, accessed: 4/5/2020.

Figure 7.17 Truss frame of Old St. Peter's Basilica on a fresco from St. Peter's Basilica, Roman, Italy.

Source: Wikipedia.*

*https://en.wikipedia.org/wiki/File:Affresco_dell%27aspetto_antico_della_basilica_costantiniana_di_san_pietro_nel_IV_secolo.jpg accessed: 4/5/2020.

elements is not as critical as it is in China. Thanks to its advantages in terms of stability and processing feasibility, the truss form is also the most popular structural form for half-timber skeleton frame for house-building, taking the *Fachwerk* (truss-frame house, Figure 7.20, a) in the German-speaking area as a reference.

Moreover, a great many iron elements also play an important role. Since the members may be in tension, woodworking joints can no longer ensure joint safety, and iron connections were developed in Roman times (Taylor 2003). In Palladio's truss-bridge design, he also paid special attention to inventing an iron connecting device (Tampone 2003). The early adaptation and long-term development of iron elements not only provided European wooden structures with a much greater load-bearing capacity, but ensured they had a vast potential for further evolution.

Given the above, it's easy to understand why, having a deep-rooted, mature technique with a greater load-bearing capacity, higher efficiency, easier operation and more convenient truss structure, there was no actual need – even when it was "invented" and recognized as a genius idea – to apply the woven arch structure in practice. And that's exactly what happened: even though this idea not only appeared in Da Vinci's secret manuscript, and was also very nearly presented in Palladio's treatise, history simply turned the page and went on to the truss.

7.4.3 Absence of the truss structure in the East Asian building cultures

There was no truss tradition in Asian building cultures before the modern influence came from the west. In China, for at least a millennium, cultural tastes have obviously favoured horizontal and vertical elements. Oblique elements are avoided in the visible structure of a building. This phenomenon becomes more obvious over time. Triangle-frame roof structures are common in the Tang Dynasty (618–907), and although they could be considered a rudimentary form of truss, there would have been too many crucial steps[11] to the full form, and this evolution never occurred. The use of triangular roof frames declined in the Song Dynasty (960–1279), and they almost disappeared in the last two dynasties (from the fourteenth century onwards). Inclined posts, where they are essential for structural stability and unavoidable in multi-storey buildings – such as the famous Guanyin Pavilion of the Dule Temple and the Sakya Pagoda of the Fogong Temple ("The Wooden Pagoda," Figure 7.18), both dating from of the Liao Dynasty (916–1125) – are hidden inside the "mezzanine floors." The high-ranking palace hall described in the official building regulations of the Song Dynasty, *Yingzao Fashi*(1103), can be seen as an ideal model of the Chinese architectural tradition: layers of huge beams placed on each other above the ceiling, framing the curved silhouette of the huge roof (Figure 7.19).

A short note must be added here, that there are two basic types of wooden frame structure in China, the above mentioned high-ranking buildings are catalogued under the *tailiang* construction, which is used in North China and for high-standard buildings, characteristic by the multiple layers of structural members. On the contrary, our bridge corridors (Figure 4.94) are catalogued under the *chuandou* construction, which is used in South China and characteristic by the wooden pillars reaching from the floor to the purlin, and that they are connected by layers of tie-beams penetrating them. Nevertheless, the favour for horizontal and vertical elements is the same in both construction types (Figure 7.20).

The cultural aversion to inclined members is less decided in Japan, where we find various forms of inclined members in roofs, especially when they are blocked by ceilings: they have an enormously significant structural function; mostly as levers, never as pulling bars.

The idea of using timber in tension does not exist in traditional East Asian architectural tradition. Although such elements might occasionally be found

---

11 E.g. the appearance of a hanging post/members in tension.

(a) (b)

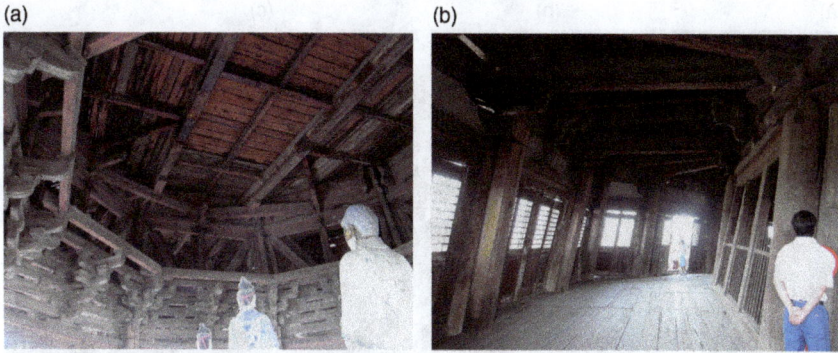

Figure 7.18 Sakya Pagoda, Fogong Temple, Ying County, Shanxi Province, North China, 2007. (a) The diagonal beams and posts are only applied on the mezzanine (hidden) floors. (b) Deformation in the upper floor of the pagoda.

Figure 7.19 Schematic representation of a high-ranking building according to the building rules in the *Yingzao Fashi* from the Song Dynasty.

Source: Liu 1984, fold page, Fig. 134-1.

in mechanical structures, they seldom appear in buildings. If certain tensile forces are present in building members, they would only appear as the result of the secondary loads or functions. For example some tie beams in wooden structures can withstand tensile forces and keep the frame steady under lateral

(a)   (b)   (c)

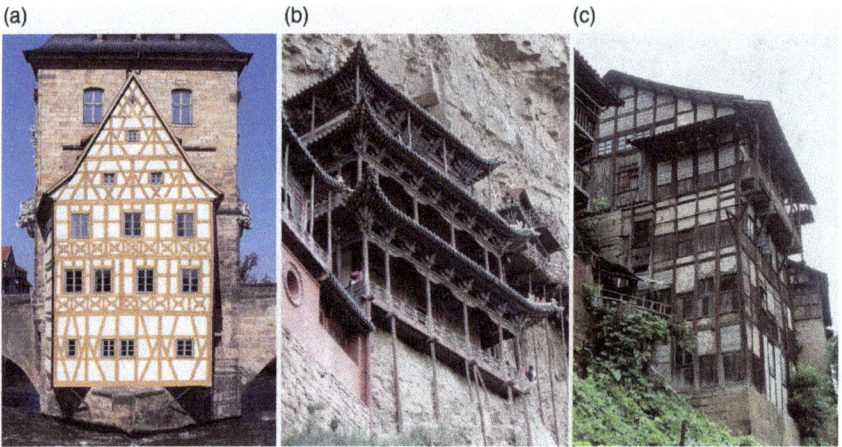

Figure 7.20 Wooden frame structure in multi-storey buildings in Germany and China. (a) A truss-frame building (*Fachwerkbau*) in Bamberg, Germany. *Source*: Photo: Klaus Zwerger. (b) Hanging Temple (Xuankong si), Shanxi, North China, *tailiang* construction.(c) Town of Fubao, Sichuan, South China, *chuandou* construction.

Figure 7.21 The two cantilevers under a bridge function in the same way as the *dougong* under a roof. Huangqiling Bridge (黄旗岭桥), Zhouning, Fujian, 2019.

forces (e.g. from wind), while the main static loads upon them remain structural weight from above or from themselves. Hanging posts are also found in pointed roofs or the centre of a tower or pagoda, but their function is not to bear the weight of the beams attached to them, as the king posts do in a truss, but rather to connect the ridge beams and other centripetal members and keep them in their correct position.

In East Asia, leverage is the predominant mechanical means for resolving the span challenge, both in buildings on land and over water, in the form of corbelling or cantilever, which are only different terms to express the same mechanical idea. Where the cantilever bridges take on the role of large-scale timber bridges (Figure 7.21), the corbelling system (*dougong*) takes care of the huge protruding eaves in the same way (Figure 7.19). Even the layered roof frame utilizes the lever effect. By means of a reversed form of cantilever, each horizontal beam layer delivers the load to the beam directly beneath it, at a position very close to the position where the latter is supported. This normal or reversed corbel form has a very colourful name in Chinese: 疊澁 (*diese*, lit. "lying unsmoothly"). This is used as a construction term in masonry work, describing the corbelling and reversed corbelling layers of brick or stone. The almost hieroglyphic image of these two characters vividly describes this conformation principle, in both masonry and timber construction.

7.4.4 Joint system

We've already emphasized the crucial role of the wood joint system in building a woven arch structure in a real construction setting (i.e. where a 'working bridge' is needed, rather than a decorative garden item), and the success of the Chinese (MZ) examples due to this feature. Now we will say that even in a broader context, too, woodworking joints have always played a more significant role in Eastern Asian architecture than in the western world.

In traditional timber construction in the western world, inclined members are not only load-bearing members, but they form triangles with each other, so they are also crucial for structural stability. A triangular frame is a rigid structure: i.e, as long as the members are fixed together, the frame will not deform, even if the individual joints are rotatable (non-rigid).

In Eastern Asia, especially in China, where inclined members are not favoured and wooden structures are composed of horizontal and vertical members only, without triangular stability to rely on, a parallelogram frame has a tendency to deform. Therefore, the individual joints have to be rigid in themselves to ensure the stability of the frame.

In other words, whereas in the European building culture, that has a truss tradition, the stability of a timber structure is to a large extent a question of geometry, in China-centred East Asian building culture, stability is to a great extent a question of jointing.

In a wooden frame, where stability is ensured by triangles, the joints between members can simply be lapped together and pinned with wooden dowels (Figure 7.23). Lap joints are relatively easy to process, since they can be cut simply, using a saw. Dowels are also an effective way to fasten together two wooden members. Since an individual joint consisting of two connecting members cannot rotate within a triangular frame, the joint does not necessarily have to be cut very precisely. Thus, a lap joint plus wooden dowels is one of the most common joint forms in traditional European wooden frame structures.[12]

---

12 As observed by the author, and also supported by Zwerger (2015).

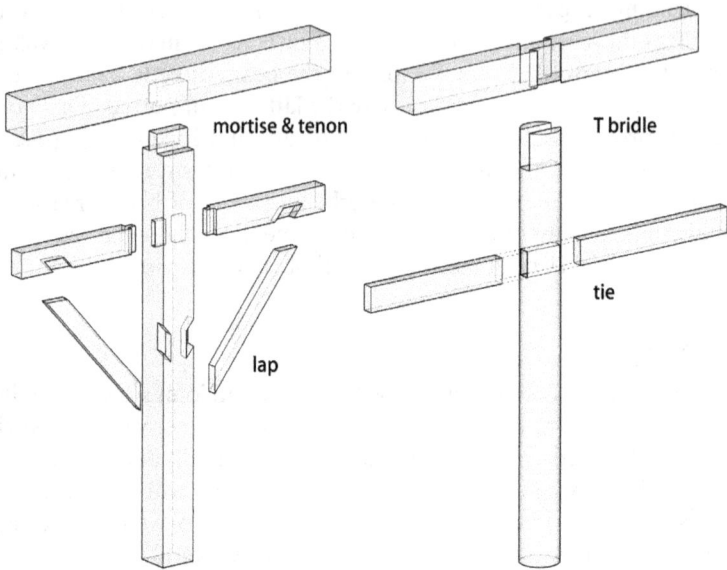

Figure 7.22 Typical joint forms between a pillar and beams in European (left) and East Asian (right) frames.

In East Asia, since the 'lap joint and wooden dowel combination' cannot stop the joints rotating, which would lead to an immediate deformation in a rectangular frame, this joint form is seldom used. Here, when a horizontal member meets a vertical member, the horizontal beam or tie beam will always go through the beam (Figure 7.22, Figure 7.24). Thus, in addition, East Asian joints are by necessity always processed not only accurately, but extremely tightly connected.[13] This is sometimes done by making the recessed (female) part of a joint a bit narrower than the projecting (male) part, and sometimes by other means. The improved rigidity of the joints ensured by these measures, in turn ensures the stability of the frame. This wooden joint tradition established a basis for the development of the MZ bridges.

7.5 Seed, soil, growing environment

We can use the growth cycle of plants as a model to illustrate the histories of the woven arch bridges. To grow, a plant requires a seed, fertile soil, and an environment with the proper moisture and warmth. In this sense, the bridge-building tradition is a plant; the idea of having a woven structure is the seed; the local environment in need of such bridges is the soil, and the people with

13 Ibid.

Figure 7.23 This centre post in Museum Cemento Rezola, Spain, shows the dominating role of lap joints and the triangle principle in European frame structures, 2019.

the right technique and motivation to put that technique into practice to build a bridge and to hand down that technique is the growing environment.

7.5.1 The seed

The woven arch structure is not 'native' to any single place. The real and only origin of this structural principle is nature. People in many different types of environment have been able to achieve this idea by observing nature, by playing about with sticks, or using branches to build a temporary shelter. As soon as it is noticed that the sticks or branches can be arranged in a way that they support and restrain each other – that here is a structure with some rigidity and load-bearing capacity – the seed has appeared.

The seed can be passed on among people and from one place to another, by simply playing with some sticks. With three stick-shaped objects, you can demonstrate the idea of a mutually supporting system, and with six you can

Figure 7.24 The relationship between the horizontal members and the pillar is a representative configuration in East Asia. *Jodo-ji jododo*, Hyogo, Japan.

Source: Zwerger 2015, Fig. 437.

build the simplest form of bridge – a simple enough game, with something magical to it.

After careful examination, it's now proven that Da Vinci's woven arch bridge designs were not influenced from abroad, but rather from his study of Caesar's Rhine Bridge (based on Caesar's narration back in the first century BC) as a result of the extensive social trend at the time for reviving Caesar's legacy. The Norwegian example also very likely originated from the locality where a series of types of bridges that might well have developed the woven arch idea existed in the same time period.

By contrast, the example described in German literature adopted the idea from another area, as indicated by its name (the "so-called Caucasian Bridge").

The origins of the Japanese examples, both the designs in Japan and in the one in the US, are unclear. They both appear in their primary forms, indicating that they are the direct results of a playful idea. The idea could have been self-generated, appeared separately or have been shared between the two sides of bridgebuilders in a way unknown to us. It could also have been imported from abroad, and if so, most probably from China.

The relationship between the Chinese examples, with their rich variety, is also of multiple possibilities. From the earliest known examples, the Rainbow Bridges in the North China Plains, to the late vernacular examples, including the strut-bridge-based versions in the Southeast (MZ area), and the cantilever-bridge-based versions in the Northwest (Gansu) and Mid-South (Hubei) – one or the other branch could be entirely independent, or rather, the seeds might have wandered over this area, in the form of a magical folk game for centuries, until they happened upon the right soil and the right climate.

7.5.2 *The soil*

If it drops in the desert, into the ocean or in the dry crack of a rock, a seed has no hope of growing roots. Only seeds that are planted in soil have a chance to grow. The soil for our bridge is the demand for bridges with the specific capabilities of a woven arch structure.

Gardens with sponsors full of curiosity can be a form of soil, like the exotic gardens in the nineteenth and early twentieth century, including H. E. Huntington's Japanese garden, and the possible exotic gardens mentioned in Prof. Meyer's book when he describes the "Caucasian bridge." However, when we move on to actual projects for the construction of bridges for real traffic purposes, the soil is critical.

The largest span capability of a woven arch bridge is about 40 m. So in the vicinity of large rivers on plains, or places where rivers are easily spanned by more convenient means – these places provide no soil for the woven arch structure. Instead, the places where real woven arch bridges are built are usually less developed mountain areas, with rich resources of the forest. It's easy to understand that adequate timber resources and smaller valleys and rivers provide the necessary conditions for such timber structures. However, limited local development in terms of culture, technique, and economy also plays an important role. If we are to believe Palladio and Da Vinci's reconstructions, the Roman military engineers were playing with a similar idea when they used some of their talents in timber construction on the battlefield in the land of the "barbarians" – in circumstances where their amazing engineering achievements in masonry construction would have been useless.

Later, in Europe, the seed of the woven arch bridge idea was made visible at least twice: in the Da Vinci manuscript and in German literature (in which the bridge was attributed to some unknown "Caucasians"), but there was no soil for the bridge. Both Italy and the German-speaking area had a tradition of building wooden truss bridges, a far more effective and convenient structural type. Therefore, there is no motivation to incorporate the woven arch idea into the real structure. This holds true for most parts of Europe, except for Norway – a poor and least-developed area in Europe until modern times.

The situation is similar in China, too. The woven arch bridges idea took root in the MZ area, a mountainous boundary area of rather delayed and limited development. Most of the inhabitants' ancestors had migrated from the lower

Yangtze River area centuries previously. That area was also famous for its numerous bridges, but, unlike the MZ area, was already very developed culturally and economically, and their advanced masonry techniques enabled them to build beautiful and poetic masonry bridges, so that would be have been no soil for the woven arch bridges either. Indeed, the places in China with woven arch bridges that have survived are all marginal areas. The only examples that don't fit this model are the long-gone Rainbow Bridges: although they were built at the centre of the empire of the time, they were built as a result of every other means at the time having been tried and having failed.

Although both countries were without a truss tradition, China had its advanced masonry bridge technique in at least some developed areas whereas Japan did not have an alternative. The masonry technique for bridges was introduced very late in Japan: the first stone arch bridge[14] was built in the seventeenth century, by Chinese craftsmen. Only the cantilever bridge and the suspension bridge were part of the Japanese large-scale bridges tradition.

The woven arch bridge, through ingenious and sophisticated in terms of structural features, is not considered to be a result of advanced technology, either historically or in the modern world. It is only given a chance in a very narrow set of conditions, when the conventional types of bridges are inadequate in terms of engineering capability, in places where more effective bridge types are not feasible.

7.5.3 *Growing environment*

A seed can lie dormant for many years until it is woken by the right amount of warmth and moisture. However, it can as easily be killed by a bout of cold in the Spring. Its growth requires a continuously temperate environment. Although the conditions of the soil are critical, the growing environment for the seed of the woven arch bridges idea to root and sprout is even more demanding, and to get it to grow into a mature plant, that's the most critical part of all.

In the various attempts, especially the playful garden cases, the creator's personal will is the deciding factor, and this in turn has its roots in social background and personal experiments – the Huntington Moon Bridge being the best example of this. However, the first group of rather personal attempts, whether they exist in literature or in physical form, did not endure, in the sense that neither was the construction knowledge passed down to successive generations of builders, nor did they establish a local tradition.

There is a second group of examples that did not endure for long. Some of them appeared in a series or in a group in a period as a phenomenon, but the idea did not take root strongly enough to grow into a lasting tradition. The Rainbow Bridges and the Norwegian example(s)[15] belong here. For the

14 The Megane Bridge in Nagasaki.
15 There may have been more than the one we found.

Rainbow Bridges (North Song) government demand and support were the deciding factors, but the building activities ceased soon after the government was forced to leave North China, so the soil of the bridges disappeared along with the dynasty.

For the Norwegian examples, we have little information apart from the images from the photo archive: that is we do not know the number, period, and geographical scope of the woven arch bridges there. The only thing we can say with some certainty is that the woven arch bridges (clamped bridges) were connected to the mining industry, as were as all the other bridge types that blossomed at the same time. The mining industry was responsible for this: with its demands on the traffic system, its economic resources, its urgent requests which left no time to call in fine craftsmanship and instead, called for experimental means – that and the gentle hilly land and gentle rivers, prepared the growth sprout of the woven arch bridge. This bridge type went out of fashion as soon as the mining activities and the concomitant bridge-building fever died down.

Actually, the bridges in the MZ area are the most unusual examples. It is only in this area that the tradition has been maintained so successfully, for centuries and generations, interrupted only briefly by modernization for a few short decades. The idea of the woven arch could have emerged at any time. If it is indeed somehow related to the Rainbow Bridges, the concept might have been introduced in the form of the chopstick game after the end of the North Song Dynasty, brought in by the immigrants from the north. Or it might have been generated spontaneously, without outside influence, either an inspiration from nature or as an improvement on the strut bridge, as was the process in Norway. In the latter case, it could theoretically have occurred at any time.

However, seen in the context of our growing environment theory, MZ mountain societies might not have been ready to afford such a sophisticated bridge-building technique until the middle of the Ming Dynasty. The early mountain residents were mainly "refugees" from the chaos caused by wars in the north. They found shelter in the mountains and fed themselves by cultivating virgin lands, gathering food in the forest, and hunting, with little need for travel. The many centuries of silver mining did not bring prosperity, but rather heavy taxes, rebellions, looting and bandits and gave rise to nightmarish legends. It was only after the government suppressed the rebellion, closed the mines, and established county governments in the remote mountains in the fifteenth century, that the mountain society underwent its first stages of economic development. The upturn of the forest economy led to an increase in the demand for roads and bridges, and led to the development of the bridge carpenter profession. However, this profession was still a career for the poor, for families and groups who could not earn their living by means of a cultivation-and-education lifestyle, but rather, had to risk their lives building bridges in dangerous conditions in order to earn a better life. When such professional bridge carpenters were working in huge groups, with a systematic, sophisticated division of labour, the bridge-building technique reached its peak.

So again, it is thanks to the remoteness of the mountain region, that it lagged behind the outside world in economic, cultural, and technological matters, as this in turn generated a centuries-long steady legacy of woven arch bridges. The MZ bridges, the embodiment of the woven arch bridge at its most developed form – and our cultural heritage – were indeed the result of creative, inventive, highly imaginative ways of dealing with the challenges of a tough environment, and of using wisdom and courage to survive in the roughest conditions. From 'mission impossible' to the highest achievement, the story of the evolution of the woven arch bridges is one of hardship and determination, like the grit in an oyster that produces a rare and beautiful pearl.

References

Bertin, Vito. 2012. *Leverworks: One Principle, Many Forms*. Beijing: China Architecture & Building Press.

Hirahara, Naomi. 2013. "From Japan to America: The Garden and the Japanese American Community." in: Li, T. June. *One Hundred Years in the Huntington's Japanese Garden: Harmony with Nature*. San Marino, Calif.: Huntington Library Press. 94–107.

Killer, Josef. 1985. *Die Werke der Baumeister Grubenmann*. Basel: Birkhäuser.

Liu, Diyu
刘涤宇. 2014. 历代《清明上河图》——城市与建筑. 同济大学出版社.

Liu, Dunzhen
刘敦桢. 1984.中国古代建筑史（第二版）.中国建筑工业出版社.

Liu, Jie
刘杰. 2017.中国木拱廊桥建筑艺术. 上海人民美术出版社.

Meyer, Franz Sales. (1895)1981. *Zimmermannsbuch*. Hannover: Th. Schäfer GmbH. 358–9.

Needham, Joseph. 1971. *Science and Civilisation in China, Vol. 4: Physics and Physical Technology, Part 3: Civil Engineering and Nautics*. Cambridge: Cambridge University Press. 162–5.

Nitto, Kazuhiko
日塔和彦. 2004. 日本の歴史の木造橋（刎橋を中心に）.木の建築フォラム / 岩国. 現代に生きる伝統技術. 75–91.

Sagae
寒河江市教育委員会. 1985.寒河江市史編纂叢書（第三十二集）: 白岩臥竜橋関係資料（工藤善兵衛家資料・鈴木修助家資料・渡辺半右衛門家資料）.

Shekhar, R. K. Chandra. 2005. *Academic Dictionary of Civil Engineering*. Delhi: Isha Books.

Tang, Huancheng
唐寰澄. 1987. 中国古代桥梁. 北京：文物出版社: 64–77.
唐寰澄. 2010.中国木拱桥. 北京：中国建筑工业出版社.

Tampone, Gennaro. 2003. "Palladio's timber bridges." In *Proceedings of the First International Congress on Construction History*: Madrid, 20–24 January 2003. Madrid: Instituto Juan de Herrera.

Tardini, Chiara. 2019. "Survey and assessment of wooden bridges: a template proposal." In *7th International Symposium on Chinese Covered Bridges*. Shanghai and Qingyuan: Secretariat of the Organizing Committee of the Symposium & the Timber Architecture Design and Research Center, 23–8.

Taylor, Rabun. 2003. *Roman Builders – A Study in Architectural Process*. New Haven and London: Cambridge University Press. 180–1.

Thönnissen, Udo. 2015. *Hebelstabwerke: Tradition und Innovation*. Zurich: gta Verlag.

Zwerger, Klaus. 2015. *Das Holz und seine Verbindunge*. Basel: Birkhäuaser.

Afterword

In September 2002, as a freshman at the Department of Civil Engineering at Tsinghua University in Beijing, my first course, "Introduction to Civil Engineering," was taught by Prof. Liu Xila (刘西拉). In his stirring introductory lecture, he showed us an image of an amazing historical arch bridge, the Rainbow Bridge from the famous *Qingming Shanghe Tu* scroll. The wooden arch bridge was so extraordinary, and the professor stated that it would not be an easy task even for modern civil engineers to make statics calculations for it.

The next month, the Eighth Structure Design Competition of the university was held in our department. Following the tradition of previous years, the topic of the competition was bridge structure under loading tests. For my part, I wanted to attempt to reproduce the structure of the Rainbow Bridge, but my classmates decided to build truss bridges or fish-bellied-beam structures: these types had been discussed in class and are proved to have optimal structural features. Working alone, and being at the beginning of my studies, I failed to complete the model. However, the fascination with the Rainbow Bridge was imprinted on my brain and has stayed with me ever since.

Four years later, with a diploma thesis on the calculation models for historical Chinese wood architecture, I received my bachelor's degree in civil engineering. Although this study[1] has been quoted dozens of times by Chinese colleagues in the decade after its publication, I always felt that it still lacked something essential.

For my master's, I turned to the field of architecture and devoted myself to the History of Chinese Architecture. At around the same time, I began to wonder why, even with the knowledge which was being taught at university, the history of my homeland was still a "foreign country."

During my studies at the Southeast University in Nanjing, I had the opportunity to audit a course in building history taught by Prof. Zhao Chen at the nearby Nanjing University. One of the leading scholars in the field of the wooden arch bridges in Southeast China (MZ bridges), the professor discussed

---

1 Liu and Yang. (刘妍, 杨军. 2007. 独乐寺辽代建筑结构分析及计算模型简化. 东南大学学报 (自然科学版), (05): 887–91.

these bridges within the context of the tectonic theory. Later, during my doctoral studies at the Technical University of Munich in Germany, I invited him to be my second mentor.

I chose woven arch bridges as the topic for my doctoral thesis. This was a consequence of my curiosity about the topic throughout the years and, more directly, due to a lucky opportunity which arose in the summer of 2009, when I graduated with my master's: I had a chance to participate in a conservation project on historical wooden arch bridges in Taishun, Zhejiang. I took this chance and was able to visit the core area of the woven arch bridges, collect first-hand material and begin to form some basic understanding of their structure.

My doctoral studies at the TU Munich were in the field of *Bauforschung* (Building Archaeology), under the supervision of Prof. Manfred Schuller, the leading building archaeologist in that field. My research aim was initially a scientific study on some one hundred woven arch bridges that survived in Southeast China, including their typology and genealogy analysis, following the route directed by Prof. Zhao Chen. This goal was soon adjusted.

The first new inspiration beyond the route of the discipline of architecture came from "local knowledge" from the "native's point of view" in the field of anthropology, although my viewing angle had not included such a theoretical system at the beginning. My idea was simple but effective: to understand the craftsmanship of the bridge carpenters, I had to become one of them and think the way they did.

I received my first (three-week) training in carpentry in the woodworking shop (*Schreinerei*) of the Deutsches Museum (Munich) in 2011, and after that in large-scale model-building from my German colleagues, following the German tradition of the study on historical timber structures. However, my most crucial woodworking experience was with the traditional Chinese carpenters.

From 2011 to 2015, I spent several months each year in the Southeast China mountains on field investigation, which at the beginning were focused on a survey of the bridges. During the fieldwork, I made the acquaintance of a number of bridge carpenters, and as a result took part in the construction of three bridges, working with three different bridge-building families. The third project was a particular opportunity, which was to build a bridge in the garden of the Nepal Himalaya Pavilion in Regensburg, Germany. In this case, I acted as a carpenter master myself, taking over the design activities when the traditional carpenters had run into trouble. Through these activities, and through interviews with a number of carpenters in the MZ area and in other regions in China, I was able to dive inside the methodology used by traditional craftsmen in timber construction.

At this stage, the 'initial question' raised on my first day at university, which was "how to make a correct calculation on the traditional structures," was no longer a main focus. A correct analysis model in the scientific system must be established on the basis of a thorough understanding of the entire set of measurements involved in the bridge-building craftsmanship. The reason for this is that the real challenge lies not in reproducing the visible structural form or practice methods, but the knowledge structure of these two disciplines. This anthropological study works as an interpreter between the two disciplines, and can provide a much broader view which extends beyond the calculation.

Although my study started as a 'Chinese' topic, several 'foreign' examples revealed themselves to me in the course of my research activities. The first was the already well-known Da Vinci bridges and soon enough, came my 'discovery' of the Japanese examples, and more European material.

When I embarked on this research journey, the only examples known outside China were Da Vinci's designs. The Chinese examples played such a leading role that it was clear that scholars from both China and the West tended to believe (from their publications and private communications) that Da Vinci's designs might be the result of cultural influence from China – even though they had no evidence to support this idea.

However, my case studies on the Italian and Japanese examples show that they are independent in themselves (whether or not there was any external influence). Soon, with more western examples in Germany and Norway found scattered in construction history, the 'universal uniqueness' or the 'shared isolation' situation of these examples became strikingly evident. These discoveries led to the topic of my research evolving into "constructional thoughts": the deep-seated thinking patterns in the respective cultures which led to different routes of development and then to a rather similar result in the form of the woven arch bridge.

To achieve a plausible comparative study, I had to go deep enough in each case study. In addition to sifting through historical literature, I needed some other tools to permit me to get the necessary depth.

Bauforschung or Building Archaeology is just such a handy tool. It's the discipline of my study in Germany, a sub-field of History of Architecture. *Bauforschung* applies archaeological methods to objects above ground, and through a thorough and in-depth investigation, it creates a 'biography' of the object. It is of particular advantage in the reconstruction of the building and rebuilding history of a structure, together with its construction methods, especially those with little or no written documentation. Using this method, we can look beyond the present appearance of the structure, and 'read' the construction problems and the measurements taken accordingly in the entire lifecycle of the building. And it's great fun! In the two case studies of *Bauforschung* in this book – the Huntington Moon Bridge and the Rulong Bridge – I felt like a detective, using my wits to find, sort and piece together the evidence to solve the cases. To quote my 'doctoral father' (*Doktorvator*): "*Bauforschung* means observing! The *Bauforscher* is a detective who collects countless bits of evidence, including those that may seem trivial at first glance, and puts them together to form a complete puzzle."[2]

2 Schuller, Manfred. The application of Bauforschung-methodology and presentation. In: De Jonge, Krista, and Koen Van Balen, eds. *Preparatory architectural investigation in the restoration of historical buildings*. Leuven: Leuven University Press (2002): pp. 31–47.

I finished my dissertation in 2016 and earned my doctoral degree the following year. Written in German, the first version of my writing focused on the technical aspects as described above. However, I felt there was still something missing, and that it wasn't from the architectural discipline.

I turned to historical anthropology for help, in the course of my fellowship at the Society of Fellows in Liberal Arts in the Southern University of Science and Technology, Shenzhen, China. Thanks to this interdisciplinary academic platform, and thanks to the various communications channels with this and other institutes, I received inspiration and suggestions from colleagues working in different disciplines, especially anthropologists, historians, and geographers. During this time I kept on re-visiting the mountains, collecting, among other forms of local literature, the genealogy books of the carpenter families in particular. When putting together piece by piece, the various types of literature combined to reveal a broader scene of the historical landscape of the mountain society.[3]

This book is the fruit of a decade of research, and even the composing period has taken more than five years. Echoing the "unique yet universal" feature of the object, the writing framework is designed to make each chapter as independent as possible. However, I have to admit that because of the length of this particular battle, in depth and strength, the chapters are not as equal as I would wish. This unevenness of the chapters parallels my personal growth, from a novice student to a confident scholar. I sometimes see reflections of myself in this research, a metaphor for my own life. My personal life experiences also inspired my unique interpretation of the study — changing not only cultures, academic fields but also genders, the ambiguity and instability of my identity resonated with the complexity of the subject.

My research on the woven arch bridges is not yet finished. Some remaining 30 MZ-area bridges still need to be surveyed, the tonnes of material buried in my hard disks need to be sorted and collated, the drawings of dozens of bridge plans need to be completed, and hundreds of hours of video need to be edited and processed and made into documentary films. In the meantime, though, I'm relieved that I can set a milestone on the path, with this book. Although so much work remains unfinished, and I anticipate that my understanding will evolve in the coming years. However, that will be a story for another decade.

---

3 The work based on historical anthropology is reproduced in a condensed form in this English version, as it will be read mainly by western readers. This part of the material and analysis is shown in a more complete form in the Chinese version of the book bearing the title "Woven arch bridge: technical and social histories" (2021). The absence of material does not affect the solidity of the conclusions.

Acknowledgements

The completion of this work owes so much to so many people, and I fear that it would be impossible to make a complete list of all the people and institutions that have helped and supported me along the way. Therefore, please know that, although some names may be missing from the list, my gratitude goes out to all that have helped me on this long journey.

First and foremost, my gratitude goes to my supervisor Professor Manfred Schuller and to my second mentor Professor Zhao Chen (赵辰), who have provided direction and academic advice along the way, supporting me with suggestions, targeted questions and factual information.

Professor Philip Caston was the one who taught me the methodology of *Bauforschung* (building archaeology), step-by-step. He, together with professors Tom Peters, Terry Miller, and Ron Knapp, provided great examples of how to survey timber bridges during our bridge research trips together.

I am particularly indebted to Professor Klaus Zwerger for information and assistance regarding western and eastern wooden constructions. Special thanks also go to Dr Ren Congcong (任丛丛) for her help on sources and understanding the Japanese examples, and to Dr Yuan Changgeng (袁长耕), for the inspiring suggestions on anthropological matters.

Special thanks to Dr Nora Eibisch, my mentor in my doctoral studies, for her generous help in navigating the many different aspects of academic life in a foreign land. Dr Eibisch was also instrumental in making it possible for me to study woodworking in the workshop (*Schreinerei*) of the Deutsches Museum. My thanks also go to the *Schreinerei* for giving me the opportunity to undertake basic training as a joiner.

My gratitude also goes to my alma mater, the Institute of Architectural History of the Southeast University in China, and the colleagues working there, especially professors Chen Wei (陈薇), Hu Shi (胡石) and Chen Jian'gang (陈建刚), for their help and support in my field investigations. My thanks also to Professor Wang Shuzhi (王树芝) in the Institute of Archaeology, Chinese Academy of Social Sciences, for the help on the dendrochronological analysis.

Profound thanks to the bridge masters and carpenters I have named in the second part of this book, especially masters Wu Fuyong (吴复勇), Wu Dagen

(吴大根), Zheng Duoxiong (郑多雄), Zhang Changzhi (张昌智) and Huang Chuncai (黄春财). They allowed me to become an 'insider,' permitting me a close-up and thorough observation of the skills and knowledge related to their construction techniques, those closely-guarded secrets which have been handed down in their families for generations.

The list continues with Mr Heribert Wirth, sponsor and owner of my project in the Nepal Himalayan Pavilion in Regensburg, who gave me the chance to set up a woven arch bridge in Germany; and to Mr David MacLaren, the curator of the Huntington Japanese Garden, and Andrew Mitchell, the architect maintaining the Moon Bridge, for their kind permission for me to carry out my investigation on the Huntington bridge example and their help in my investigation.

For their help and information in my field investigations in the MZ area, I would like to thank the cultural workers from the counties of Qingyuan, Taishun, Jingning, Longquan, Lianduqu, Zhejiang Province and the counties of Pingnan, Shouning, Zhouning, Zhenghe, Fujian Province. In this group, I would like to express my particular appreciation to Mr Su Xudong (苏旭东) from Pingnan County, Mr Zheng Yong (郑勇) from Zhouning County, Mr Gong Difa (龚迪发) and Mr Gong Jian (龚健) from Shouning County, Mr Xue Yiquan (薛一泉) and Mr Ji Haibo (季海波) from Taishun County. In addition, a number of scholars also shared their information on the MZ bridge study with me, including professors Liu Jie (刘杰), Yang Yan (杨艳) and Chen Zhenguo (陈镇国), and engineer Yao Hongfeng (姚洪峰). Special thanks also to the members of the referred families in Chapter 6, who take care of their treasured genealogy books of their families and kindly allowed me to look through and take photos. My thanks also go to the 'bridge-fans' Mr He Nanda (何南大) and Mr Ying Jiakang (应嘉康), who added their information.

Next on my list is a great (in both senses of the word) group of people (the names of some of whom I recorded and the names of others which I didn't have the opportunity to learn), to whom I am indebted: these are the villagers I met on my research journeys, who offered me help of all kinds, including free accommodation, transportation, food, and assistance, even in the middle of nowhere. Their warmth and kindness to a stranger will always be treasured in my heart.

Most of the time I travelled alone, and I therefore also feel very grateful to those who joined me, even just briefly, on my field trips in the mountains. Top of this list is Yu Yannan (于燕楠) who carried out the investigation on the Rulong Bridge together with me. Others who have joined me on a field trip at one point or another include Zhou Miao (周淼), Dong Shuyin (董书音), Gao Wenjuan (高文娟), Yang Yang (杨扬), Yang Xi (杨熹), Meng Xianchuan (孟宪川), Wang Xinyu (王新宇) Wu Chao (吾超), documentary director Qiu Ping (邱萍) and her team, and later my students, Chen Shuo (陈硕), Wen Han (文涵), Liang Xiaorui (梁晓蕊), Du Mengzeshan (杜孟泽杉), Lin Yu (林宇), and Li Xueqi (李雪琦) from Nanjing University. Most of them even helped in the survey and documentation. In addition, there are many other friends – from both the real and the online world– whose encouragement inspired me on my trips.

I am grateful for the assistance of Li Jiayong (李嘉泳) and Gu Xueping (顾雪萍), who helped me with some drawings of the bridges, and my student assistants Liao Wanning (廖琬凝) and Zhang Shaohua (张少华) from the Southern University of Science and Technology, who helped me by copying and collating some of the local literature.

On writing and language issues, my special thanks to Ms Francesca Brizi for the careful and helpful copy-editing of this English version of the book. Furthermore, many friends and colleagues provided valuable help and review input when I was writing the preliminary version of this book (in the shape of my dissertation in German), including Dr Nora Eibisch, Dr Alexandra Harrer, Dr Barbara Berger, and Mr Wei Shaochen (韦劭辰).

I'd also express my gratitude to the following institutes and personal for permitting me to use their photos or images in this book, including the Huntington Library, Pasadena Museum of History, Anno Musea i Nord-Østerdalen, Rørosmuseet, Palace Museum, Veneranda Biblioteca Ambrosiana, Biblioteca Nacional de España, Shouning Museum, Mr Tang Hao (唐浩), Mr Gong Jian, Professor Hu Shi, Dr Zhou Miao, Mr Ying Jiakang, Mr Wu Chao, Professor Kazuhiko Nitto, and Professor Klaus Zwerger.

During the long and arduous journey, my research has been financed by the travel funds of the TU Munich Graduate School (2013); the Research Funding (2012–4) by the Key Laboratory of Urban and Architectural Heritage Conservation of the Ministry of Education, Southeast University in China; the Laura Bassi Award at the TU Munich (2015–6), and the research fund (2018–9) by the Society of Fellows in the Liberal Arts, Southern University of Science and Technology in China.

Finally, my love and my most grateful thanks to my family, who have supported me with the warmest understanding for all these years.

List of referred historical MZ bridges

Bridge	Address	Bridge	Address
Beijian Bridge	Town of Sixi, Taishun County, Zhejiang Province.	北涧桥	浙江省泰顺县泗溪镇
Chixi Bridge	Chixi Village, Township of Chengyuan, Zhenghe County, Fujian Province.	赤溪桥	福建政和县澄源乡赤溪镇赤溪村
Dabao Bridge	Xiaodong Village, Township of Kengdi, Shouning County, Fujian Province	大宝桥	福建省寿宁县坑底乡小东村
Dachikeng Bridge	Dachikeng Village, Township of Dajun, Jingning County, Fujian Province.	大赤坑桥	浙江省景宁县大均乡大赤坑村
Dan Bridge	Menghulin Village, Township of Kengdi, Shouning County, Fujian Province.	单桥	福建省寿宁县坑底乡猛虎林村
Denglong Bridge	Bapu Village, Town of Qibu, Zhouning County, Fujian Province.	登龙桥	福建省周宁县七步镇八蒲村
Dongkeng Xia (Downstream) Bridge	Dongkeng Village, Town of Dongkeng, Jingning County, Zhejiang Province	东坑下桥	浙江省景宁县东坑镇东坑村
Duoting Bridge	Dianban Village, Town of Rixi, Jin'an, Fujian Province	多亭桥	福建省福州晋安区日溪镇嶂山村委店坂自然村
Fushou Bridge	Xixi Village, Town of Xixi, Shouning County, Fujian Province.	福寿桥	福建省寿宁县犀溪乡犀溪村
Helong Bridge	Shenghuang Village, Town of Shenghuang, Minqing County, Fujian Province.	合龙桥	福建省闽清区省黄乡省黄村

(continued)

Bridge	Address	Bridge	Address
Hengli Bridge	Qingyuan County(not exist)	亨利桥	浙江省庆元县（不存）
Houlong Bridge	Houlong Village, Township of Limen, Zhouning County, Fujian Province	后垄桥	福建省周宁县礼门乡后垅村
Houshan Bridge	Houshan Village, Township of Lingyao, Zhenghe County, Fujian Province	后山桥	福建省政和县岭腰乡后山村
Huangqiling Bridge	Huangqiling Village, Township of Limen, Zhouning County, Fujian Province	黄旗岭桥	福建省周宁县礼门乡黄旗岭村
Jiaolong Bridge,	Dalixi Village, Township of Chengyuan, Zhenghe County, Fujian Province.	蛟龙桥	福建省政和县澄源乡大梨溪村
Jielong Bridge,	Zhangkeng Village, Town of Dongkeng, Jingning County, Zhejiang Province	接龙桥	浙江省景宁县东坑镇章村
Kuaizi Bridge,	Ankou Village, Township of Shuiji, Jianyang, Fujian Province.	筷子桥	福建省建阳区水吉镇安口村
Lanxi Bridge	Xiyang Village, Township of Wudabao, Qingyuan County, Zhejiang Province	兰溪桥	浙江省庆元县五大堡乡西洋村
Lanxia Bridge,	Lanxia Village, Lanxia County, Shunchang County, Fujian Province,	岚下桥	浙江省顺昌县岚下乡岚下村
Lingjiao Bridge	Xiongdai Village, Township of Jiadi, Jingning County, Zhejiang Province	岭脚桥	浙江省景宁县家地乡弓岱村
Liren Bridge	Youxi Village, Township of Qinyang, Shouning County, Fujian Province	里仁桥	福建省寿宁县芹洋乡尤溪村
Longjin Bridge	Houlong Village, Township of Pingcheng, Pingnan County, Fujian Province	龙津桥	福建省屏南县屏城乡后垅村
Longjin Bridge	Liuyuan Village, Township of Tingping, Minhou County, Fujian Province	龙津桥	福建省闽侯县廷坪乡流源村
Longtan Bridge	Longtan Village, Township of Yangyuan, Zhenghe County, Fujian Province	龙滩桥	福建省政和县杨源乡龙滩村
Luoling Bridge	Yangyuan Village, Township of Yangyuan, Zhenghe County, Fujian Province	落岭桥	福建省政和县杨源乡杨源村

(continued)

Bridge	Location		
Meichong Bridge	Meichong Village, Town of Yingchuan, Jingning County, Zhejiang Province	梅崇桥	浙江省景宁县英川镇梅崇村
Niao Bridge	Ximen Village, Town of Songyuan, Qingyuan County, Zhejiang Province	袅桥	浙江省庆元县松源镇西门村
Rulong Bridge	Yueshan Village, Township of Jushui, Qingyuan County, Zhejiang Province.	如龙桥	浙江省庆元县举水乡月山村
Qiancheng Bridge	Tangkou Village, Township of Tangkou, PingnanCounty, Fujian Province	千乘桥	福建省屏南县棠口乡棠口村
Shengxian Bridge	Xianfeng Village, Township of Xixi, Shouning County, Fujian Province.	升仙桥	福建省寿宁县犀溪乡仙峰村
Shizhukeng Bridge	Shizhukeng Village, Township of Limen, Zhouning County, Fujian Province.	石竹坑桥	福建省周宁县礼门乡石竹坑村
Shunde Bridge	Yangshun Village, Town of Pingnan, Longquan County,Zhejiang Province	顺德桥	浙江省龙泉市屏南镇洋顺村
Tingxia Bridge	Louxia Village, Town of Shanyang, Gutian County, Fujian Province	亭下桥	福建省古田县杉洋镇楼下村
Tongle Bridge	Township of Lingbei, Taishun County, Zhejiang Province	同乐桥	浙江省泰顺县岭北乡
Wan'an Bridge	Changqiao Village, Town of Changqiao, Pingnan County, Fujian Province	万安桥	福建省屏南县长桥镇长桥村
Wenxing Bridge	Kengbian Village, Town of Xiaocun, Taishun County, Zhejiang Province.	文兴桥	浙江省泰顺县筱村镇坑边村
Wodu Bridge	Yuanjia'ao Village, Fenghua, Ningbo, Zhejiang Province	卧渡桥	浙江省宁波市奉化区袁家岙村
Wohung Bridge	Kenggen Village, Township of Zhangcun, Qingtian County, Zhejiang Province.	卧虹桥	浙江省青田县章村乡坑根村
Xiaban Bridge	Wangdacuo Village, Township of Yangyuan, Zhenghe County, Fujian Province	下坂桥	福建省政和县杨源乡王大厝村
Xian'gong Bridge	Town of Aoyang, Shouning County, Fujian Province	仙宫桥	福建省寿宁县鳌阳镇
Xianju Bridge	Xianju Village, Township of Xianren Taishun County, Zhejiang Province	仙居桥	浙江省泰顺县仙稔乡仙居村

Bridge	Address	Bridge	Address
Xiaodong (Upstream) Bridge	Xiaodong Village, Town of Kengdi, Shouning County, Fujian Province.	小东上桥	福建省寿宁县坑底乡小东村
Xidong Bridge	Town of Sixi, Taishun County, Zhejiang Province.	溪东桥	浙江省泰顺县泗溪镇
Xuzhou Bridge	Liaocuo Village, Township of Zhuoyang, Gutian County, Fujian Province	徐州桥	古田县卓洋乡廖厝村
Yangmeizhou Bridge	Yangmeizhou Village, Township of Kengdi, Shouning County, Fujian Province	杨梅洲桥	福建省寿宁县坑底乡杨梅洲村
Yangxitou Bridge	Yangxitou Village, Township of Xiadang, Shouning County, Fujian Province	杨溪头桥	福建省寿宁县下党乡杨溪头村
Yongge Bridge	Shancha Village, Town of Shangyong, Dehua County, Quanzhou, Fujian Province.	永卓桥	福建省德化县上涌镇山茶自然村
Yongzhen Bridge	Xushan Village, Town of Bohai, Jingning County, Zhejiang Province.	永镇桥	浙江省景宁县渤海镇徐山村
Yueyuan Bridge	Yang'an Village, Town of Xiayang, Yanping County, Fujian Province.	月圆桥	福建省延平区峡阳镇洋安村
Yuqing Bridge	Wuyishan city, Fujian Province	余庆桥	福建省武夷山市
Zhangkeng Bridge	Zhangkeng Village, Township of Qinyang, Shouning County, Fujian Province.	张坑桥	福建省寿宁县芹洋乡张坑村

Index

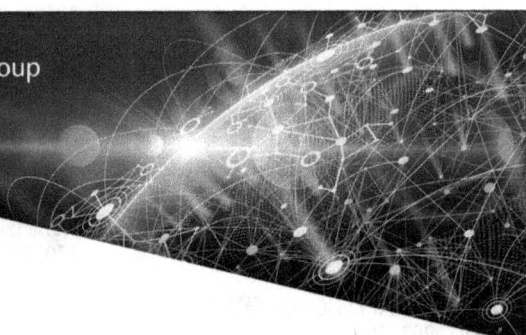

For Product Safety Concerns and Information please contact our EU
representative GPSR@taylorandfrancis.com
Taylor & Francis Verlag GmbH, Kaufingerstraße 24, 80331 München, Germany

41598CB00079B/3768